Lecture Notes in Computer Science 15213

The series Lecture Notes in Computer Science (LNCS), including its subseries Lecture Notes in Artificial Intelligence (LNAI) and Lecture Notes in Bioinformatics (LNBI), has established itself as a medium for the publication of new developments in computer science and information technology research, teaching, and education.

LNCS enjoys close cooperation with the computer science R & D community, the series counts many renowned academics among its volume editors and paper authors, and collaborates with prestigious societies. Its mission is to serve this international community by providing an invaluable service, mainly focused on the publication of conference and workshop proceedings and postproceedings. LNCS commenced publication in 1973.

Luca Maria Aiello · Tanmoy Chakraborty ·
Sabrina Gaito
Editors

Social Networks Analysis and Mining

16th International Conference, ASONAM 2024
Rende, Italy, September 2–5, 2024
Proceedings, Part III

Springer

Editors
Luca Maria Aiello
IT University of Copenhagen
Turin, Italy

Tanmoy Chakraborty
Indian Institute of Technology Delhi
New Delhi, Delhi, India

Sabrina Gaito
Università degli Studi di Milano
Milan, Italy

ISSN 0302-9743 ISSN 1611-3349 (electronic)
Lecture Notes in Computer Science
ISBN 978-3-031-78547-4 ISBN 978-3-031-78548-1 (eBook)
https://doi.org/10.1007/978-3-031-78548-1

This Springer imprint is published by the registered company Springer Nature Switzerland AG
The registered company address is: Gewerbestrasse 11, 6330 Cham, Switzerland

Preface

With ASONAM 2024, the International Conference on Advances in Social Network Analysis and Mining marked its 16th anniversary as the flagship, premier and leading venue in the rapidly growing domain of social network analysis and mining. It has emerged into one of the most well-established and successful conferences. It is a great pleasure that the acceptance rate has stabilized below 20% for full papers since ASONAM was organized in Istanbul in 2012. Indeed, this reflects maturity and stability in terms of number of submissions, acceptance rate and participation, and ASONAM has earned its permanent position among top-tier international conferences.

This year, we moved from ACM/IEEE as sponsors to having Springer as the Sponsor and publisher of the proceedings. Authors of all papers presented at ASONAM and the co-located events are invited to submit expanded versions of their manuscripts to the prestigious SNAM journal, NetMAHIB journal or the LNSN series, which are characterized by their high visibility and fast processing of submissions. Special thanks to Springer Nature for their various publication venues which have been well integrated with ASONAM to the benefit of both parties.

We gathered over four days to witness interesting and exciting research achievements by various authors who present full, short, poster or demo papers covering a wide spectrum of research contributions to the foundations and applications of social networks. This would not have been possible without the dedication of a large team of motivated research leaders working closely together for twelve months to put together the attractive and intensive scientific program. Their great achievements contributed much to the visibility of ASONAM. I would like to heartily thank them all.

We do not forget in particular the generous support received from the operational organizing team who spent considerable time and effort handling daily issues and activities, answering emails, updating the Websites, etc. Special thanks to Min-Yuh Day, Panagiotis Karampelas, Tansel Ozyer, Mehmet Kaya, Deniz Bestepe, Diaylo Steiman and Jalal Kawash, who worked hard to produce the proceedings, communicate with participants/authors and handle the registration; also special thanks go to the local arrangements team from the University of Calabria. Indeed, without their highly appreciated effort it would have been really very hard to maintain the quality of the social program and keep the trend of providing rich meals and breaks during the conference and the excursion trip organized.

Thank you to all organizers including general chairs, program chairs and the chairs of various tracks and workshops, to participants, to authors who submitted papers, to program committee members and to the reviewers who invested their valuable time and effort to provide timely and comprehensive reviews. We encourage researchers and practitioners to submit again next year to get the opportunity and privilege to present their work at ASONAM 2025.

September 2024

Luca M. Aiello
Tanmoy Chakraborty
Sabrina Gaito

Organization

Steering Chair

Reda Alhajj	University of Calgary, Canada

Honorary Chair

Frans N. Stokman	University of Groningen, Netherlands

General Chairs

Andrea Tagarelli	University of Calabria, Italy
Roberto Interdonato	CIRAD, UMR Tetis, France
Jon Rokne	University of Calgary, Canada

Program Committee Chairs

Luca M. Aiello	IT University of Copenhagen, Denmark
Tanmoy Chakraborty	IIT Delhi, India
Sabrina Gaito	University of Milan, Italy

Industry Track Chairs

Francesco Gullo	University of L'Aquila, Italy
Gianmarco De Francisci Morales	CENTAI, Italy

Workshops Chairs

I-Hsien Ting	National University of Kaohsiung, Taiwan
Rajesh Sharma	University of Tartu, Estonia
Lucio La Cava	University of Calabria, Italy

Multidisciplinary Track Chairs

Ester Zumpano	University of Calabria, Italy
Shirin Nilizadeh	University of Texas at Arlington, USA
Carmela Comito	ICAR-CNR, Italy

PhD Forum and Posters Track Chairs

Huzefa Rangwala	George Mason University, USA
Alessia Antelmi	University of Turin, Italy
Domenico Mandaglio	University of Calabria, Italy

Demos and Exhibitions Chairs

Elio Masciari	University of Naples, Italy
Tansel Ozyer	Ankara Medipol University, Turkey

Tutorial Chairs

Pasquale De Meo	University of Messina, Italy
Shady Shehata	YOURIKA Labs, Canada
Davide Vega	University of Uppsala, Sweden

Publicity Chairs

Shang Gao	Jilin University, China
Buket Kaya	Firat University, Turkey
Kashfia Sailunaz	University of Calgary, Canada

Publication Chairs

Min-Yuh Day	National Taipei University, Taiwan
Panagiotis Karampelas	Hellenic Air Force Academy, Greece

Registration Chairs

Jalal Kawash	University of Calgary, Canada
Mehmet Kaya	Firat University, Turkey

Web Chair

Deniz Bestepe	Istanbul Medipol University, Turkey

Additional Reviewers

Abdessamad Benlahbib	FSDM, Morocco
Abdessamad Imine	Loria, France
Abiola Akinnubi	COSMOS-UALR, USA
Adnan Hoq	University of Notre Dame, USA
Aisling Third	Open University, UK
Akira Matsui	Yokohama National University, Japan
Alessandro Visintin	University of Padua, Italy
Alexander Rodriguez	Georgia Institute of Technology, USA
Amrit Poudel	University of Notre Dame, USA
Anastasios Giovanidis	Centre National de la Recherche Scientifique, France
Anatoliy Gruzd	Toronto Metropolitan University, Canada
Anggy Eka Pratiwi	Indian Institute of Technology Jodhpur, India
Ankan Mullick	IIT Kharagpur, India
Anurag Singh	National Institute of Technology Delhi, India
Arlei Silva	Rice University, USA
Ashwin Shreyas Mohan Rao	University of Southern California, USA
B. Aditya Prakash	Georgia Tech, USA
Bailu Jin	Cranfield University, UK
Bijaya Adhikari	University of Iowa, USA
Billy Spann	University of Arkansas at Little Rock, USA
Bing He	Georgia Institute of Technology, USA
Bohan Jiang	Arizona State University, USA
Casey Doyle	Sandia National Laboratories, USA
Charalampos Chelmis	University at Albany State University of New York, USA
Christine Largeron	Université de Lyon, France
Constantine Dovrolis	Georgia Institute of Technology, USA
Courtland Vandam	Massachusetts Institute of Technology, USA

David Skillicorn	Queen's University, Canada
Debanjan Datta	Virginia Tech, USA
Dong Wang	University of Illinois Urbana-Champaign, USA
Eduard Dragut	Temple University, USA
Ehsan Ul Haq	Hong Kong University of Science and Technology, China
Etienne Gael Tajeuna	Laval University, Canada
Fattane Zarrinkalam	University of Guelph, Canada
Fernando Terroso-Saenz	Catholic University of Murcia, Spain
Frank Liu	Southern Illinois University, USA
Fujio Toriumi	University of Tokyo, Japan
George Panagopoulos	École Polytechnique, France
Gita Sukthankar	University of Central Florida, USA
Hadassa Daltrophe	Shamoon College of Engineering, Israel
Hamid R. Rabiee	Sharif University of Technology, Iran
Hanjia Lyu	University of Rochester, USA
Hasan Davulcu	Arizona State University, USA
Hitkul Jangra	Indraprastha Institute of Information Technology, Delhi, India
Huimin Zeng	University of Illinois at Urbana-Champaign, USA
Humayun Kabir	Microsoft, USA
Isabel Murdock	Carnegie Mellon University, USA
Jiaming Cui	Georgia Institute of Technology, USA
Jiamou Liu	University of Auckland, New Zealand
Jiten Sidhpura	Sardar Patel Institute of Technology, India
Jose Luis Fernandez-Marquez	University of Geneva, Switzerland
Julio Cesar Soares dos Reis	Federal University of Viçosa, Brazil
Keith Burghardt	University of Southern California, USA
Kenji Yokotani	Tokushima University, Japan
Keyan Guo	University at Buffalo, USA
Kijung Shin	Korea Advanced Institute of Science and Technology, South Korea
Kshiteesh Hegde	Western Digital, USA
Lanyu Shang	University of Illinois Urbana-Champaign, USA
Lara Quijano-Sanchez	Universidad Autónoma de Madrid, Spain
Lu-An Tang	NEC Labs America, USA
Mainuddin Shaik	University of Arkansas at Little Rock, USA
Mehrdad Jalali	Karlsruhe Institute of Technology, Germany
Michael Smit	Dalhousie University, Canada
Mirela Riveni	University of Groningen, The Netherlands
Muhammad Abulaish	South Asian University, India
Nayoung Kim	Arizona State University, USA

Neha Gondal	Boston University, USA
Nicholas Botzer	University of Notre Dame, USA
Nikhil Muralidhar	Stevens Institute of Technology, USA
Niloofar Yousefi	University of Arkansas at Little Rock, USA
Nishant Vishwamitra	Clemson University, USA
Nur Dean	Farmingdale State College, USA
Orchid Chetia Phukan	Indraprastha Institute of Information Technology, India
Raed Alharbi	University of Florida, USA
Rafael Elias De Lima Escalfoni	CEFET-RJ, Brazil
Rajesh Sharma	University of Tartu, Estonia
Rajiv Ramnath	Ohio State University, USA
Sajedul Talukder	University of Alabama at Birmingham, USA
Sangeeta Lal	Keele University, UK
Sankita Patel	SVNIT, India
Sharma Chakravarthy	University of Texas at Arlington, USA
Shreya Ghosh	Pennsylvania State University, USA
Shubham Gupta	Indian Institute of Technology Jodhpur, India
Siyi Guo	University of Southern California, USA
Sriram Pemmaraju	University of Iowa, USA
Subhodip Biswas	Virginia Polytechnic Institute, USA
Suman Kundu	Indian Institute of Technology Jodhpur, India
Tanvir Amin	Google, USA
Theresa Migler	California Polytechnic State University, USA
Tim Weninger	University of Notre Dame, USA
Tobias Hecking	German Aerospace Center, Germany
Toshiharu Sugawara	Waseda University, Japan
Trenton Ford	University of Notre Dame, USA
Tuan Le	New Mexico State University, USA
Ulrik Brandes	ETH Zürich, Switzerland
Wael Khreich	American University of Beirut, Lebanon
Wang-Chien Lee	Pennsylvania State University, USA
William Power	Temple University, USA
Xinwei Deng	Virginia Tech, USA
Xinyang Zhang	University of Illinois Urbana-Champaign, USA
Xueying Liu	Virginia Polytechnic Institute and State University, USA
Yang Zhang	University of Illinois at Urbana-Champaign, USA
Yifan Ding	University of Notre Dame, USA
Ying Zhao	Naval Postgraduate School, USA
Yiqiao Jin	Georgia Institute of Technology, USA
Yoshiharu Ichikawa	Keio University/NHK, Japan

Young-Woo Kwon	Kyungpook National University, South Korea
Yue Zhang	Amazon, Inc., USA
Yueqing Liang	Illinois Institute of Technology, USA
Zhenming Liu	College of William and Mary, USA
Zhenrui Yue	University of Illinois Urbana-Champaign, USA
Zhihao Hu	Virginia Tech, USA

Contents – Part III

Research

Across Platforms and Languages: Dutch Influencers and Legal Disclosures on Instagram, YouTube and TikTok ... 3
Haoyang Gui, Thales Bertaglia, Catalina Goanta, Sybe de Vries, and Gerasimos Spanakis

Exploring the Capability of ChatGPT to Reproduce Human Labels for Social Computing Tasks ... 13
Yiming Zhu, Peixian Zhang, Ehsan-Ul Haq, Pan Hui, and Gareth Tyson

Fuzzy and Overlapping Communities Detection: An Improved Approach Using Formal Concept Analysis ... 23
Martin Waffo Kemgne, Christophe Demko, Karell Bertet, and Jean-Loup Guillaume

Understanding Engagement Dynamics with (Un)Reliable News Publishers on Twitter ... 36
Alireza Mohammadinodooshan and Niklas Carlsson

PUREmotion: Understanding the Impact of Highway Construction on People's Wellbeing ... 48
Omar Hammad, Aniya Khalili Hollo, Nicholas Clements, Shelly Miller, Shivakant Mishra, and Esther Sullivan

Opinion and Self-confidence in Influence Networks: A Coupled Dynamics Model ... 58
Emmanuel Kravitzch, Vineeth S. Varma, Antoine O. Berthet, and Yezekael Hayel

The Emergence of Threads: The Birth of a New Social Network ... 69
Peixian Zhang, Yupeng He, Ehsan-Ul Haq, Jiahui He, and Gareth Tyson

SocFedGPT: Federated GPT-Based Adaptive Content Filtering System Leveraging User Interactions in Social Networks ... 79
Sai Puppala, Ismail Hossain, Md Jahangir Alam, and Sajedul Talukder

Anonymous Dissent in the Digital Age: A YouTube Dislikes Dataset ... 89
Sujan Dutta, Mallikarjuna Tupakula, Sumeet Kumar, and Ashiqur R. KhudaBukhsh

PhD

From Prison Gang to Transnational Criminal Organization. A Social Network Analysis of the Tren de Aragua 103
Laura Homann and Oscar Alvaro-Montes

Methodology for Identifying Social Groups Within a Transactional Graph 117
Maxence Morin, Baptiste Hemery, Fabrice Jeanne, and Estelle Pawlowski-Cherrier

Evaluating Deep Graph Network Performance by Augmenting Node Features with Structural Features 131
Mohamad Abushofa, Amir Atapour-Abarghouei, Matthew Forshaw, and A. Stephen McGough

Beyond the Surface: Navigating Complex Systems via ABMs and Hypergraphs 148
Daniele De Vinco

Which Reveals Ideology Better? Comparing Self-Presentation and Public Rhetoric in the Facebook Climate Debate via Embeddings Analysis 164
Luigi Arminio and Luca Rossi

Multidisciplinary

Five Patterns of Vaccine Misinformation on Telegram 181
Fabio Malini, Francis Sodré, Athus Cavalini, Gabriel Herkenhoff, and Fabio Goveia

MLCDG: Multi-Level Contrastive Graph Clustering in Dynamic Graphs 197
Mohamed Mahmoud Amar, Mohamed Bouguessa, and Abdoulaye Banire Diallo

Dissecting the Advocacy Discourse Behind the #StopAsianHate Movement on X/Twitter 211
Yuze Sha, Nicholas Micallef, and Yan Wu

Evaluating LLMs Capabilities Towards Understanding Social Dynamics 230
Anique Tahir, Lu Cheng, Manuel Sandoval, Yasin N. Silva, Deborah L. Hall, and Huan Liu

Two-Stage Stance Labeling: User-Hashtag Heuristics with Graph Neural Networks 245
Joshua Melton, Shannon Reid, Gabriel Terejanu, and Siddharth Krishnan

Enhancing School Success Prediction with FRC and Merged GNN 262
Melike Yildiz Aktas, Aadyant Khatri, Mariam Almutairi, Lulwah Alkulaib, and Chang-Tien Lu

Balancing Efficiency and Quality in LLM-Based Entity Resolution on Structured Data 278
Navapat Nananukul and Mayank Kekriwal

Analyzing Characteristics of Nontrivial Information Diffusion via Implicit Links on Social Media 294
Yuto Tamura, Sho Tsugawa, and Kohei Watabe

Insights About Radicalism on the Digital Era: A Sociological Approach 307
Victoria Ferro, Athus Cavalini, Fabio Goveia, and Fabio Malini

Investigating Gender Euphoria and Dysphoria on TikTok: Characterization and Comparison 322
S. J. Dillon, Yueqing Liang, H. Russell Bernard, and Kai Shu

PRAGyan - Connecting the Dots in Tweets 338
Rahul Ravi, Gouri Ginde, and Jon Rokne

Identifying Cyberbullying Roles in Social Media 355
Manuel Sandoval, Mohammed Abuhamad, Patrick Furman, Mujtaba Nazari, Deborah L. Hall, and Yasin N. Silva

Exploration of Hugging Face Models by Heterogeneous Information Network and Linking Across Scholarly Repositories 371
Muhammad Asif Suryani, Saurav Karmakar, and Brigitte Mathiak

Author Index 387

Research

Across Platforms and Languages: Dutch Influencers and Legal Disclosures on Instagram, YouTube and TikTok

Haoyang Gui[1], Thales Bertaglia[1](✉), Catalina Goanta[1], Sybe de Vries[1], and Gerasimos Spanakis[2]

[1] Utrecht University, Utrecht, The Netherlands
{h.gui,t.f.costabertaglia,e.c.goanta,s.a.devries}@uu.nl
[2] Maastricht University, Maastricht, The Netherlands
jerry.spanakis@maastrichtuniversity.nl

Abstract. Content monetization on social media fuels a growing influencer economy. Influencer marketing remains largely undisclosed or inappropriately disclosed on social media. Non-disclosure issues have become a priority for national and supranational authorities worldwide, who are starting to impose increasingly harsher sanctions on them. This paper proposes a transparent methodology for measuring whether and how influencers comply with disclosures based on legal standards. We introduce a novel distinction between disclosures that are legally sufficient (green) and legally insufficient (yellow). We apply this methodology to an original dataset reflecting the content of 150 Dutch influencers publicly registered with the Dutch Media Authority based on recently introduced registration obligations. The dataset consists of 292,315 posts and is multi-language (English and Dutch) and cross-platform (Instagram, YouTube and TikTok). We find that influencer marketing remains generally underdisclosed on social media, and that bigger influencers are not necessarily more compliant with disclosure standards.

Keywords: influencer marketing · legal disclosures · social media measurement · YouTube · Instagram · TikTok

1 Introduction

Social media is undergoing fundamental changes due to the presence of users who rely on monetization, known as influencers or content creators. Influencers engage in various monetization business models, the most popular being *influencer marketing* consists of brands hiring influencers to deliver advertising services in exchange for money, goods and/or services. Such ads tend to look like content rather than advertising. As a result, influencer marketing remains largely undisclosed or inappropriately disclosed on social media [6,12].

Despite an exponential interest in influencer studies across various computer science communities in the past years [6,10,19], the resulting body of academic work in this field has faced three main problems. *First* is the problem of the

L. M. Aiello et al. (Eds.): ASONAM 2024, LNCS 15213, pp. 3–12, 2025.
https://doi.org/10.1007/978-3-031-78548-1_1

evasiveness of influencer definitions and classifications. In academic literature, influencers are defined either in terms of size [19], network influence [8] or based on manual curation by researchers [6]. These approaches remain unrelated to legal standards.

Second, not all monetized posts can be objectively identified. Thus, measuring hidden advertising generally suffers from an inherent degree of subjectivity in the perception of which content is monetized. *Third*, laws worldwide establish abstract disclosure obligations but often do not include practical standards. This leads researchers to propose their own (non-legal) disclosure standards.

This study proposes a transparent methodology for measuring influencer disclosure compliance based on legal standards. We focus on the Netherlands, where both authorities and the advertising industry have been very active in setting clear disclosure standards. We introduce a novel dataset of influencers registered with the Dutch Media Authority based on a legal registration obligation imposed in 2022 by Dutch media law [2]. We collect and analyze a multi-language and cross-platform dataset to measure and characterize the advertising disclosures by Dutch influencers. Our research makes several contributions. First, it provides a comprehensive, multi-language (English and Dutch), cross-platform (Instagram, YouTube, and TikTok) measurement of influencer marketing disclosures based on legal standards. Second, it proposes and applies an original disclosure taxonomy that distinguishes between legally sufficient (green) disclosures and legally insufficient (yellow) disclosures. Finally, it identifies a sub-dataset of affiliate marketing based on a simple and effective methodology and uses it to measure different disclosure practices across different platforms, languages, and sizes of influencers.

2 Related Work

Research on content monetization has primarily focused on: monetization effectiveness [11,16,18], influencer marketing strategies [1], the impact of disclosures and regulation [6,7,9,12], and the detection of undisclosed sponsored content [3,10,19]. In this context, [19] compiled a dataset of 35,000 posts and 99,000 stories from Instagram, categorizing influencers by their audience size and employing deep neural networks to distinguish between disclosed and undisclosed sponsored posts. [10] compiled a large dataset of 1.6 million Instagram posts and employed network features, including brand mentions and connections between posts, to train deep learning models for detecting hidden advertisements. Additionally, [4] investigated the reliability of human annotators in detecting undisclosed ads, highlighting the implications of such inconsistencies for machine learning models. [12] also applied web measurement methods and identified only 10% AM content as disclosed out of 3,472 YouTube videos and 18,273 Pinterest pins. While these studies offer substantial insights, they exhibit a notable gap in connecting computational findings with legal standards within specific jurisdictions.

3 Content Monetization and Legal Disclosures

Influencer Marketing and Dutch law. Based on the contractual transaction models, influencer marketing practices include *Endorsements*, where money is exchanged for advertising services; *Barters*, which involve goods or services being provided in return for advertising services; and *Affiliate Marketing* (AM), where each sale results in a referral commission [13]. In the Netherlands, media and consumer law determine applicable disclosure standards. Laws are generally vague and principle-based. However, self-regulatory organizations such as the Dutch Advertising Organization (Stichting Reclame Code) have proposed more specific rules, such as which hashtags should be clear enough for disclosure purposes. These rules are included in the Dutch Advertising Code, which theoretically must be aligned with Dutch law. In this study, we therefore focus on the more specific rules of the Dutch Advertising Organization to computationally model legally required disclosures. In parallel, the Dutch Media Authority is a state organization that has adopted specific national guidelines relating to identifying influencers. As a result, starting with 1 July 2022, Dutch influencers must register in the Video-Uploader Registry if they: (a) have more than 500k followers on Instagram, YouTube or TikTok; (b) make regular video content (at least 24 videos in the past 12 months); (c) make revenue based on the content; and (d) are registered with the Dutch Chamber of Commerce.

A Legal Framework for Measuring Influencer Marketing. These legal developments allow us to propose a simple and effective approach to measuring disclosures and overcome the research gaps identified above. First, the Dutch Video-Uploader Registry provides a means to identify influencers accurately based on legal criteria. This public registry, mandated by the government, includes influencers who have formalized their monetization activities through registration, offering a formal list that avoids definitional subjectivities. Second, we focus on the legal standards for disclosure as outlined in the Dutch Advertising Code. We categorize disclosures into green disclosures, which follow legal standards (e.g., specific hashtags and words in Dutch and their English translations), and yellow disclosures, which are more inconspicuous and commonly used by influencers (e.g., #ambassador, #partner). Lastly, we propose a method for identifying affiliate marketing (AM) as a benchmark for hidden advertising.

4 Methodology

Data Collection and Cleaning. Between August and October 2023, we collected textual data from the Dutch Video-Uploader Registry. We focus on text data as monetization disclosures remain largely communicated in writing. 209 registrations were officially made by 1 July 2023. However, this number included not only influencers but also other online media companies. We filtered out all the non-influencer accounts through annotations made by the research team, leading to a total of 150 influencers. Out of these 150 influencers, 133 are active on Instagram, 141 on YouTube, 131 on TikTok and 105 are on all three platforms. We used each of the respective platform's API (Instagram's Crowdtangle [5], YouTube Data API v3 [17] and TikTok Research API [15] to collect all the available data of the respective influencers. Due to API limitations or bugs (especially for the TikTok Research API), we could only retrieve data from 132 influencers from Instagram, 136 from YouTube and 127 from TikTok.

The collected data features a total of 300,199 posts. We used `lingua-py` [14] to identify the language of each post. The resulting dataset reflects 292,315 posts recognized as either English or Dutch text. The relevant text data consists of 122,913 Instagram posts from 2011 to 2023, 128,444 YouTube video descriptions from 2007 to 2023 and 48,842 TikTok video descriptions from 2016 to 2023)

Detecting Legal Disclosures. We identify disclosures as follows. *Green Disclosures* are legal disclosures made in compliance with the Dutch Advertising Code. The Code specifies that platform toggles must be used (e.g., the *Paid partnership*) and that word disclosures must be positioned at the beginning of the text. We consider disclosure words in the first five words of each post (after tokenization and removing all punctuation) to be compliant. While all platforms in our study use the disclosure toggle, we only managed to collect disclosure toggle information from Instagram. *Yellow Disclosures* are disclosures which are not legally sufficient but are still used by influencers. We identify them based on a list created using observations from the dataset and expert insights from the author team.

Detecting Affiliate Marketing. Based on AM textual cues, we compiled a list with the co-occurrence of these terms based on dataset observations. Our set of co-occurrence terms includes variations of these relevant words. When words co-occur in one post together, content can be categorized as AM. We checked the accuracy of this approach by manually annotating 10% of 13,917 AM posts across the dataset, where we only found 2 false positives.

5 Findings

We focus on three main research questions: First, what are the practices of Dutch influencers with respect to complying with legal standards? Second, how do Dutch micro- macro- and mega-influencers influencers disclose content on different platforms? Third, What is the engagement difference between disclosed and non-disclosed content across different platforms and influencer sizes?

Legal Disclosure Practices. Overall, the amount of content voluntarily disclosed by influencers (green and yellow disclosures aggregated) shows that registered influencers only flag a marginal amount of their content as being monetized (6.53%) and, therefore, needing disclosure. Table 1 shows a general breakdown of the overall dataset and a distribution of disclosure practices and AM content across three platforms and two languages. Besides Dutch, the influencers also post content in English (43.5%).

Within the disclosed content category, we note a very low usage of green disclosures in general, with YouTube English having the lowest proportion (0.011%), where yellow disclosures are exclusively used (12.840%). One possible explanation is that green disclosures require strict positioning, so influencers

Table 1. Percentage of disclosed and AM content by each platform and language

	Instagram English	Instagram Dutch	Youtube English	Youtube Dutch	TikTok English	TikTok Dutch
Percentage of disclosures	**2.834%**	**3.708%**	**12.851%**	**10.146%**	**1.120%**	**3.013%**
Green disclosure	1.049%	0.800%	0.011%	0.226%	0.143%	0.388%
Yellow disclosure	1.785%	2.908%	12.840%	9.920%	0.977%	2.624%
Percentage of AM	**2.841%**	**0.460%**	**3.182%**	**12.777%**	**0.278%**	**0.141%**
Green disclosed AM	0.076%	0.055%	0.000%	0.027%	0.004%	0.004%
Yellow disclosed AM	0.164%	0.178%	1.294%	0.640%	0.046%	0.038%
Undisclosed AM	2.602%	0.227%	1.888%	12.110%	0.227%	0.098%

may be placing them at the end of the text. On YouTube, this is additionally problematic since text on the platform tends to be longer than Instagram or TikTok posts. Overall, this finding reveals a preference of Dutch influencers for using popular disclosure cues that do not comply with Dutch law.

Table 1 illustrates the overall amount of AM content in the dataset per platform and language (total 4.76%), as well as how much AM is disclosed using green and yellow disclosures (total 0.49%). While green and yellow disclosures only allow us to track disclosures that were voluntarily made by influencers, they do not reveal non-disclosed advertising. Using the AM sub-dataset as a benchmark, it is possible to identify hidden advertising as non-disclosed AM (total 4.27%). Moreover, the green disclosure of AM content is meagre across all platforms and languages (even the highest is just 0.076% on Instagram English). Except for YouTube English, most AM content from the other venues remains undisclosed (especially for YouTube Dutch, with 12.110% of undisclosed AM content).

Moving to disclosure positions, we calculate the position within the sentence (in # of words) where the first disclosure word is shown. Although sentence length varies across different platforms, none of them has a median number lower than the first five words. Moreover, Instagram and YouTube have relatively different medians in English and Dutch, whereas the difference between TikTok's English and Dutch is small.

While all platforms in our study use the disclosure toggle, we could only collect disclosure toggle information from Instagram. Figure 1 presents the distribution of different disclosure types in Instagram data. Green disclosures are divided into three categories: *Words & position*, which refers to the right words at the beginning of the text (first five words); *Toggle*, which indicates the use of the platform toggle in the platform interface; and *Toggle, words, & position*, which involves using the right words at the beginning of the text along with the platform toggle.

We find that disclosures are used insufficiently across both English (more than 60%) and Dutch (around 80%). Overall, there are more toggle disclosures in English than in Dutch, and for both languages, there is an insignificant amount of legal disclosures placed sufficiently early in the text.

Influencer Size and Disclosures. We further investigate whether influencers with more followers disclose more monetized content and if this disclosure is legally sufficient. We determine size using the number of followers on the day

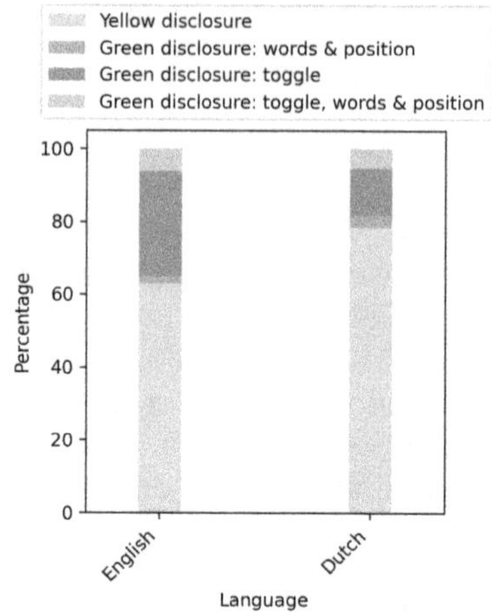

Fig. 1. Instagram disclosure composition.

when the data was collected. We divide the dataset into three audience size categories: *micro-influencers* (less than 500K followers), *macro-influencers* (more than 500K but less than 1M followers), and *mega-influencers* (more than 1M followers). Here, we hypothesise that the bigger the following, the more professional the influencer is. We start by looking into the general distribution of influencers by size. Table 2 shows the distribution of different sizes of influencers across platforms. The second and third rows in Table 2 present the disclosure distribution for each platform. The results show that macro-influencers from YouTube disclose most advertising across all platforms using yellow disclosures. This corresponds with findings from Table 1, showing the higher prevalence of disclosures and AM content on YouTube compared to the other two platforms. Moreover, macro- and mega-influencers have a similar distribution of disclosure content on Instagram and TikTok, generally disclosing more than micro-influencers.

Table 2. Overview of disclosure and AM by different influencer size. # denotes the absolute number and % the proportions of the corresponding disclosure type.

	Instagram			YouTube			TikTok		
	Micro	Macro	Mega	Micro	Macro	Mega	Micro	Macro	Mega
# Influencers	73	35	24	65	43	28	58	35	32
% Green Disclosure	5.51	7.62	14.90	0.52	0.71	0.09	0.72	5.97	6.17
% Yellow Disclosure	18.03	31.05	22.89	7.01	73.88	17.78	10.60	36.11	40.43
% Green AM	1.17	0.69	2.12	0.15	0.03	0.00	0.00	1.01	1.01
% Yellow AM	2.33	4.24	4.08	0.75	4.34	4.35	0.00	8.08	12.12
% Non-disclosed AM	7.68	5.67	72.02	2.52	77.87	9.97	12.12	41.41	24.24

As suggested by Table 1, almost no green disclosures are found, and the majority (around 90%) of AM content stays undisclosed. These findings suggest that more disclosures originate from influencers with a large audience, whether in terms of overall disclosures or AM specifically. To further investigate this pattern, we analyze the top five influencers on each platform with the most AM disclosures. We then measure the proportion of disclosed AM among all AM created by each of the selected influencers, showing their compliance with disclosures. Figure 2 presents the results.

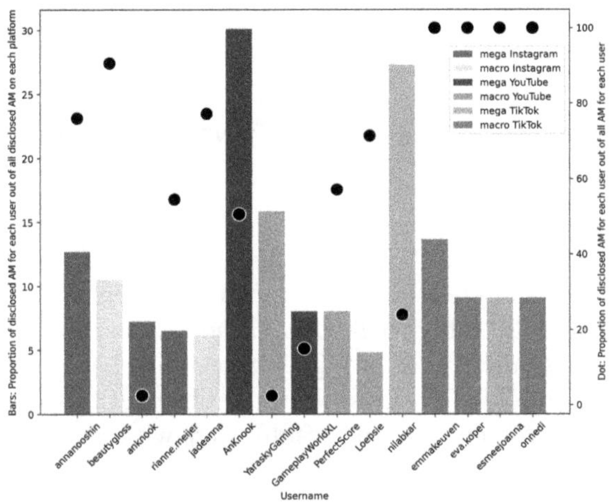

Fig. 2. Top 5 accounts with the most disclosed AM on each platform and the corresponding disclosure rate for their own AM.

The left y-axis indicates the scale of the bars, showing the proportion of disclosed AM for each examined user out of all disclosed AM posts on each platform. As a result, the 5 influencers with the highest proportions of disclosed AM on all platforms are either macro- or mega-influencers, but none are micro-influencers. Instagram's accounts are more representative than the other two platforms, and it would not be reasonable to infer that influencers from YouTube and TikTok are more likely to disclose AM because of the skewed distribution.

Finally, the right y-axis of Fig. 2 shows the proportion of AM that is disclosed for each user (indicated by dots). Four accounts from TikTok disclose all their AM content, showing their high compliance. However, none of them have more than five AM posts in total, which makes the result not representative. In comparison, the results from Instagram and YouTube show that macro-influencers tend to disclose more AM content than mega-influencers from the same platform. These findings do not support the hypothesis that the bigger the influencers are, the more compliant they tend to be.

Engagement and Disclosures. To understand how disclosures affect engagement, we conduct a series of comparative experiments on AM posts from all three platforms. For each post, we define engagement as the sum of the number of likes and comments. Figure 3 shows box plots of audience engagement in AM posts for different disclosure word positions. The engagement score is normalized by the Z-score so that the results between different platforms are comparable.

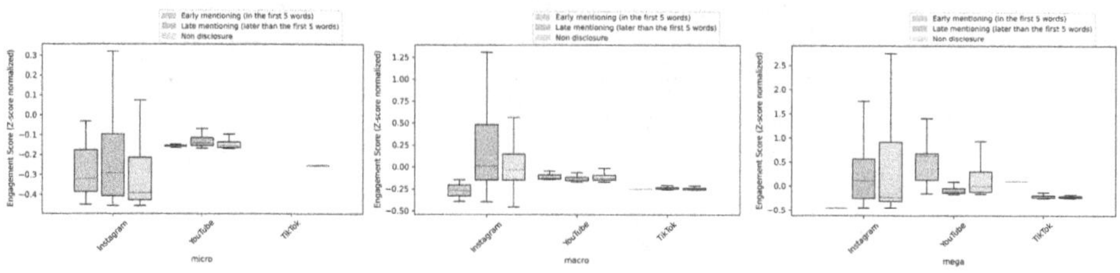

Fig. 3. Box plots of engagement for AM by different disclosure word positions in each influencer category

This plot suggests that for TikTok few disclosures are found positioned in the first five words of the sentence for all sizes of influencers, which corresponds to the findings from Table 2. A higher median in *late disclosure* can be observed on Instagram, while *early disclosure* is higher than *non disclosure* for mega influencers on YouTube and TikTok. Overall, these observations suggest that disclosures can benefit engagement but the results vary with the different positioning of disclosure words.

Lastly, we extend the experiment of the composition of different disclosures on Instagram from Fig. 1 and explore differences in engagement. In Fig. 4, "Green disclosures: word & position" are only visible in micro-influencers due to its few occurrences. Except for it and "Green disclosures: toggle, words & position" in mega-influencers, green disclosures tend to perform better than yellow disclosures regarding the median. The variance of green disclosures is also better than other practices in micro- and mega-influencers. Moreover, in micro- and macro-influencers, "Green disclosure: toggle, words & position" also performs better regarding the median than those only using toggle. The more compliant the AM posts are, the higher engagement they tend to attract. However, findings from mega-influencers contradict this assumption, as those using both toggle and words & position perform the worst.

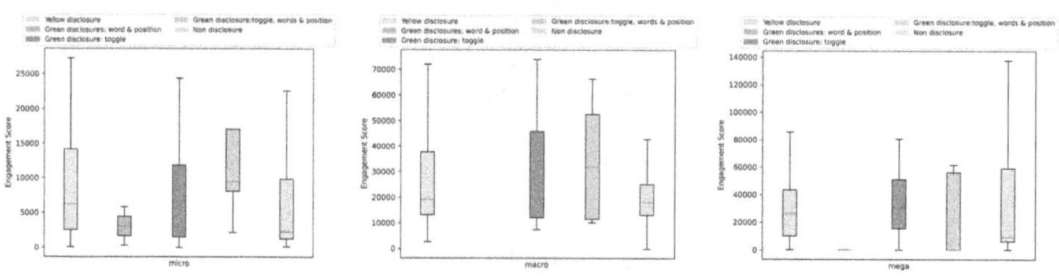

Fig. 4. Engagement by different disclosure strategies of AM on Instagram

6 Discussion and Future Research

This paper presents granular information on how disclosures are done on social media using Dutch law as a starting point to measure legal compliance. Our analysis shows that the general volume of disclosed content is astonishingly low. The content voluntarily disclosed by influencers, whether with green or yellow disclosures, amounts to a mere 6.53% out of the overall dataset. According to our results, in the case of affiliate marketing, only up to 10% is disclosed, leaving 90% of influencer marketing undisclosed. This result aligns with the low disclosure rates found in previous research on English YouTube and Pinterest affiliate marketing by [12], which was, on average, around 10%.

The growing popularity of content monetization has led to an ecosystem where influencers must be present on multiple platforms and often create content for different language audiences. It is important to understand the particularities of content creation on each of these platforms. Further research should investigate platform-specific disclosure affordances.

Limitations Although we used the TikTok Research API, our data retrieval was incomplete due to API problems. We reported the issue to TikTok and used the partial data we retrieved. Data incompleteness is often seen more in earlier data points than in later ones.

Acknowledgments. This research has been supported by funding from the ERC Starting Grant HUMANads (ERC-2021-StG No 101041824).

References

1. Alassani, R., Göretz, J.: Product placements by micro and macro influencers on instagram. In: International Conference on Human-computer Interaction, pp. 251–267. Springer (2019)
2. media authority, D.: Video-uploader registreren (2022). https://www.cvdm.nl/voor-mediamakers/video-uploaders/video-uploader-registreren/
3. Bertaglia, T., Heisig, L., Kaushal, R., Iamnitchi, A.: Instasynth: Opportunities and challenges in generating synthetic instagram data with chatgpt for sponsored content detection. In: Proceedings of the International AAAI Conference on Web and Social Media, vol. 18, pp. 139–151 (2024)
4. Bertaglia, T., Huber, S., Goanta, C., Spanakis, G., Iamnitchi, A.: closing the loop: testing chatgpt to generate model explanations to improve human labelling of sponsored content on social media. In: Longo, L. (ed.) Explainable Artificial Intelligence, pp. 198–213 (2023)
5. CrowdTangle: CrowdTangle API documentation (2019). https://github.com/CrowdTangle/API/wiki/Home
6. Ershov, D., Mitchell, M.: The effects of influencer advertising disclosure regulations: Evidence from instagram. In: Proceedings of the 21st ACM Conference on Economics and Computation, pp. 73–74. EC '20, Association for Computing Machinery (2020), event-place: Virtual Event, Hungary
7. Goanta, C., Ranchordás, S.: The Regulation of Social Media Influencers. Edward Elgar Publishing (2020)

8. Han, J., et al.: FITNet: Identifying fashion influencers on twitter. Proc. ACM Hum.-Comput. Interact. **5** (2021)
9. James, T.: The real sponsors of social media: How internet influencers are escaping FTC disclosure laws. Ohio St. Bus. LJ **11**, 61 (2017), publisher: HeinOnline
10. Kim, S., Jiang, J.Y., Wang, W.: Discovering undisclosed paid partnership on social media via aspect-attentive sponsored post learning. In: Proceedings of the 14th ACM International Conference on Web Search and Data Mining, pp. 319–327. WSDM '21, Association for Computing Machinery (2021)
11. Lee, J.A., Sudarshan, S., Sussman, K.L., Bright, L.F., Eastin, M.S.: Why are consumers following social media influencers on instagram? exploration of consumers' motives for following influencers and the role of materialism. Int. J. Advert. **41**(1), 78–100 (2022), publisher: Taylor & Francis
12. Mathur, A., Narayanan, A., Chetty, M.: Endorsements on social media: An empirical study of affiliate marketing disclosures on YouTube and pinterest. Proc. ACM Human-Comput. Interact. **2**, 1–26 (2018)
13. Parliament, E., Union, D.G.f.I.P.o.t., Michaelsen, F., Collini, L.: The impact of influencers on advertising and consumer protection in the single market. Publications Office of the European Union (2022)
14. Stahl, P.M.: lingua-py (2024-04-30). https://github.com/pemistahl/lingua-py, original-date: 2021-07-13T09:52:34Z
15. TikTok: Research API | TikTok for developers (2023). https://developers.tiktok.com/products/research-api/
16. Wibawa, R.C., Pratiwi, C.P., Larasati, H.: The role of nano influencers through instagram as an effective digital marketing strategy. In: Conference Towards ASEAN Chairmanship 2023 (TAC 23 2021), pp. 233–238. Atlantis Press (2021)
17. YouTube: YouTube data API documentation. https://developers.google.com/youtube
18. Zak, S., Hasprova, M.: The role of influencers in the consumer decision-making process. In: SHS Web of Conferences, vol. 74, p. 03014. EDP Sciences (2020)
19. Zarei, K., et al.: Characterising and detecting sponsored influencer posts on instagram. In: 2020 IEEE/ACM International Conference on Advances in Social Networks Analysis and Mining (ASONAM), pp. 327–331 (2020)

Exploring the Capability of ChatGPT to Reproduce Human Labels for Social Computing Tasks

Yiming Zhu, Peixian Zhang, Ehsan-Ul Haq, Pan Hui(✉), and Gareth Tyson

The Hong Kong University of Science and Technology (Guangzhou), Guangzhou, China
yzhucd@connect.ust.hk

Abstract. Harnessing the potential of large language models (LLMs) like ChatGPT can help address social challenges through inclusive, ethical, and sustainable means. In this paper, we investigate the extent to which ChatGPT can annotate data for social computing tasks, aiming to reduce the complexity and cost of undertaking web research. To evaluate ChatGPT's potential, we re-annotate seven datasets using ChatGPT, covering topics related to pressing social issues like COVID-19 misinformation, social bot deception, cyberbully, clickbait news, and the Russo-Ukrainian War. Our findings demonstrate that ChatGPT exhibits promise in handling these data annotation tasks, albeit with some challenges. Across the seven datasets, ChatGPT achieves an average annotation F1-score of 72.00%. Its performance excels in clickbait news annotation, correctly labeling 89.66% of the data. However, we also observe significant variations in performance across individual labels. We believe that this research opens new avenues for analysis and can reduce barriers to engaging in social computing research.

Keywords: ChatGPT · Crowdsourcing · Social Data Annotations

1 Introduction

Crowd-sourced human intelligence is commonly used for text data annotation [9]. Such annotations are then used for training various models, including stance detection [8], hate speech detection [11], and bot detection [7]. While unsupervised methods are being introduced for classification tasks, such methods usually require large data samples. Thus, social computing research still relies heavily on human annotations. This, however, creates significant barriers for less well-funded research labs. For example, the annotation of a 10,000-post social media dataset by three human annotators would take approximately five hours. With a rate of $25 per individual worker, this would cost hundreds of dollars [5]. When performing comparative analyses across multiple datasets, these costs can easily escalate to thousands of dollars.

Recently, the release of ChatGPT has uncovered a range of cases where large language models (LLMs) can help substitute human intelligence [24]. Several works compare the use of ChatGPT to human methods [22]. For instance,

L. M. Aiello et al. (Eds.): ASONAM 2024, LNCS 15213, pp. 13–22, 2025.
https://doi.org/10.1007/978-3-031-78548-1_2

researchers have investigated the use of ChatGPT for automatic misinformation detection [20], and even generating academic writing [2]. In this paper, we explore the potential of using ChatGPT for five text-based social computing annotation tasks. We seek to understand whether ChatGPT has the potential to reproduce human-generated annotations. ChatGPT's annotations can highlight its usefulness against crowd-sourced annotations. To achieve this, we first use ChatGPT to label seven annotation datasets on five distinct social problems – COVID-19 controversies (3x), social bot deception, cyberbully, clickbait news, and Russo-Ukrainian War stance detection. We compare ChatGPT's labels with the human assigned labels on those datasets. Our results show that ChatGPT *does* have the potential to perform data annotation tasks. Performance is highest for the clickbait headlines dataset, with ChatGPT correctly annotating 89.66% of headlines. In contrast, performance is worst for the COVID-19 hate speech task, which only correctly annotates 52.24% of posts. Closer inspection reveals that performance varies substantially across individual labels. We observe significant gaps (over 25%) exist between labels' accuracy on five out of seven datasets. For instance, in social bot detection, while ChatGPT can identify 81.1% of human tweets, it only identifies 45.5% of bot tweets.

We hope that this work can open up new lines of analysis and can act as a basis for future research into the use of ChatGPT for human annotation tasks. Our contributions are as follows:

1. We evaluate the efficacy of ChatGPT at replacing human annotators for five important social computing data tasks, spanning seven datasets.
2. We show that ChatGPT *can* replace human annotators, yet this varies across tasks and datasets. Performance is highest for clickbait headlines detection (89.66% accuracy), and lowest for hate speech detection (52.24% accuracy).

2 Related Work

Recent research has looked at using ChatGPT for annotating social computing data. Huang et al. [12] report that ChatGPT is able to correctly annotate 80% of the implicit hateful tweets from the `LatentHatred` dataset [6]. In addition, the authors show that ChatGPT's explanations can reinforce human annotators' perception of the target text in explaining why tweets would be annotated as hateful or not. Our study also examines how well ChatGPT performs in annotating hate speech. However, we do not limit the annotation to a binary decision for whether tweets are hateful or not, and further include a neutral label. Note, existing literature has highlighted the importance of this, stating that tweets with neutral expressions can defense the spread of hateful content [15].

Similar to us, others have used ChatGPT for performing stance detection [1]. Zhang et al. [22] evaluate ChatGPT's performance on detecting political stance on two prevalent datasets, `SemEval-2016` [16] and `P-Stance` [14]. The authors report that ChatGPT can outperform most state-of-art stance detection models in zero-shot settings, suggesting ChatGPT's potential to handle stance annotation tests. Major challenges still remain though. Aiyappa et al. [1] show that

ChatGPT's performance varies across different model versions. The authors also point out that such variance is due to the possibility of data leakage, where past prompts are used for training the next ChatGPT generation. Nonetheless, it remains unclear how well ChatGPT performs in annotating individuals' stances in a broader context, such as those who are in favor or against a particular issue. Our analysis also finds further challenges, e.g., that ChatGPT has a tendency to overestimate neutral stances.

In addition to the aforementioned themes, others have experimented with ChatGPT to perform tasks such as sentiment analysis and fake news detection [3]. However, most of this literature only focuses on a single annotation task. In contrast, we seek to perform a comparative analysis across different data annotation tasks aiming to solve social problems.

3 Methodology

3.1 Overview of Annotation Tasks and Datasets

We follow a comparative approach to analyze the differences in human annotations and ChatGPT annotations by utilizing seven different datasets. We select seven annotation datasets covering five social problems that are commonly used in academic research: ***(i)*** COVID-19 controversies (vaccine arguments, anti-Asian hate speech, and COVID-19 fake news) [11,17,19], ***(ii)*** social bot deception [7], ***(iii)*** cyberbully [13], ***(iv)*** clickbait news [4], and ***(v)*** Russo-Ukrainian War [10,18,23]. For these seven datasets, we then attempt to recreate the human annotations using ChatGPT.

Our dataset selection strategy is based on the following requirements: ***(i)*** The datasets must be in English to avoid differences in language provision, and ***(ii)*** The datasets must be annotated by human annotators, as we wish to compare the human annotations with ChatGPT. We list our targeted annotation tasks and datasets statistics in Table 1.

3.2 ChatGPT Annotation

For each annotated dataset, we try to recreate the same annotations using ChatGPT. We utilize OpenAI API, configured with module `gpt-3.5-turbo-0631`, to annotate each target dataset. We rely on an official prompt example for classification tasks from OpenAI.[1]

In the official document, most prompts are imperative sentences starting with a verb. As such, we choose "Classify", which is frequently used in annotation work. We use this verb to design our prompt. According to the official template, we find that starting a new row with a word describing the subject and object in the prompt is effective. Thus, we follow this pattern, injecting the subject and objective of the annotation task here. Another benefit of this template is that it is flexible to inject text input to annotate and specify a desired format for ChatGPT to respond with its annotation.

[1] https://platform.openai.com/docs/guides/completion/prompt-design.

Table 1. A summary of selected datasets and ChatGPT's annotation. The column "ChatGPT annotation" shows the volume of ChatGPT's responses matching any candidate label of the dataset, with the percentage representing its proportion to dataset size. The column "Label (Human/ChatGPT)" details the proportion of each label's size to dataset size, annotated by human or ChatGPT respectively.

Dataset	Size	ChatGPT annotation	Labels (Human/ChatGPT)
Vaccine Stance [19]	5,926	5,832 (98.41%)	Anti-vaccine (21.56%/10.62%) Pro-vaccine (39.72%/26.86%) Neutral (38.72%/62.52%)
COVID-19 Hate Speech [11]	2,290	2,280 (99.56%)	Hate (18.73%/34.30%) Counterhate (22.58%/23.73%) Neutral (58.69%/41.97%)
COVID-19 Fake News [17]	10,700	10,540 (98.50%)	Real news (52.34%/64.00%) Fake news (47.66%/36.00%)
Social Bot [7]	16,824	16,235 (96.50%)	Human (54.08%/69.01%) Bot (45.92%/30.99%)
Anti-LGBT Cyberbully [13][1]	4,299	4,193 (97.53%)	Cyberbully (29.22%/49.10%) Not cyberbully (70.78%/50.90%)
Clickbait Headlines [4]	32,000	31,084 (97.14%)	Clickbait (50.00%/40.54%) Not clickbait (50.00%/59.46%)
Russo-Ukrainian Stance [18]	1,460	1,321 (90.48%)	Pro-Ukraine (63.90%/48.52%) Pro-Russia (36.10%/51.48%)

[1] https://www.kaggle.com/datasets/kw5454331/anti-lgbt-cyberbully-texts

Based on this template, we modify this example into a generalized prompt template applicable to our seven distinctive datasets. The template can be adjusted to be applied for annotating text in different context as shown as follows: "*Classify the text about {Topic} with a label from [Label 1, Label 2, ...]. Text: "{text to classify}". Desired format: <label_for_ classification>*", where *{Topic}* refers to the topic or background of the text; The *[label1, label2, ...]* refers to the set of candidate labels for ChatGPT to annotate the text; and *{text to classify}* refers to the text input for ChatGPT to produce label. In addition, the plain-language index *Desired format* indicates that ChatGPT should only respond using the label without any other text. We then apply this template to generate a ChatGPT prompt according to the dataset's original annotation strategy. Following this, we pass all data to ChatGPT for annotation. When ChatGPT responds to such a prompt, it is necessary to parse the response and extract its annotation. We consider a response parsable only if it follows the desired format – only providing a label without any other text. Thus, we extract ChatGPT's annotation by matching its response to any candidate labels for the dataset.

In all, we only encounter an average of 3.13% ($SD = 2.99\%$) responses per-dataset that fail to be parsed. Note, a small number (0.1%) of failed cases are because ChatGPT states there is not enough information for it to make a decision. For example, a failed response from Russo-Ukrainian Stance dataset states "*Cannot classify the given text with the label [Pro-Russia, Pro-Ukraine] as it does not provide any relevant information about the topic.*"

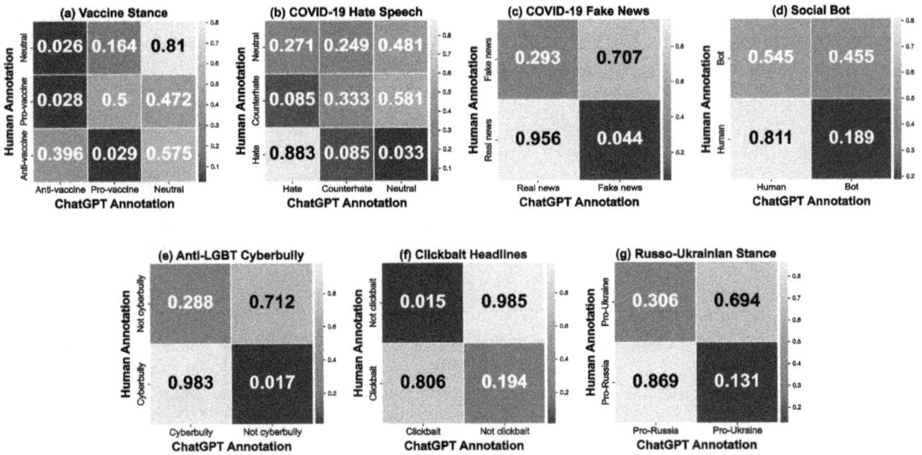

Fig. 1. The confusion matrices of ChatGPT's annotations for the seven annotation datasets. The y-axis label for a given row shows human label of text, and values in each cell show the percentage of those text annotated in the corresponding x-axis label by ChatGPT.

We emphasize that there are many ways in which our methodology could be expanded and refined. Our future work will involve exploring alternative forms of prompt formulation and response mining.

4 Results and Analysis

To evaluate the performance of ChatGPT, we compare ChatGPT's annotation against the original human annotations contained within each dataset. We treat the original human annotations as the gold standard that ChatGPT must predict. As such, we treat ChatGPT as a prediction engine, which we can then evaluate using traditional classifier metrics. We use weighted F1-score to evaluate ChatGPT's performance on data annotation. Given a dataset, a higher F1-score indicates that ChatGPT provides annotations *more* similar to humans.

4.1 Results Summary

Table 1 presents statistics for the ChatGPT annotation results. The annotated datasets contain 73,493 text items. ChatGPT annotates 71,484 (97.27%) items. For the remainder, 2% of responses do not match any candidate label in the dataset, and 0.73% of responses fail due to API errors. This confirms that ChatGPT can generate easily extractable annotation labels in desired format.

Table 2 presents the weighted recall, precision, and F1-score for ChatGPT's predictions for each dataset. For the seven datasets, ChatGPT achieves an average weighted F1-score of 72.00% ($SD = 13.90\%$). This suggests that, *as a data annotator, ChatGPT has the potential to generate annotations similar*

Table 2. A summary of ChatGPT's overall and label-wise annotation performance on selected datasets. The prefix "w-" notes that the measurements' calculations are weighted average by the number of human-annotation for the label.

Dataset	w-Recall	w-Precision	w-F1	Label	Recall	Precision	F1
Vaccine Stance	59.81%	66.11%	59.17%	Anti-vaccine	39.65%	80.13%	53.05%
				Pro-vaccine	50.02%	74.03%	59.70%
				Neutral	80.99%	50.25%	62.02%
COVID-19 Hate Speech	52.24%	55.61%	51.88%	Hate speech	88.27%	48.08%	62.25%
				Counterhate speech	33.33%	31.79%	32.54%
				Neutral speech	67.19%	31.79%	56.03%
COVID-19 Fake News	83.75%	85.55%	83.43%	Real news	95.63%	78.18%	86.03%
				Fake news	70.70%	93.65%	80.57%
Social Bot	64.96%	65.33%	63.70%	Human	81.12%	64.16%	71.65%
				Bot	45.54%	66.75%	54.14%
Anti-LGBT Cyberbully	79.08%	87.17%	80.03%	Cyberbully	98.28%	58.43%	73.29%
				Not cyberbully	71.17%	99.02%	82.81%
Clickbait Headlines	89.66%	90.92%	89.56%	Clickbait	80.57%	98.17%	88.50%
				Not clickbait	98.53%	83.86%	90.60%
Russo-Ukrainian Stance	75.85%	79.83%	76.26%	Pro-Ukraine	69.35%	90.02%	78.34%
				Pro-Russia	86.91%	62.50%	72.71%

to humans. However, its performance varies across different domains. ChatGPT performs well on COVID-19 Fake News, Anti-LGBT Cyberbully, Clickbait Headlines, and Russo-Ukrainian Stance datasets (weighted F1-score $> 75\%$). In contrast, ChatGPT performs poorly on Vaccine Stance, COVID-19 Hate Speech, and Social Bot datasets (weighted F1-score $< 65\%$).

We next explore how ChatGPT performs on different labels in each annotation dataset. Figure 1 presents the confusion matrices for the seven datasets. For each matrix, the y-axis refers to the ground truth human labels, and the x-axis refers to ChatGPT's labels. The value in a cell, row i and column j, presents the proportion of text with label i that are annotated with label j by ChatGPT. For all annotation tasks, the proportions of correctly annotated tweets by ChatGPT vary per distinct label. In all, *five out of seven* datasets present such an imbalance, where a gap of more than 25% exists between labels' accuracy. These results suggest that *for a given annotation task, ChatGPT's accuracy varies heavily across different labels.*

4.2 Task Analysis and Implications

Next, we dive into each annotation task to present implications according to ChatGPT's performance. To assist the following analyses on ChatGPT's label-wise performance, we also present ChatGPT's recall, precision, and F1-score for each annotation label in Table 2.

Rank 1st: Clickbait Headlines. For Clickbait Headlines, ChatGPT's overall performance ranks *first* out of seven annotation tasks by a weighted F1-score of 89.56%. ChatGPT correctly identifies 89.66% clickbait or non-clickbait news headlines. This is the highest accuracy among the annotation datasets. In addition, the weighted precision rate of 90.92% suggests that ChatGPT can effectively annotate clickbait news headlines, while only introducing a small number of false positives. In addition, ChatGPT attains recall, precision, and F1-score exceeding 80% across all labels. Specifically, the corresponding confusion matrix (Fig. 1(f)) shows that ChatGPT can correctly identify 80.6% of clickbait and 98.5% of non-clickbait headlines. In conclusion, *ChatGPT shows better overall performance on Clickbait Headlines than the other six annotation datasets.*

Rank 2nd: COVID-19 Fake News. For COVID-19 Fake News, ChatGPT's overall performance ranks *second* out of seven annotation tasks, with a weighted F1-score of 83.43%. ChatGPT correctly distinguish 83.75% of news as real or fake. However, ChatGPT's performance across different news varies when measured by recall and precision. For real news, ChatGPT achieves a high recall of 95.63%, but a lower precision of 78.18%. For fake news, ChatGPT only attains a recall of 70.7%, but a higher precision of 93.65%. Such a pattern is caused by the existence of many false positives of real news annotated by ChatGPT. Moreover, corresponding confusion matrix (Fig. 1(c)) shows that, while ChatGPT can identify 95.6% real news, 29.3% fake news are misidentified as real. *These results suggest that, for COVID-19 Fake News, ChatGPT is able to annotate real news with high accuracy, but also has a tendency to misidentify fake news as real.* As fake news detection mainly relies on news content when training models, we conjecture *such a limitation is caused by a lack of comprehensive COVID-19 fake news data for training ChatGPT.*

Rank 3rd: Anti-LGBT Cyberbully. For Anti-LGBT Cyberbully annotation, ChatGPT's overall performance ranks *third* out of seven tasks, with a weighted F1-score of 80.03%. Overall, ChatGPT correctly annotates 79.08% posts. However, when annotating cyberbullying posts, ChatGPT achieves 98.28% recall, but only attain 58.43% precision. Such a pattern indicates the existence of false positives among cyberbullying posts annotated by ChatGPT. The confusion matrix supports this by showing that ChatGPT only correctly annotates 71.2% non-cyberbully posts and 28.8% non-cyberbullying posts are misidentified as cyberbully. To summarize, *ChatGPT is better at annotating cyberbully posts (using our prompt) when compared against other posts containing non-cyberbully expressions. However, ChatGPT often misclassifies non-cyberbullying posts as cyberbully.*

Rank 4th: Russo-Ukrainian Stance. For the Russo-Ukrainian Stance detection dataset, ChatGPT's overall performance ranks *fourth* out of seven tasks, with a weighted F1-score of 76.26%. Note, the invasion of Ukraine took place after the training date of our ChatGPT version. Overall, ChatGPT correctly annotates 75.85% of tweets' stance. We find that, for the pro-Ukraine tweets, ChatGPT reports a high precision of 90.02%, with a low recall of 69.35%. In contrast, for the pro-Russia tweets, ChatGPT reports a high recall of 86.91%, but with a low precision of 62.50%. This means that, when ChatGPT annotates pro-Ukraine tweets, it is usually correct (high precision), but the same is not true for annotating a tweet pro-Russia (low precision). The corresponding confusion matrix (Fig. 1(e)) shows that 30.6% pro-Ukraine tweets are mis-annotated as pro-Russia by ChatGPT. In summary, *when ChatGPT labels a tweet as expressing a pro-Ukraine stance, it is usually correct.* Yet, this comes at the cost of a low recall rate. In contrast, *ChatGPT has more false positives when labeling tweets as pro-Russian, but does have a higher recall.*

Rank 5th: Social Bots. For the Social Bot dataset, ChatGPT's overall annotation performance ranks *fifth* out of seven annotation datasets, with a weighted F1-score of 63.70%. Overall, ChatGPT correctly annotates 64.96% of tweets as posted by humans or bots. For precision, ChatGPT only attains 64.16% for human tweets and 66.75% for bot tweets. This suggests ChatGPT's ability to reproduce precise annotations is still limited on social bot detection, i.e., ChatGPT often reports false positives for both human and bot tweets. For recall, ChatGPT attains 81.12% for human tweets but only 45.54% for bot tweets. Such a significant difference implies that, while ChatGPT manages to identify most human tweets, it often fails to distinguish bot tweets from human ones. The corresponding confusion matrix supports this by showing that 54.5% of bot tweets are mis-annotated, compared to only 18.9% human tweets. In summary, *ChatGPT has the potential to act as an annotator to distinguish most human content from bot content. Yet, ChatGPT's capability is limited when identifying content generated by social bots. In our case, ChatGPT often accidentally annotates social bot tweets as human-generated. As a result, its annotation for human tweets can involves lots of false positives.*

Rank 6th: Vaccine Stance. For the Vaccine Stance dataset, ChatGPT's overall performance ranks *sixth* out of seven datasets, with a weighted F1-score of 59.17%. Overall, ChatGPT correctly annotates 59.81% of tweets' stance towards COVID-19 vaccine. ChatGPT is more precise when detecting tweets expressing anti-vaccine (precision = 80.13%) and pro-vaccine stance (precision = 74.03%), compared to tweets with neutral expression (precision = 50.25%). In the meantime, ChatGPT's recall varies across these three labels. While ChatGPT achieves a high recall of 80.99% for neutral tweets, it attains only 39.65% for anti-vaccine tweets and 50.02% for pro-vaccine tweets. This indicates that ChatGPT is conservative when annotating tweets' stance towards COVID-19 vaccine and often mis-labels tweets' stance as neutral. According to the corresponding confusion matrix (Fig. 1(a)), ChatGPT mis-annotates 57.5% of anti-vaccine tweets and 47.2% pro-vaccine tweets as neutral tweets. Therefore, *ChatGPT performs poorly in labeling tweets' stance on the COVID-19 vaccine. This is because many anti-vaccine or pro-vaccine tweets are mislabeled as neutral by ChatGPT.*

Rank 7th: COVID-19 Hate Speech. For the COVID-19 Hate Speech dataset, ChatGPT's performance ranks *seventh* out of seven, with a weighted F1-score of 51.88%. Overall, ChatGPT correctly annotates 52.24% of tweets as expressing or countering anti-Asia hate. Specifically, for hate speech, ChatGPT attains a F1-score of 62.25% with a high recall of 88.27%. For counterhate speech, ChatGPT attains a very low F1-score of 32.54%, with a recall of 33.33%. This suggests that ChatGPT's performance varies when annotating hate and counterhate content on COVID-19. Importantly, ChatGPT's precision for all three labels is lower than 50%, confirming its annotation for each label involves many false positives. As highlighted by the corresponding confusion matrix (Fig. 1(b)), ChatGPT mis-labels 27.1% of neutral tweets as hate tweets, and 24.9% neutral tweets as counterhate tweets. Meanwhile, it mis-annotates 58.1% of counterhate tweets as neutral. In summary, *ChatGPT's annotations are inaccurate for the COVID-19 Hate Speech task. ChatGPT often mis-annotates neutral content as hate speech, and fails to distinguish counterhate speech from neutral content.*

5 Conclusion and Discussion

This study has investigated ChatGPT's potential to act as a data annotator for social computing data. We have compared ChatGPT's performance on seven datasets against the original human annotations. ChatGPT is often able to reproduce human labels, achieving an average F1-score of 72.00%. That said, we discover significant variations in its performance across different domains.

Our work has a number of limitations, which form the basis of our future work. *First*, this study only examines ChatGPT's annotation performance with a small number of datasets covering five social issues. We would like to further inspect ChatGPT's potential as a annotator on other domains (e.g., named-entity recognition), or on other social problems (e.g., political polarization). *Second*, we acknowledge that we only use a single prompt for annotation, and there are many variants that could be experimented with. Prompt design is a major theme of future work, which we believe we yield better results. We are keen to explore how state-of-art prompt-tuning methods, like few-shot prompting and Chain-of-Thought technique [21], can assist with prompt formulation.

References

1. Aiyappa, R., An, J., Kwak, H., Ahn, Y.Y.: Can we trust the evaluation on chatgpt? arXiv preprint arXiv:2303.12767 (2023)
2. Aydın, Ö., Karaarslan, E.: Openai chatgpt generated literature review: Digital twin in healthcare. Available at SSRN 4308687 (2022)
3. Bang, Y., et al.: A multitask, multilingual, multimodal evaluation of chatgpt on reasoning, hallucination, and interactivity. arXiv preprint arXiv:2302.04023 (2023)
4. Chakraborty, A., Paranjape, B., Kakarla, S., Ganguly, N.: Stop clickbait: Detecting and preventing clickbaits in online news media. In: 2016 IEEE/ACM International Conference on Advances in Social Networks Analysis and Mining (ASONAM), pp. 9–16. IEEE (2016)
5. Díaz, M., et al.: Crowdworksheets: Accounting for individual and collective identities underlying crowdsourced dataset annotation. In: Proceedings of the 2022 ACM Conference on Fairness, Accountability, and Transparency, pp. 2342–2351 (2022)
6. ElSherief, M., et al.: Latent hatred: A benchmark for understanding implicit hate speech. arXiv preprint arXiv:2109.05322 (2021)
7. Fagni, T., Falchi, F., Gambini, M., Martella, A., Tesconi, M.: Tweepfake: About detecting deepfake tweets. PLoS ONE **16**(5), e0251415 (2021)
8. Glandt, K., Khanal, S., Li, Y., Caragea, D., Caragea, C.: Stance detection in covid-19 tweets. In: Proceedings of the 59th Annual Meeting of the Association for Computational Linguistics and the 11th International Joint Conference on Natural Language Processing (Long Papers), vol. 1 (2021)

9. Haq, E.U., Lu, Y.K., Hui, P.: It's all relative! a method to counter human bias in crowdsourced stance detection of news articles **6**(CSCW2) (nov 2022). https://doi.org/10.1145/3555636. https://doi.org/10.1145/3555636
10. Haq, E.U., Tyson, G., Lee, L.H., Braud, T., Hui, P.: Twitter dataset for 2022 russo-ukrainian crisis. arXiv preprint arXiv:2203.02955 (2022)
11. He, B., Ziems, C., Soni, S., Ramakrishnan, N., Yang, D., Kumar, S.: Racism is a virus: Anti-asian hate and counterspeech in social media during the covid-19 crisis. In: Proceedings of the 2021 IEEE/ACM International Conference on Advances in Social Networks Analysis and Mining, pp. 90–94 (2021)
12. Huang, F., Kwak, H., An, J.: Is chatgpt better than human annotators? potential and limitations of chatgpt in explaining implicit hate speech. arXiv preprint arXiv:2302.07736 (2023)
13. Kennedy, C.J., Bacon, G., Sahn, A., von Vacano, C.: Constructing interval variables via faceted rasch measurement and multitask deep learning: a hate speech application. arXiv preprint arXiv:2009.10277 (2020)
14. Li, Y., Sosea, T., Sawant, A., Nair, A.J., Inkpen, D., Caragea, C.: P-stance: A large dataset for stance detection in political domain. In: Findings of the Association for Computational Linguistics: ACL-IJCNLP 2021, pp. 2355–2365 (2021)
15. Mathew, B., et al.: Thou shalt not hate: Countering online hate speech. In: Proceedings of the International AAAI Conference on Web and Social Media, vol. 13, pp. 369–380 (2019)
16. Mohammad, S., Kiritchenko, S., Sobhani, P., Zhu, X., Cherry, C.: Semeval-2016 task 6: Detecting stance in tweets. In: Proceedings of the 10th International Workshop on Semantic Evaluation (SemEval-2016), pp. 31–41 (2016)
17. Patwa, P., et al.: Fighting an infodemic: Covid-19 fake news dataset (2020)
18. Peixian, Z., Ehsan-Ul, H., Yiming, Z., Pan, H., Gareth, T.: Echo chambers within the russo-ukrainian war: The role of bipartisan users. In: 2023 IEEE/ACM International Conference on Advances in Social Networks Analysis and Mining (ASONAM) (2023). https://doi.org/10.1145/3625007.3627475
19. Poddar, S., Mondal, M., Misra, J., Ganguly, N., Ghosh, S.: Winds of change: Impact of covid-19 on vaccine-related opinions of twitter users. In: Proceedings of the International AAAI Conference on Web and Social Media. vol. 16, pp. 782–793 (2022)
20. Sallam, M., et al.: Chatgpt output regarding compulsory vaccination and covid-19 vaccine conspiracy: A descriptive study at the outset of a paradigm shift in online search for information. Cureus **15**(2) (2023)
21. Wei, J., Wang, X., Schuurmans, D., Bosma, M., Xia, F., Chi, E., Le, Q.V., Zhou, D., et al.: Chain-of-thought prompting elicits reasoning in large language models. Adv. Neural. Inf. Process. Syst. **35**, 24824–24837 (2022)
22. Zhang, B., Ding, D., Jing, L.: How would stance detection techniques evolve after the launch of chatgpt? arXiv preprint arXiv:2212.14548 (2022)
23. Zhu, Y., Haq, E.u., Lee, L.H., Tyson, G., Hui, P.: A reddit dataset for the russo-ukrainian conflict in 2022. arXiv preprint arXiv:2206.05107 (2022)
24. Zhu, Y., Yin, Z., Tyson, G., Haq, E.U., Lee, L.H., Hui, P.: Apt-pipe: A prompt-tuning tool for social data annotation using chatgpt. In: Proceedings of the ACM on Web Conference 2024, pp. 245–255 (2024)

Fuzzy and Overlapping Communities Detection: An Improved Approach Using Formal Concept Analysis

Martin Waffo Kemgne(✉), Christophe Demko, Karell Bertet, and Jean-Loup Guillaume

L3i (Laboratoire Informatique, Image Et Interaction), La Rochelle Université, La Rochelle, France
{martin.waffo_kemgne,christophe.demko,karell.bertet, jean-loup.guillaume}@univ-lr.fr

Abstract. This paper introduces an innovative method for identifying fuzzy and overlapping communities in graphs using Formal Concept Analysis (FCA). While several initial works have explored the application of FCA to non-overlapping community detection, this paper aims to extend one of these approaches to the detection of overlapping communities. We conduct experiments on various benchmark graphs, including both synthetic and real networks. The performance of the proposed method is evaluated using well-known metrics and compared with other algorithms in the literature.

Keywords: Overlapping communities · Fuzzy communities · FCA · graphs

1 Introduction

In an era marked by an exponential proliferation of data, modeling complex systems through networks or graphs has become fundamental for understanding the underlying structures that govern the dynamics of social, biological, and technological systems. These networks, which model connections between elements or actors, reflect the intrinsic complexity of the systems they represent, ranging from protein interactions to social networks and technological infrastructures.

At the heart of this exploration lies the notion of community structure, a pervasive feature in networks that manifests the tendency of nodes to cluster into groups or communities, where internal links are numerous while inter-community links are scarcer. The study of community structures within complex networks has garnered significant attention in the field of network science, providing deep insights into the organization, dynamics, and function of various real systems [22].

Traditionally, community detection algorithms have sought to partition networks into distinct groups or communities, with the underlying assumption that

L. M. Aiello et al. (Eds.): ASONAM 2024, LNCS 15213, pp. 23–35, 2025.
https://doi.org/10.1007/978-3-031-78548-1_3

each node is a member of only one community [12]. Although useful, this approach often fails to capture the nuanced relationships inherent in many complex networks where nodes can belong to multiple communities, reflecting a more realistic scenario of overlapping or fuzzy community memberships [30].

While Formal Concept Analysis (FCA) [14] has been effective in detecting non-overlapping communities [2,11,13,20], its application to fuzzy and overlapping communities remains unexplored. In this paper, we propose an adapted FCA approach for the detection of fuzzy and overlapping communities. Building upon our previous work, we introduce a refinement step to enhance community detection accuracy. We conduct extensive experiments on various benchmark graphs, including synthetic and real networks, comparing the performance of our method with existing approaches.

The structure of this paper is organized as follows: Sect. 2 provides an overview of previous work on community detection. We discuss existing FCA-based approaches and their limitations in Sect. 3. Our proposed approach and motivation are detailed in Sect. 4. To evaluate the consistency and performance of our method, we present general simulations in Sect. 5. Finally, we conclude in Sect. 6 by summarizing our contributions and outlining future research directions.

2 Related Works

Community structure analysis in networks is an increasingly studied area within network science, focusing on the detection of both overlapping and non-overlapping communities.

2.1 Preliminary Definitions

Let $G = (V, E)$ be a graph with $n = |V|$ and $m = |E|$.

Clique: A clique in a graph is a subset of nodes $M \subseteq V$ where every pair of nodes is interconnected. A clique is maximal (MC) if adding any other node disrupts the clique condition, indicating it is not contained within a larger clique.

Partition: A partition divides the node set V of graph $G = (V, E)$ into disjoint subsets, or communities, where $\bigcup_{i=1}^{k} V_i = V$ and $V_i \cap V_j = \emptyset$ for all distinct i and j. Each node is exclusively in one community.

Fuzzy Partition: Differing from a classical partition, a fuzzy partition permits nodes to belong to multiple communities, represented by a membership degree α_{ic} ranging from 0 to 1. Here, 0 indicates no membership, and 1 denotes full membership. A membership matrix $A = [\alpha_{ic}]$ defines this structure, allowing nodes to exhibit varying affiliations, thus more accurately reflecting complex community dynamics.

2.2 Detection of Non-Overlapping Communities

The detection of non-overlapping communities in networks has been extensively studied, leading to the development of various algorithms. Pioneering methods include Newman and Girvan's algorithm [23] and the widely-used Louvain algorithm by Blondel et al. [4]. Santo Fortunato provided a comprehensive review of over fifty community detection methods [12], highlighting approaches based on modularity optimization, such as Newman and Girvan's [15]. However, optimizing modularity is NP-hard [5], leading to the use of heuristic methods like the Louvain algorithm [4]. Despite their popularity, modularity-based methods suffer from resolution limit issues [12] and other drawbacks [25]. Alternative approaches, such as the Infomap algorithm [27], Label Propagation algorithm [26], and Stochastic Block Models (SBM) [17], offer effective community detection without relying on modularity. For a detailed comparison of methods, Fortunato's review is a valuable resource [12].

Despite all the methods previously mentioned, there exists one that we have not discussed, which is the focus of this paper. These methods rely on the notion of cliques and FCA to detect communities [2,11,13]. These methods, while effective in identifying community structures, overlook the possibility that a node may belong to multiple communities simultaneously. In the following section, we will explore in depth the methods and techniques dedicated to the detection of overlapping communities.

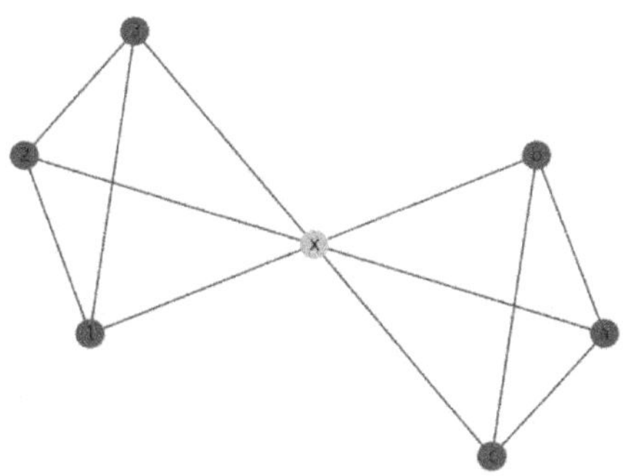

Fig. 1. Graph with Two Fuzzy Communities and a Vertex with Equal Membership Degree to Both Communities

2.3 Detection of Overlapping Communities

While traditional methods have been effective in identifying non-overlapping communities, they often fall short in capturing the intricate interactions found in various domains. For instance, social networks, interdisciplinary research, ecology, and the medical field all exhibit scenarios where entities belong to multiple categories simultaneously. Recognizing these limitations, recent research has shifted towards detecting overlapping communities. Early approaches include those based on the line graph [1,10], as well as methods involving seed extension or contraction [29].

Another method, among the most well-known not involving the *line graph*, is the CPM (*Clique Percolation Method*) algorithm [24]. Several other methods for detecting overlapping communities exist, particularly fuzzy communities [16] where, in addition to being overlapping, a degree of membership to a community is associated with each node. To this end, several modified versions of modularity have been proposed by different authors [21,28] to extend the modularity of Newman and Girvan [15], but the one we will use in this article is that of Shen et al. [28]. In the following section, we will present in detail those methods based on FCA.

3 Detection of Communities Based on FCA

FCA [14] is a method of data analysis and knowledge representation that deals with information in the form of hierarchies of concepts constructed from binary relations. It uses two important formalisms: the notion of a formal context and a formal concept. FCA has been used in several studies for the detection of non-overlapping communities, providing a rigorous framework based on lattice theory to identify groups of elements sharing common attributes [2,11,13]. In FCA, community detection can essentially be summarized as the search for *hidden* attributes shared by certain objects. This section presents the approaches used in [2,11,13]. Historically, Freeman was the pioneer in 1996 in using FCA to analyze social networks and identify communities by focusing on maximal overlapping cliques [13]. However, this approach excluded actors from bridging cliques. Falzon, in 2000, proposed an improvement by including these actors in the analysis without excluding them [11]. Ali in 2014 further developed this idea by incorporating all actors in the network, even those not belonging to the maximal cliques [2]. However, its application to detecting overlapping and fuzzy communities remains underexplored. Our objective is to develop an FCA-based method that overcomes these limitations by revealing hidden overlapping fuzzy communities.

3.1 Preliminary Definitions

Formal Context: A formal context, denoted as $\mathbb{K} = (G, M, I)$, comprises a finite set of objects G, a finite set of attributes M, and an incidence relation $I \subseteq G \times M$ indicating which objects possess which attributes.

Galois Connection Operators: Formal Concept Analysis introduces two operators, $\alpha : 2^G \rightarrow 2^M$ and $\beta : 2^M \rightarrow 2^G$, establishing a Galois connection between the sets 2^G and 2^M. These operators are defined as follows:

- $M \supseteq \alpha(A) = \{a \in M | \forall o \in A, (o, a) \in I\}$ for $A \subseteq G$,
- $G \supseteq \beta(B) = \{o \in G | \forall a \in B, (o, a) \in I\}$ for $B \subseteq M$.

Formal Concept: A formal concept, denoted as (A, B), consists of subsets $A \subseteq G$ and $B \subseteq M$ satisfying $\alpha(A) = B$ and $\beta(B) = A$. Here, A represents the extension and B the intension of the concept.

Formal concepts, ordered by the relation $(A_1, B_1) \geq (A_2, B_2) \Leftrightarrow A_1 \supseteq A_2 \Leftrightarrow B_2 \supseteq B_1$, form a complete lattice.

3.2 Freeman and Falzon's Methods for Community Detection

As mentioned earlier, Falzon's method [11] is an improvement of Freeman's method [13], which takes into account the actors from bridging cliques in the detection phase.

Definition of the Notion of Overlap. We generalize Freeman's [13] definition of overlapping node sets to c-overlapping node sets as follows.

The binary relation *c-overlap*, $o_c \subseteq L_k \times L_k$, is defined as:

(i) $(n_i, n_i) \in o_c$—each node set overlaps with itself, hence o_c is reflexive.
(ii) $|n_i \cap n_j| > c \Rightarrow (n_i, n_j) \in o_c$—two node sets having more than c elements in common overlap, hence o_c is symmetric.
(iii) $(n_i, n_j) \in o_c$ and $(n_j, n_k) \in o_c \Rightarrow (n_i, n_k) \in o_c$—$o_c$ is transitive.

Note that for $c = 0$, the *0-overlap* coincides with *overlap*. It is clear that o_c is an equivalence relation and partitions L_k into subsets (disjoint for $c = 0$) that form the basis of the level k groups, which are the communities.

Principle of Freeman's Method. Consider a graph $G = (V, E)$ and let $C = \{C_1, C_2, \ldots, C_n\}$ be the set of maximal cliques in G that contain at least three vertices. The choice of cliques of size at least three is intuitive, as triangles, which are cliques of size three, exemplify the principle that *the friends of my friends are my friends.* The first step involves constructing the context $\mathbb{K} := (V, C, I)$ where I is the binary relation defined by: for every $x \in V$ and $M \in C$, $I(x, C_i) = 1$ if the node x belongs to the maximal clique C_i and $I(x, C_i) = 0$ otherwise, $\forall i \in \{1, 2, ..., n\}$.

The second step is to construct the concept lattice associated with the above context. Then, to detect communities, Freeman [13] relies on the notion of overlapping cliques in the Galois lattice. He identifies, from the Galois lattice of the previously constructed context, the set of nodes of the lattice where at least two paths leading from their current position in the lattice to the infimum are not of the same length, which he calls bridging cliques. He then proceeds to eliminate the edges departing from these nodes to obtain disjoint groups. These groups represent the communities. To obtain them, it is sufficient to apply the overlap relation on the remaining maximal cliques after the elimination of the bridging cliques; these maximal cliques are at the first level of the lattice. The equivalence classes for this relation represent the communities (by uniting the elements or maximal cliques of the same equivalence class).

However, in large networks, there can often be multiple bridging MCs, and often, by eliminating them, some network actors are also eliminated. This is why Falzon [11] proposed another formalism to address this limit and take into account the actors from bridging cliques.

Principle of Falzon's Method. T The step of constructing the context is similar to that described in subsection 3.2.

The second step involves constructing the concept lattice associated with the previous context. The construction of the concept lattice, is based on the algorithm that Falzon [11] proposed for building concept lattice.

In the final phase, Falzon [11] introduces a community detection method utilizing the properties of maximal cliques in a concept lattice. Her algorithm identifies community structures by generating groups G_k at each lattice level, which represent equivalence classes formed based on overlaps among lattice elements. This approach effectively detects communities, including nodes in bridging cliques, offering an improvement over Freeman's method [13], which excludes such nodes. The method has been applied to several graphs and has yielded good results; see [11] for more details. Despite the various improvements to this method, its application to fuzzy communities remains unexplored. This is the subject of the following section.

4 Proposed Approach

While traditional methods based on FCA, such as Falzon's method [11], have been effective in identifying non-overlapping communities, they are unable to detect overlapping communities. To illustrate this, consider the graph shown in Fig. 1, where it is intuitive that there are at least two communities. Falzon's method, when applied to the same graph, fails to detect these communities, concluding that there is no possible decomposition into communities. Given this observation, our objective is to propose a new method that addresses these shortcomings and is thus capable of detecting overlapping fuzzy communities.

4.1 Principle of the Method

The principle of the method is similar to that of Falzon [11].

Let $G = (V, E)$ be a graph and $C = \{C_1, C_2, ..., C_n\}$ the set of maximal cliques in the graph. The first step involves constructing the context $\mathbb{K} := (V, C, I)$ following the same principle as in subsection 3.2. Thus, the step of constructing the lattice is identical to that described in [11]. The second step is the generation of communities.

First Case: If for all $C_i, C_j \in C$ such that $i \neq j$, $C_i \cap C_j = \emptyset$, then the set of communities is $\pi = C = \{C_1, C_2, ..., C_n\}$.

Second Case: Conversely, if there exists $C_i, C_j \in C$ such that $i \neq j$ and $C_i \cap C_j \neq \emptyset$. First determine the cardinality of each non-empty element of the lattice. Let c be the smallest cardinality; $c \geq 1$ by definition. Let $\mathcal{B}$ be the concept lattice previously constructed.

Then, apply Falzon's algorithm to our lattice $\mathcal{B}$, where the overlap function has been modified by the *c-overlap*. Thus, the communities to be obtained are overlapping. Therefore, to refine the method introduced in our first work, we incorporate the following two small steps.

Assume that we initially detected m communities noted $\pi_1 = \{G_1, G_2, \ldots, G_m\}$.

First, we identify all nodes that have not been classified, i.e., those that do not belong to any maximal clique of size at least 3. We denote this set of unclassified nodes by $U = V \setminus (\bigcup_{i=1}^{m} G_i)$.

Next, for each node $n \in U$, we generate $G(a)$ as the set of $G_i \in \pi_1$ with which a has a link, meaning that a is adjacent to at least one node of G_i.

After that, we identify nodes a that are linked to only one community; this corresponds to nodes for which $G(a)$ contains a single element. If $G(a)$ contains a single element and a is not adjacent to any other unclassified node $b \in U$ with $b \neq a$, then we add node a to the single community in $G(a)$ with which it is linked. Thus, all unclassified nodes that are linked to a single community and are not linked to any other node outside this community are integrated into this community. This reduces the number of unclassified nodes, and we obtain modified new communities $\pi_2 = \{G'_1, G'_2, \ldots, G'_m\}$, where each G'_i corresponds to the community G_i augmented by one or more new nodes.

Secondly, the new unclassified nodes are now $U' = V \setminus (\bigcup_{i=1}^{m} G'_i)$. To augment meaningful communities, we proceed as follows: We identify the connected components of the subgraph G' of G defined by the graph $G' = (V \setminus U, E)$. Connected components of size at least 3 will be considered as new communities, which we note π_3. The final communities will be $\pi = \pi_2 \cup \pi_3$.

To compute the membership degree α_{xK_i} of element x to community K_i, given a family $\pi = \{K_1, K_2, ..., K_m\}$ of communities, use the formula: $\alpha_{xK_i} = \frac{\sum_{y \in K_i} w(x,y)}{\sum_{z \in V} w(x,z)}$. where $w(x, y)$ represents the edge weight between x and y.

Illustrative Examples: Our method identifies two communities in Fig. 1: {{1,2,3,x}, {a,b,c,x}}, unlike Falzon's method, which fails to detect any.

4.2 Advantages of Our Approach

Our approach provides several advantages over existing methods: It detects overlapping communities for deeper insights into network interactions, ensures determinism for consistent and reproducible results, simplifies calculations with a non-parametric framework, and effectively addresses the resolution limit problem by utilizing cliques instead of modularity.

5 General Simulations

In this section, we compare the performance of our *FALO* (Falzon Overlap) method against established techniques for detecting overlapping communities. We selected the following methods due to their recognized effectiveness: *k-clique* [24], an early method in community detection; *SLPA* [31], known for its precision; *DEMON* [7], favored for large networks [8]; *UMSTMO* [3], a recent non-parametric approach; and *Core Expansion (COREE)* [6], which tackles group fragmentation. This selection provides a robust evaluation of *FALO*.

Remark: In our simulations, we choose the cover with the highest extended modularity [28] in our dendrogram as the best option.

5.1 Description of Data

In synthetic network evaluations, we used the LFR benchmark, known for simulating realistic networks [18]. For Real-world Data Sets, We analyzed several diverse real networks to evaluate algorithm performance: **Zachary's Karate Club Network**, **Football Network**, **Dolphin Social Network** and **Jazz Musicians Network**.

5.2 Evaluation Metrics

In our network analysis, we employ three key metrics: extended modularity (EQ) [28], Overlapping Normalized Mutual Information (ONMI) [19], and F1-score[9]. ONMI and F1-score necessitate ground truth data and are thus utilized solely for synthetic networks.

5.3 Performance Evaluation

For these experiments, the algorithms k-clique, DEMON, SLPA, UMSTMO, and COREE were sourced from the Python library cdlib, and the code for FALO can be found on our GitLab repository at https://gitlab.univ-lr.fr/mwaffo01/falo-algorithm-for-overllaping-community-detection. We used the default values for all the algorithms that offer adjustable parameters. The average result was computed based on 10 independent runs. All experiments were performed on a desktop computer equipped with a 12th generation Intel(R) Core(TM) i9-12900H processor at 2.50 GHz and 32 GB of RAM.

Evaluations on Synthetic Networks. In evaluations on synthetic networks using the LFR benchmark, we set standard parameters following Lancichinetti et al. [18]. These parameters include: $n = 1000$ nodes, average node degree $\hat{k} = 20$, maximum degree $k_{\max} = 50$, minimum community size $c_{\min} = 20$, maximum community size $c_{\max} = 100$, power law exponents $\tau_1 = -2$ and $\tau_2 = -1$, number of overlapping community memberships $Om = 2$, and number of overlapping nodes $On = 200$ with mixing parameter $\mu = 0.1$. To comprehensively assess algorithm behavior, we varied parameters μ, Om, On, and $\hat{k}$. Furthermore, considering that the plots associated with the three metrics exhibit similar patterns for each parameter, we have chosen to include only one metric for each parameter due to space constraints, with the exception of μ, for which images of all associated metrics are presented. Nonetheless, we provide conclusions related to all metrics for each parameter, even if the associated plots are not included.

Effect of μ (mixing parameter): In our study, μ ranged from 0.1 to 0.8, with other parameters fixed as in Sect. 5.3. As shown in Fig. 2, all detection methods' performance declined with increasing μ, indicating the growing difficulty of community detection with higher mixing. FALO and SLPA exhibited remarkable stability and outperformed other algorithms for all the metrics.

Effect of Om (number of memberships of overlapping nodes): We varied Om from 2 to 8 while keeping other parameters constant. As depicted in Fig. 3a, most algorithms exhibited declining performance with increasing Om, reflecting the increased complexity introduced by multiple memberships. But in summary, SLPA demonstrated robustness across all levels of membership overlap, while FALO showed strong early performance, albeit with diminishing returns as Om increased.

Effect of On (number of overlapping nodes): We studied the impact of On, by varying it from 200 to 800. FALO and SLPA demonstrated notable performance across all metrics, while DEMON and COREE performed comparably, as depicted in Fig. 3b.

Effect of $\hat{k}$ (Average Degree): We explored how varying ($\hat{k}$) from 10 to 25 impacts algorithm performance. Overall, FALO and SLPA emerged as the most effective algorithms, especially in networks with varying densities, while UMSTMO consistently showed lower performance; from the Fig. 3c.

FALO vs SLPA: Previous simulations have demonstrated the effectiveness of both *FALO* and *SLPA*, with *SLPA* showing slight superiority. However, considering that *SLPA* is parametric, depending on two parameters, r and t, and we have used the default parameters for general simulations, it is pertinent to question what would happen if we vary these parameters. To investigate this, we adjusted t from 5 to 25. The analysis from Figs. 4 indicates scenarios where *FALO* outperforms *SLPA* under certain parameterizations, specifically when $5 \leq t \leq 15$.

Evaluations on Real Networks. We assessed *FALO*'s performance against other algorithms on five real-world networks, using the EQ metric for effectiveness in the absence of ground truth data.

The Table 1 summarizes the performance of different algorithms on five real-world datasets, measured using the EQ metric. *FALO* and *SLPA* consistently demonstrate strong performance across the datasets, with *SLPA* achieving the highest EQ score in the Football, Polbooks and Jazz datasets. In contrast, *FALO* achieves the highest EQ score in the Karate and Dolphins indicating its effective-

Table 1. Performance of algorithms on real-world data sets. Values in **bold** indicate the best performance, while values in *italic* indicate the second best performance.

	FALO	Kclique	DEMON	SLPA	UMSTMO	COREE
Karate	**0.306**	0.185	0.145	0.175	0.106	*0.302*
Football	0.420	0.190	0.285	**0.561**	0.074	*0.436*
Dolphins	**0.405**	0.361	0.153	*0.393*	0.039	0.290
Polbooks	*0.441*	0.436	0.113	**0.467**	0.081	0.292
Jazz	*0.282*	0.002	0.060	**0.326**	0.000	0.183

ness in identifying overlapping communities. Overall, *FALO* and *SLPA* emerge as top-performing algorithms across the datasets.

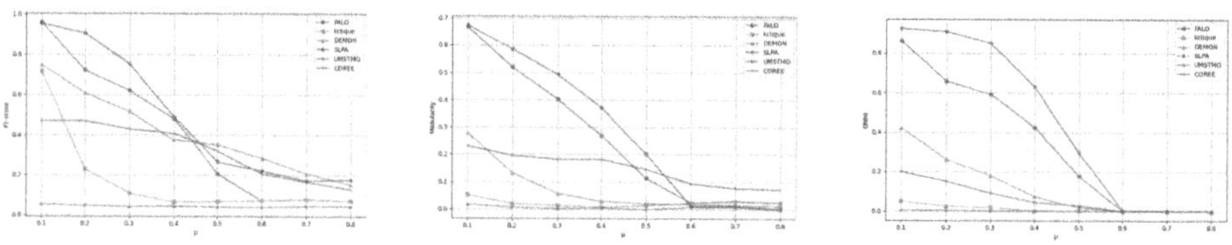

Fig. 2. Comparative Analysis of Algorithm Performance on LFR Networks with Parameter μ Varying from 0.1 to 0.8

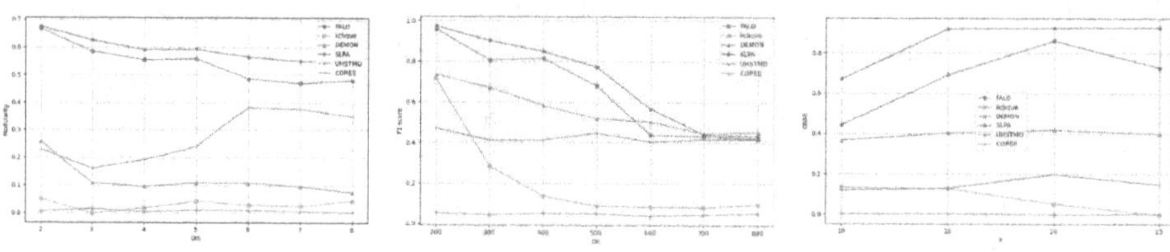

(a) Modularity for Om (b) F1-Score for On (c) NMI for $\hat{k}$

Fig. 3. Comparative Analysis of Algorithm Performance on LFR benchmark with parameter O_m Varying from 2 to 8, O_n from 200 to 800, and $\hat{k}$ from 10 to 25.

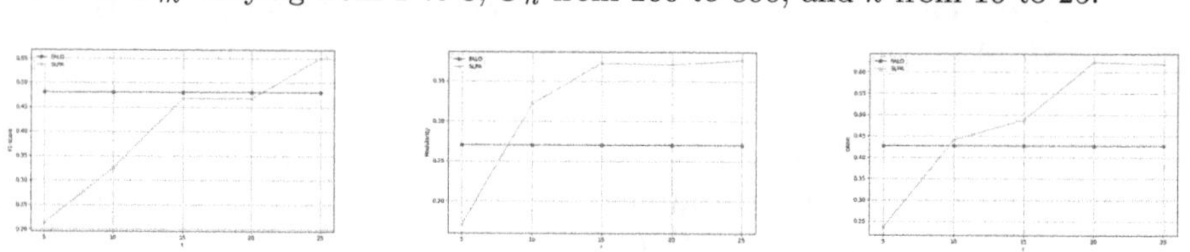

Fig. 4. Comparative Analysis between FALO and SLPA, with Parameter t Varying from 5 to 25

6 Conclusion

This paper introduces a novel method leveraging FCA for detecting fuzzy and overlapping communities in complex networks. The approach, validated through experiments on synthetic and real-world networks, proves its effectiveness. Our method offers enhanced capabilities over traditional approaches by accommodating overlapping communities. This research contributes to a deeper comprehension of community structures, with potential applications in various domains, including artificial intelligence for tasks like clustering.

Future research directions include addressing the challenge of assigning nodes not belonging to any maximal clique to communities systematically. Additionally, further analysis of the algorithmic complexity is warranted for comprehensive insights.

Acknowledgments. This research work is supported by the National Research Agency through the project "SmartFCA - ANR-21-CE23-0023: Formal Concept Analysis : an intelligent tool for the analysis of complex data."

References

1. Ahn, Y.Y., Bagrow, J.P., Lehmann, S.: Link communities reveal multiscale complexity in networks. Nature **466**(7307), 761–764 (2010)
2. Ali, S.S., Bentayeb, F., Missaoui, R., Boussaid, O.: An efficient method for community detection based on formal concept analysis. In: Foundations of Intelligent Systems: 21st International Symposium, ISMIS 2014, Roskilde, Denmark, June 25-27, 2014. Proceedings 21,D pp. 61–72. Springer (2014). https://doi.org/10.1007/978-3-319-08326-1_7
3. Asmi, K., Lotfi, D., El Marraki, M.: Overlapping community detection based on the union of all maximum spanning trees. Library Hi Tech **38**(2), 276–292 (2020)
4. Blondel, V.D., Guillaume, J.L., Lambiotte, R., Lefebvre, E.: Fast unfolding of communities in large networks. J. Stat. Mech: Theory Exp. **2008**(10), P10008 (2008)
5. Brandes, U., Delling, D., Gaertler, M., Görke, R., Hoefer, M., Nikoloski, Z., Wagner, D.: On finding graph clusterings with maximum modularity. In: Graph-Theoretic Concepts in Computer Science: 33rd International Workshop, WG 2007, Dornburg, Germany, June 21-23, 2007. Revised Papers 33, pp. 121–132. Springer (2007). https://doi.org/10.1007/978-3-540-74839-7_12
6. Choumane, A., Awada, A., Harkous, A.: Core expansion: a new community detection algorithm based on neighborhood overlap. Soc. Netw. Anal. Min. **10**, 1–11 (2020)
7. Coscia, M., Rossetti, G., Giannotti, F., Pedreschi, D.: Demon: a local-first discovery method for overlapping communities. In: Proceedings of the 18th ACM Sigkdd International Conference on Knowledge Discovery and Data Mining, pp. 615–623 (2012)
8. Coscia, M., Rossetti, G., Giannotti, F., Pedreschi, D.: Uncovering hierarchical and overlapping communities with a local-first approach. ACM Transact. Knowl. Discov. Data (TKDD) **9**(1), 1–27 (2014)

9. Epasto, A., Lattanzi, S., Paes Leme, R.: Ego-splitting framework: From non-overlapping to overlapping clusters. In: Proceedings of the 23rd ACM SIGKDD International Conference on Knowledge Discovery and Data Mining, pp. 145–154 (2017)
10. Evans, T.S., Lambiotte, R.: Line graphs, link partitions, and overlapping communities. Phys. Rev. E **80**(1), 016105 (2009)
11. Falzon, L.: Determining groups from the clique structure in large social networks. Social Netw. **22**(2), 159–172 (2000)
12. Fortunato, S.: Community detection in graphs. Phys. Rep. **486**(3–5), 75–174 (2010)
13. Freeman, L.C.: Cliques, galois lattices, and the structure of human social groups. Social Netw. **18**(3), 173–187 (1996)
14. Ganter, B., Wille, R.: Formal concept analysis: mathematical foundations. Springer Science & Business Media (2012). https://doi.org/10.1007/978-3-642-59830-2
15. Girvan, M., Newman, M.E.: Community structure in social and biological networks. Proc. Natl. Acad. Sci. **99**(12), 7821–7826 (2002)
16. Gregory, S.: Fuzzy overlapping communities in networks. J. Stat. Mech: Theory Exp. **2011**(02), P02017 (2011)
17. Holland, P.W., Laskey, K.B., Leinhardt, S.: Stochastic blockmodels: First steps. Social Netw. **5**(2), 109–137 (1983)
18. Lancichinetti, A., Fortunato, S., Radicchi, F.: Benchmark graphs for testing community detection algorithms. Phys. Rev. E **78**(4), 046110 (2008)
19. McDaid, A.F., Greene, D., Hurley, N.: Normalized mutual information to evaluate overlapping community finding algorithms. arXiv preprint arXiv:1110.2515 (2011)
20. Missaoui, R., Messaoudi, A., Ibrahim, M.H., Abdessalem, T.: Detecting communities in complex networks using formal concept analysis. In: Advances in Knowledge Discovery and Management: Vol. 9, pp. 77–105. Springer (2022). https://doi.org/10.1007/978-3-030-90287-2_5
21. Nepusz, T., Petróczi, A., Négyessy, L., Bazsó, F.: Fuzzy communities and the concept of bridgeness in complex networks. Phys. Rev. E **77**(1), 016107 (2008)
22. Newman, M.E.: The structure and function of complex networks. SIAM Rev. **45**(2), 167–256 (2003)
23. Newman, M.E., Girvan, M.: Finding and evaluating community structure in networks. Phys. Rev. E **69**(2), 026113 (2004)
24. Palla, G., Derényi, I., Farkas, I., Vicsek, T.: Uncovering the overlapping community structure of complex networks in nature and society. Nature **435**(7043), 814–818 (2005)
25. Peixoto, T.P.: Hierarchical block structures and high-resolution model selection in large networks. Phys. Rev. X **4**(1), 011047 (2014)
26. Raghavan, U.N., Albert, R., Kumara, S.: Near linear time algorithm to detect community structures in large-scale networks. Phys. Rev. E **76**(3), 036106 (2007)
27. Rosvall, M., Bergstrom, C.T.: Maps of random walks on complex networks reveal community structure. Proc. Natl. Acad. Sci. **105**(4), 1118–1123 (2008)
28. Shen, H., Cheng, X., Cai, K., Hu, M.B.: Detect overlapping and hierarchical community structure in networks. Physica A **388**(8), 1706–1712 (2009)
29. Wang, X., Jiao, L., Wu, J.: Adjusting from disjoint to overlapping community detection of complex networks. Phys. A **388**(24), 5045–5056 (2009)

30. Xie, J., Kelley, S., Szymanski, B.K.: Overlapping community detection in networks: the state-of-the-art and comparative study. ACM Comput. Surveys (csur) **45**(4), 1–35 (2013)
31. Xie, J., Szymanski, B.K., Liu, X.: Slpa: Uncovering overlapping communities in social networks via a speaker-listener interaction dynamic process. In: 2011 IEEE 11th International Conference on Data Mining Workshops, pp. 344–349. IEEE (2011)

Understanding Engagement Dynamics with (Un)Reliable News Publishers on Twitter

Alireza Mohammadinodooshan(✉) and Niklas Carlsson

Linköping University, Linköping, Sweden
{alireza.mohammadinodooshan,niklas.carlsson}@liu.se

Abstract. According to the Pew Research Center, a majority of X (formerly Twitter) users in the U.S. (55%) regularly consume news through the platform, exceeding the ratio for all other major social media platforms. Still, the current literature falls short in providing insights into the relative interaction patterns seen for different classes of news on this platform. To address this gap, this study provides a large-scale analysis of user interactions with different news classes, emphasizing both the bias and reliability of the publishers. To this end, we have compiled a robust dataset comprising more than 75 million tweets posted over 56 months by 2,041 labeled U.S. news publishers. Using this dataset, we study the engagement patterns across news categories, identifying several statistically significant variances. Understanding these dynamics is crucial for developing informed strategies for news dissemination, audience targeting, and content moderation. Accordingly, this study offers data-driven insights to support such strategy development.

1 Introduction

Social media have transformed people from passive consumers to active participants in news dissemination [12]. In fact, through their interactions (i.e., likes, shares, and comments), users play a key role in the successful dissemination of news, including unreliable content, making it crucial to understand these interaction patterns. While prior research has extensively studied user interaction patterns with different news publishers on Facebook [4] and Reddit [20], the dynamics of these interactions on X remain relatively underexplored; despite a higher portion of X's U.S. user base (55%) regularly accessing news via its service than seen on competing services (e.g., Facebook (43%), Reddit (38%) [19]).

To address this gap, we present a large-scale analysis of user interactions with a wide range of U.S. news publishers (outlets) on X that addresses the following research question (**RQ**): *How do user engagement patterns on X differ across news publishers from varying bias and reliability classes?*

Specifically, we gathered and analyzed a dataset containing 75 million tweets from 2,041 publishers, spanning 56 months (June 2018 to Feb. 2023), along with the 5.2 billion interactions associated with these tweets. Like some other recent

L. M. Aiello et al. (Eds.): ASONAM 2024, LNCS 15213, pp. 36–47, 2025.
https://doi.org/10.1007/978-3-031-78548-1_4

studies (e.g., [4]), we categorize U.S. publishers based on both their bias ("Left", "Left-Center", "Least Biased", "Right-Center", and "Right") and reliability ("Unreliable" or "Reliable"). Using the large labeled dataset, we then provide a comprehensive study into how users engage with more or less reliable news across the political spectrum. Here, we explore key factors that shape user engagement, including the interplay between a publisher's bias, reliability, followership, content views, and different interaction types. We next outline our main contributions.

First, the paper provides insights into differences in the level of user engagement across content types, both at an aggregate level (Sect. 4.2) and when statistically analyzed (Sect. 4.3). This analysis clarifies which news publishers attract more interactions, the patterns in content volume, and the influence of followership size on publishers' success dynamics. As a concrete example, our empirical insights reveal that across all biased publisher categories, *Unreliable* publishers consistently exhibit higher interactions compared to their *Reliable* counterparts. However, a deeper examination of the follower base nuances refines these findings, emphasizing that only *Unreliable* publishers from the most extreme bias categories outperform *Reliable* counterparts in terms of engagement.

Second, by making comparisons based on what type of interactions (i.e., likes, retweets, replies, and quotes) users most engage in with different content, we provide insights into subtle differences in how users interact with different outlets (Sect. 6). A noteworthy observation is that right-leaning publishers are more effective in prompting dialogue-based interactions (i.e., replies) compared to other publishers. Moreover, users tend to engage in shallow interactions like "liking" with *Unreliable* publishers, whereas deeper forms of engagement, such as quoting and replying, are more likely to occur with *Reliable* news sources.

Third, by leveraging that X provided research access to view counts (impressions) from Dec. 2022, the paper presents a pioneering analysis of impressions and engagement rates (i.e., the portion of impressions translating into user interactions) across news classes (Sect. 7). Our analysis here shows that *Unreliable* publishers, regardless of bias, excel in capturing user attention per view. While the first two dimensions of our study complement prior research (done on other platforms than X), this third dimension introduces the first analysis of its kind.

While our target audience primarily is other researchers, we note that our findings also may help publishers refine their strategies, assist moderators in safeguarding information integrity, and provide policymakers with valuable input to address the challenges posed by *Unreliable* publishers.

Outline: After discussing the related works (Sect. 2), we present our methodology (Sect. 3), results and discussions (Secs. 4-7), and our conclusions (Sect. 8).

2 Related Works

This work aligns with works focused on user engagement with news content of varying political biases and levels of reliability. As the most similar work to our research in this group, we can mention [4]. Considering both bias and reliability dimensions and collecting an 8-month panel of 7.5M posts, they try to analyze the users' engagement with different classes of news on Facebook. Similar to what we have recorded for X, their results show that posts from *Unreliable* news outlets receive consistently higher median engagement than *Reliable*. Similar to [4], a study of 416 news organizations on Facebook from 2012 to 2017 found that the most engaging sources often are the most ideologically extreme [8], a trend that part of our results confirmed on X. Weld et al. [20] also extended this line of inquiry to Reddit by examining the effects of bias and reliability there.

In terms of reliability, the seminal work by Vosoughi et al. [18] also stands out, demonstrating the faster and broader spread of unreliable news compared to the truth. However, during the COVID-19 pandemic, Altay et al. [3] observed an upsurge in news consumption, with credible outlets experiencing increased traffic. On the other hand, and investigating the impact of bias, users' bias towards sharing right-leaning news on X has been reported by [6].

When it comes to X, separate from the influence of political bias and reliability, some researchers have explored user engagement with news publishers more broadly. One line of research here is Aldous et al.'s research [1,2], whose studies have a limited scope in terms of news publishers studied (8 and 53 news publishers, respectively). Another related work here is [5], where the authors study the user interactions with 232 news sources through 432K tweets, explicitly mentioning or retweeting those sources.

Our research differentiates itself from prior studies by providing the largest scale study (2,041 news sources, more than 75M tweets, and 5.2B interactions) on X, conducting a combined analysis of political bias and reliability, and providing a pioneering analysis of impressions and engagement rates for different news classes. Finally, we should mention that while our focus is on the U.S. news ecosystem, studies like [17] have considered other news ecosystems.

3 Methodology

Dataset Compilation: The dataset used for our analysis was created using a multi-step approach involving six primary steps. Our study commenced with compiling a comprehensive list of U.S. news publishers sourced from Media Bias Fact Check (MBFC)[10], an independent organization that evaluates the bias and reliability of media sources.

Second, using MBFC's labels and bias-meter icons assigned to each publisher, we identified the bias class for each publisher. For publishers missing one of the five labels of interest ("Left", "Left-Center", "Least Biased", "Right-Center", and "Right"), we utilized Robertson et al.'s bias scores [14] to train five kernel density estimators and determine the bias class of those publishers.

Third, we focused on classifying publishers as either *Unreliable* or *Reliable*, using the iffy index (https://iffy.news), which relies on MBFC labels and is widely employed in previous works [7]. Fourth, we identified and linked the publishers to their corresponding X accounts, mainly by visiting their websites.

Fifth, we collected all tweets posted by these accounts, along with associated user interactions (likes, retweets, quotes, and replies) from June 2018 to the end

Table 1. Summary statistics categorized by bias and reliability type. Here (B = Billion, M = Million, K = Kilo) and (Rel. = Reliable, Unrel. = Unreliable)

Bias	Left		Left-Center		Least Biased		Right-Center		Right		Total
Reliability	Rel	Unrel	Rel	Unrel	Rel	Unrel	Rel	Unrel	Rel	Unrel	
Outlets	227	15	519	10	773	9	235	9	118	126	2,041
Followers	177.0M	19.8M	437.1M	13.8M	117.4M	1.1M	75.1M	13.7M	16.7M	73.4M	945.0M
Tweets	5.9M	637.9K	19.8M	134.7K	33.3M	118.4K	9.9M	367.7K	2.6M	2.3M	75.0M
Interactions	1.2B	287.2M	1.6B	101.2M	475.8M	6.1M	233.1M	92.4M	314.6M	938.2M	5.2B

of Jan. 2023. We refer to this collection of data as the "Interactions dataset." For tweets published after Dec. 15, 2022, we were also able to gather impressions data, which we call the "Impressions dataset." While API access limitations prevented complete data collection of 97 accounts, we were able to complete the collection for 2,041 accounts. By prioritizing statistical analysis over aggregated analysis, we mitigate the impact of this limitation on our overarching conclusions.

Finally, our analysis required the follower count of publishers' Twitter accounts, including how these numbers had changed over time. We here utilized the Internet Archive's Wayback Machine to manually obtain historical snapshots and reconstruct a timeline of the follower growth for each account.

Statistical Tests: While the ANOVA family of tests at first might seem fitting for our purposes, they face challenges due to the heavy-tailed distributions under study (and non-normal errors also after log-transform) and correlations between bias and reliability factors. We thus used the non-parametric Kruskal-Wallis test to compare the medians and overall distributions across groups without assuming normality. In our analysis, p-values below 0.01 are considered significant. For similar reasons, Spearman's rank is used for correlation analysis.

4 High-Level Category Comparisons

4.1 Dataset Overview

Table 1 presents an overview of our dataset, categorized by bias and reliability dimensions. For each bias and reliability category (columns), the table shows (split over four rows): (1) the number of outlets in that category, (2) the combined number of followers as of Feb. 2023, (3) the aggregate number of tweets in the dataset, and (4) the combined sum of all interaction types in the dataset, including likes, retweets, quotes, and replies. In total (last column in the table), our dataset comprises 5.2B interactions from 75M tweets across 2,041 news outlets, engaging a combined following of 945M users. This extensive engagement underscores the significant reach and influence of news outlets on X.

4.2 Aggregate Category Comparison

Focusing first on the *Reliable* publishers, a distinct pattern unfolds: outlets affiliated with the left party (aggregating *Left* and *Left-Center*) have the largest fol-

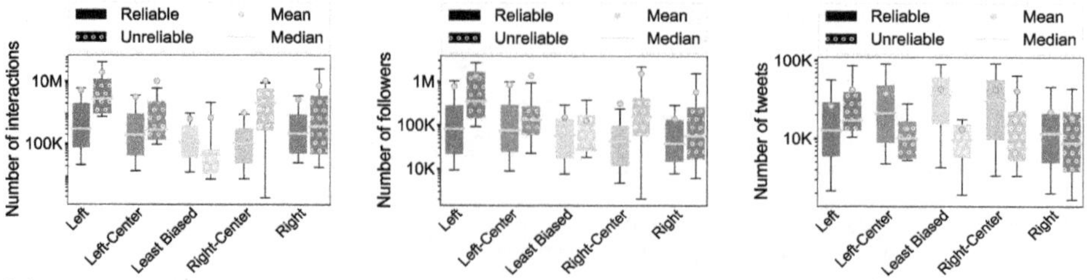

(a) Number of interactions (b) Number of followers (c) Number of tweets

Fig. 1. Distributions comparisons across different outlets (interactions dataset).

lower base at 614.1M (177M+437.1M), with a total of 2.8B (1.2B+1.6BM) interactions. Interestingly, the numerous *Least Biased* outlets do not exhibit the same level of engagement, indicating a user preference for engaging with outlets expressing a clear bias rather than those perceived as neutral. This inference is reinforced when examining the average interactions per outlet: 3.8M/outlet for left-party (*Left* and *Left-Center*), 0.6M/outlet for *Least Biased*, and 1.5M/outlet for right-party (*Right* and *Right-Center*), confirming that the same trend holds for the right-party-affiliated outlets. In addition to seeing the lowest average interaction rate per outlet, the *Least Biased* group is noteworthy for its substantial disparity in the number of *Unreliable* versus *Reliable* publishers (9 vs. 773).

Furthermore, looking at the interactions per outlet for the *Unreliable* outlets reveals substantial differences between biased classes and *Least Biased*: 15.5M/outlet for the 25 left-party outlets, 0.67M/outlet for *Least Biased*, and 7.6M/outlet for the 135 right-party outlets. These observations imply that neutral publishers may not yield the same influence as biased ones.

Considering the *Unreliable* publishers, the *Right* class outperforms other classes in both outlet count and total engagement, with 126 outlets amassing 73.4M followers and 938.2M interactions. Interestingly, the *Left-Unreliable* publishers though fewer, still attract a significant 287.2M user interactions and more interactions per outlet, indicating stronger individual engagement despite fewer outlets (for both *Reliable* and *Unreliable* spheres). These observations warrant further statistical analysis, which attributes equal weight to each outlet. We next provide such per-publisher-based analysis.

4.3 Per-publisher-based Comparison

To account for variations within each category and more fairly weight the outlets (publishers) within each category, we next compare their relative distribution statistics. Figure 1 shows box plots for the number of (a) interactions, (b) followers, and (c) tweets per outlet across the ten outlet categories. Here, the interquartile range (IQR), shown as a box, spans from the 25th to the 75th percentile, and the whiskers specify the 10th to the 90th percentile. Finally, the median is shown using a horizontal line and the mean using a marker. We note that the y-axis is plotted on a log scale, and the averages are substantially larger than the median values, indicating that the distributions are highly skewed and heavy-tailed.

Higher engagement with *Unreliable*: Comparing the total number of interactions (Fig. 1a), with the exception for the *Least Biased* class, the group of *Unreliable* outlets of each bias sees higher distribution statistics than the corresponding *Reliable* outlets. This suggests that neutral news may be consumed and interacted differently with than biased news.

Biased news more engaging: When considering the bias, biased publishers consistently achieve higher interaction levels compared to the *Least Biased* class. This difference is statistically significant for both the right and left parties, irrespective of reliability labels. Within the *Unreliable* class, it is unexpected that *Right-Center* party outperforms *Right.* Several factors contribute here, including the larger number of publishers in the *Right* class, intensifying competition, while the total user engagement potential in this group is limited. Another explanation for the higher engagement levels of *Left Unreliable* and *Right-Center Unreliable* may be their larger follower base (Fig. 1b), which we will discuss next.

Key Takeaway: For all four biased classes, the *Unreliable* outlets receive significantly more interactions than their *Reliable* counterparts.

Not all followers contribute equally to interactions: Comparing Figs. 1b and 1a reveals a correlation between the two metrics. While accounts with more followers are expected to have greater interactions due to increased visibility, the *Right-Center Unreliable* group shows exceptionally higher interaction levels that surpass follower count expectations. To understand the uneven levels of interactions per follower, Sect. 5 explores normalized interaction levels.

Left-biased outlets delivering *Reliable* content has most followers: In addition to the above correlations, we notice a decreasing trend in followers from left bias to right bias among *Reliable* publishers. This suggests that while inherent variations exist in follower counts across biases; the reliability dimension may not be the primary factor influencing follower counts. Instead, bias plays a more significant role. It is worth noting that while this trend is statistically significant, mean values influenced by outliers do not follow this pattern.

Outlets spreading *Unreliable* content often achieve high engagement with relatively fewer tweets: Examining the number of tweets in Fig. 1c reveals a divergent pattern. Most *Unreliable* classes, excluding *Left*, exhibit notably lower tweet output. Combined with our previous observation that these classes often see higher interaction levels, this implies that for these groups, the tweet frequency does not have a straightforward correlation with the interaction levels. Perhaps the nature of content, selective attention, and audience capacity play more significant roles here.

Higher tweet levels for the *Least Biased Reliable* outlets: For the *Reliable* classes, the *Least Biased* outlets exhibit a notably higher volume of tweets than their more politically biased counterparts. The trend for the *Reliable* classes, that the more biased publishers generate fewer tweets than the less biased groups, is significant among all five groups (all six possible pairwise comparisons matching this statement have p-value < 0.001). Note that in these comparisons, we compared all four biased groups against the *Least Biased* class, and for each party, the more biased one against the more centered one.

The heightened tweet levels of the *Least Biased* outlets publishing *Reliable* content see less user engagement: This phenomenon may be attributed to *Least Biased* outlets attempting to cover a broader spectrum of users and topics, potentially leading to content fatigue among followers. Additionally, the less polarizing nature of reliable and unbiased content may inherently attract less engagement compared to more controversial or partisan material. However, the number of followers also significantly influences interaction levels, potentially obscuring these effects. Therefore, and to provide a refined understanding, in the next section, we examine interaction levels normalized by follower count.

5 Followers and Interactions

The prior section suggested a correlation between the number of followers of an outlet and its interaction levels. To understand this link, we conduct a correlation analysis to quantify its impact. Here, a strong correlation is observed across all classes, quantified by a Spearman coefficient of 0.84 (p-value < 0.001), emphasizing a significant positive relationship between followers and interactions.

Second, we compare the (normalized) distributions of the interactions per follower seen across the outlets. Here, we make an extra effort to (1) consider that the number of followers changed over the measurement period, and (2) give equal weight to each post made during the measurement period. For this purpose, rather than simply calculating the ratio between the total interactions (Fig. 1a) and the number of followers (Fig. 1b), for each outlet, we calculated a weighted per-post-based metric as $\sum_{i=1}^{N} \frac{I_i}{f_i}$, where I_i is the number of interactions received for post i (out of the N posts observed for that outlet), and f_i is the outlet's (estimated) follower count at the time of that tweet post. Note that the above type of weighting is particularly important for our study periods, during which the follower base of many of these outlets has been found to be highly dynamic (especially during the COVID period [11]).

The box plot with this normalized metric (Fig. 2) reveals that some differences between classes diminish when considering per-follower interactions. Notably, the *Right-Center* class shows minimal distinctions between *Unreliable* and *Reliable* classes. While statistically significant median differences exist among *Reliable* classes, they lack a discernible pattern as in raw interaction (Fig. 1a) and follower counts (Fig. 1b). However, a consistent trend persists in *Unreliable* categories, emphasizing that political bias remains a key factor in interactions, even when normalized for follower count. Notably, extreme bias classes (*Left* and *Right*) within *Unreliable* garner higher per-follower interactions, highlighting challenges in content moderation. Despite having a larger audience, these outlets excel at engaging individuals, reflecting a higher interaction commitment per follower. We provide a deeper analysis of this trend in Sect. 7.

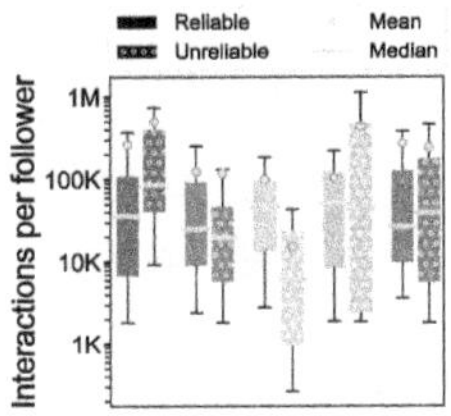

Fig. 2. Interactions per follower

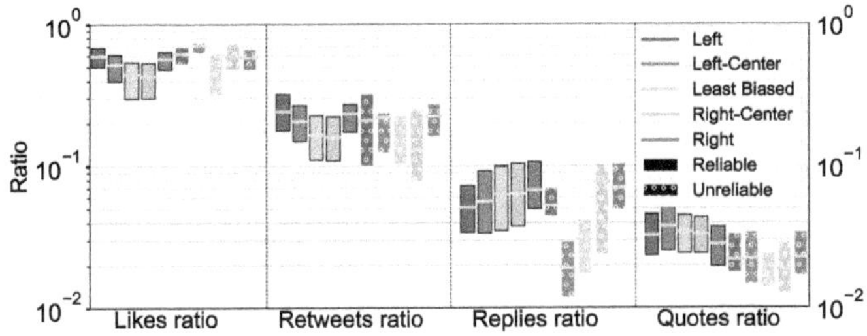

Fig. 3. Distribution of interaction types ratios

Key Takeaway: Within the *Left* and *Right* political extremes, *Unreliable* publishers exhibit statistically significant higher interactions per follower, despite having higher levels of followers.

6 Interaction Types Analysis

Different interaction types (likes, retweets, replies, and quotes) reflect varying levels of user engagement, from simple agreement to active discussion or critique. Motivated by these differences, we present a fine-grained analysis of user interactions across different outlet classes. Figure 3 shows the distribution ratios of interaction types for each category, calculated as the proportion of each interaction type relative to total interactions per publisher. For similar arguments as before, for each publisher, we calculate the mean of these ratios over all its posts (not taking the ratio of the totals) and exclude tweets without any interactions. Here, a preference for less effort-intensive interactions is observed, where we see a logarithmic decrement in the interaction levels seen across various actions: likes (69% of all interactions), retweets (19%), replies (8%), and quotes (4%). With likes dominating interactions, the likes ratio mirrors many trends observed earlier. These patterns are often reflected in retweet ratios as well.

Replies: The replies ratio offers a fresh perspective different than those seen for likes and retweets. In particular, the right-party classes exhibit a higher replies ratio than the left-party classes, suggesting a greater inclination for users to engage in conversations initiated by right-leaning publishers. For the *Reliable* classes, we observe a significant gradient: beginning with the *Left* class, which presents the lowest replies ratio, there is a marked and statistically significant increase through the *Left-Center*, *Least Biased*, and *Right-Center*, culminating with the *Right* class. This trend suggests that as we move toward the right of the political spectrum, user interactions become increasingly interactive. This higher replies ratio of right-party publishers repeats in the *Unreliable* sphere also.

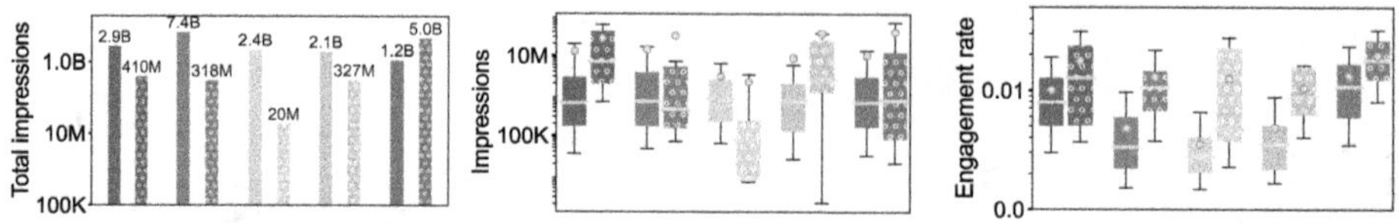

(a) Total impressions (b) Impressions distribution (c) Eng. rate distribution

Fig. 4. Impressions and engagement rates in the impressions dataset.

Quote Usage: As we progress from the simpler interactions of likes and retweets to the more involved replies, the advantage of *Unreliable* over *Reliable* counterparts diminishes, although certain groups deviate from this trend. For quotes, this trend intensifies. Here, *Unreliable* classes consistently exhibit lower ratios across all classes compared to *Reliable* groups. This implies that, despite widespread sharing and liking, *Unreliable* is quoted less frequently, suggesting users may exercise greater selectivity or thoughtfulness when engaging with content requiring additional commentary or context.

> **Key Takeaway:** The replies ratio favors the right party, and the quotes ratio favors *Reliable* publishers.

7 Impressions Study

We next study user engagement rates associated with the different classes. Here, we take advantage of X, having made available impression statistics for the last 48 d of our data collection period (Dec. 15, 2022 - Jan. 31, 2023).

Impressions Statistics: The impressions dataset includes 1.9M tweets, constituting 2.5% of the main dataset. These tweets have amassed 136.6M interactions, with likes making up the majority at 72% (99.0M likes) and an (impressive) total of 22.0B impressions (distributed as per Fig. 4a). In the *Reliable* classes, *Left-Center* dominates with 7.4B views, while in the *Unreliable* classes, *Right* leads with 5.0B impressions, correlating with its higher account count.

Number of Impressions per Outlet: To normalize for class size, we examine the distribution of interactions per outlet for each class in Fig. 4b. Aside from a marginal distinction between *Least Biased* and *Right-Center* classes (p-value=0.05), the *Reliable* classes show no statistically significant differences in impression levels. This suggests a level playing field in content exposure among these classes. In contrast, *Unreliable* groups align with trends observed in the interactions study (Fig. 1a). Biased news outlets generally command higher impressions than the *Least Biased* class. This disparity can be attributed, in part, to the marked differences in interaction levels between *Unreliable* classes, reflected in impression counts. X's algorithm, recently published [16], incorporates interaction metrics as a core component, corroborated by a Spearman correlation coefficient of 0.90 between impressions and interactions, emphasizing the algorithm's reliance on interactions to shape content visibility.

Engagement Rate: Given the above correlation, it becomes pertinent to examine the interaction levels relative to impressions. Following X's [15] and various other scholarly works convention [9], we refer to this metric as the "engagement rate". While this normalized metric has long been argued to be the most important barometer of user engagement [13], it has previously not been possible to be studied on X. We are, therefore, happy to present a head-to-head compassion of the engagement rates seen across the different classes. Figure 4c summarizes these results, allowing us to make several notable observations. First, across all political biases, *Unreliable* groups consistently achieve significantly higher engagement rates compared to their *Reliable* counterparts, amplifying trends observed in interaction numbers with more pronounced deviations. Interestingly, *Unreliable* surpasses the engagement rate of *Reliable* (even within the *Least Biased* class), with all comparative values being statistically significant (p-values < 0.01), except for the *Left* class (p-value $= 0.08$). This draw to interacting with *Unreliable* publishers can be attributed to the provocative nature of the content they share, which often matches their followers' preferences, as well as the more isolated and condensed networks they form.

Conversely, within *Reliable* classes, significant engagement rate differences are observed between classes. This highlights a consistent pattern where polarized classes (*Left* or *Right*) exhibit higher engagement rates than their respective center-aligned classes (*Right-Center* and *Left-Center*), which, in turn, outperform the *Least Biased* class. The same pattern is observed for *Unreliable* classes. This suggests that content with a clear political stance, particularly aligning with the more extreme ends of the spectrum, tends to be more engaging than neutral content, possibly influenced by the journalistic techniques and focused topics employed by these groups.

Key Takeaway: Regardless of political bias and compared to *Reliable* publishers, *Unreliable* publishers consistently achieves higher engagement rates. Furthermore, publishers with more (left- or right-) pronounced political biases receive greater engagement rates.

8 Conclusions

This paper presents a large-scale analysis of interactions between X users and news publishers, offering a comprehensive view of engagement levels and rates' dynamics across various publishers, when considering their bias, reliability, and follower base. As an example, studying the engagement rate (0.62% on average, indicating roughly one interaction per 160 content views), a statistically significant variance among publisher classes was observed. Alarmingly, the class of *Unreliable* publishers consistently achieves higher engagement rates than *Reliable* publishers. More than reliability, we also consider the influence of political bias. For instance, for the *Reliable* class, an ascending trend in reply ratios was observed from left to right-leaning publishers, suggesting that political alignment may significantly affect user engagement, especially in dialogue-centric interactions. The analysis extends beyond the roles of reliability and bias to also consider the follower count's impact on the engagement with different publishers.

In summary, this paper contributes to scholarly and practical discussions on social media engagement, capturing engagement dynamics with diverse news publishers. It offers valuable guidance for publishers, strategists, and policymakers dedicated to fostering a well-informed, critically engaged online audience.

References

1. Aldous, K., An, J., Jansen, B.: View, like, comment, post: Analyzing user engagement by topic at 4 levels across 5 social media platforms for 53 news organizations. In: Proceedings of the ICWSM (2019). https://doi.org/10.1609/icwsm.v13i01.3208
2. Aldous, K.K., An, J., Jansen, B.: Measuring 9 emotions of news posts from 8 news organizations across 4 social media platforms for 8 months. Trans. Soc. Comput. **4** (2022). https://doi.org/10.1145/3516491
3. Altay, S., Kleis Nielsen, R., Fletcher, R.: Quantifying the "infodemic": People turned to trustworthy news outlets during the 2020 coronavirus pandemic. J. Quant. Descrip. Digit. Media. **2** (2022). https://doi.org/10.51685/jqd.2022.020
4. Edelson, L., Nguyen, M.K., Goldstein, I., Goga, O., McCoy, D., Lauinger, T.: Understanding engagement with U.S. (mis)information news sources on Facebook. In: Proceedings of the IMC (2021). https://doi.org/10.1145/3487552.3487859
5. Glenski, M., Weninger, T., Volkova, S.: How humans versus bots react to deceptive and trusted news sources. In: Proceedings of the ASONAM 2018 (2018)
6. González-Bailón, S., d'Andrea, V., Freelon, D., De Domenico, M.: The advantage of the right in social media news sharing. PNAS Nexus **1** (2022)
7. Hanley, H.W.A., Kumar, D., Durumeric, Z.: A golden age: Conspiracy theories' relationship with misinformation outlets, news media, and the wider internet. ACM Hum.-Comput. Interact. **7** (2023). https://doi.org/10.1145/3610043
8. Hiaeshutter-Rice, D., Weeks, B.: Understanding audience engagement with mainstream and alternative news posts on Facebook. Digital Journalism **9** (2021)
9. Hwong, Y.L., Oliver, C., Van Kranendonk, M., Sammut, C., Seroussi, Y.: What makes you tick? the psychology of social media engagement in space science communication. Comput. Human Behav. **68** (2017)
10. Media Bias Fact Check: https://mediabiasfactcheck.com/ (2023)
11. Mohammadinodooshan, A., Carlsson, N.: Temporal dynamics of user engagement with U.S. news sources on Facebook. In: Proceedings of the SNAMS (2022)
12. Mukerjee, S., Yang, T., Peng, Y.: Metrics in action: how social media metrics shape news production on Facebook. J. Commun. **73** (2023)
13. Niciporuc, T.: Comparative analysis of the engagement rate on Facebook and Google Plus social networks. In: Proceedings of International Academic Conferences. No. 0902287, International Institute of Social and Economic Sciences (2014)

14. Robertson, R.E., Jiang, S., Joseph, K., Friedland, L., Lazer, D., Wilson, C.: Auditing partisan audience bias within google search. Hum.-Comput. Interact. **2** (2018)
15. Twitter Help Center: (2023). https://help.twitter.com/en/managing-your-account/using-the-post-activity-dashboard
16. Twitter's Algorithm: (2023). https://github.com/twitter/the-algorithm
17. Vicario, M.D., Gaito, S., Quattrociocchi, W., Zignani, M., Zollo, F.: News consumption during the Italian referendum: A cross-platform analysis on Facebook and Twitter. In: Proceedings of the DSAA (2017). https://doi.org/10.1109/DSAA.2017.33
18. Vosoughi, S., Roy, D., Aral, S.: The spread of true and false news online. Science **359** (2018). https://doi.org/10.1126/science.aap9559
19. Walker, M., Matsa, K.E.: News consumption across social media in 2021. Pew Research Center (2021)
20. Weld, G., Glenski, M., Althoff, T.: Political bias and factualness in news sharing across more than 100,000 online communities. In: Proceedings of the ICWSM (2021)

PUREmotion: Understanding the Impact of Highway Construction on People's Wellbeing

Omar Hammad[1(✉)], Aniya Khalili Hollo[2], Nicholas Clements[2], Shelly Miller[2], Shivakant Mishra[2], and Esther Sullivan[3]

[1] King Fahd University of Petroleum and Minerals, Dhahran, Saudi Arabia
omarjh@kfupm.edu.sa
[2] University of Colorado, Boulder, Boulder, USA
[3] University of Colorado, Denver, Denver, USA

Abstract. Planned disruptions such as highway construction typically have negative impact on the health & wellbeing of people living in the surrounding communities due to increased air and noise pollution as well as inconveniences such road closures. Further, these communities typically tend to have low socio-economic status with high unemployment rate and limited resources (Environmental Justice Communities). The goal of this paper is to understand this impact via a smartphone app called PUREmotion providing ecological momentary assessment. We describe the deployment of PUREmotion in a four-month study with people living in the vicinity of a major construction project (C70 project) in North Denver, Colorado. The PUREmotion app is built based on the results of several focus groups and three rounds of usability study with people from these communities. We deployed the PUREmotion app over two time periods of six weeks each with about 100 community participants in each time period. In this paper, we report on our experience of these deployments and provide a detailed analysis of the data we collected to assess the impact of construction on health and wellbeing of the people. Our findings show that living near such construction projects has a direct (negative) impact on people's wellbeing due to air pollution, bad odor, increased noise, and disruptions in daily commutes.

Keywords: Sociotechnical Systems · Wellbeing · Environmental Justice

1 Introduction

The Central 70 project (C70) was a major highway construction along a 10-mile stretch of the I-70 interstate in northeast Denver. Four communities that live in the vicinity of this construction are Globeville, Elyria-Swansea, Clayton and Cole (GESC) communities. These communities have low socio-economic status with high unemployment rate and limited resources (Environmental Justice Communities), as well as a longstanding history of environmental pollution. The Social Justice Environmental Quality - Denver (SJEQ) project (https://www.

L. M. Aiello et al. (Eds.): ASONAM 2024, LNCS 15213, pp. 48–57, 2025.
https://doi.org/10.1007/978-3-031-78548-1_5

sjeqdenver.com/) seeks to gain a greater understanding of the impacts of C70 on the health and wellbeing of the four communities, devise appropriate intervention mechanisms to alleviate any negative impacts, and provide decision-makers with valuable data concerning the potential consequences of such disruptions.

In this paper, we focus on understanding the impacts of C70 on the health and wellbeing of the people living in the vicinity of the project via the socio-technical system that we have developed. In particular, we report on the results of deploying a smartphone-based diary app called PUREmotion that we have designed and developed. PUREmotion has been created with the intention of collecting pertinent user data over a prolonged time period in order to ascertain the impact of C70 construction on the health and wellbeing of the inhabitants of four environmental justice communities. We have deployed PUREmotion to collect *just-in-time* data regarding the users' current activities, their perceptions of noise and air quality, their emotional state, any respiratory issues they are encountering, and the effects on their daily commutes. The app was deployed over two different time periods—Cohort 1: January–March 2022 and Cohort 2: May–July 2022. For each of the two cohorts, we recruited about 100 participants from the four communities. Participants were required to use the app at least once a day for 30 days during their respective cohorts. We got approval from Institutional Review Board for conducting this study.

Our research question is: To what extent does the data collected from the PUREmotion app help us understand the impact of C70 construction on the health and wellbeing of the four environmental justice communities? From our deployment of PUREmotion over the two cohorts, we collected a total of 4,565 survey entries. Our key findings include a clear relationship between poor environmental conditions (air quality, noise and odor) resulting from the construction and the level of people's happiness. Further, we found a direct positive correlation between people's level of happiness and their distance from the construction activity. As we get closer to the construction, the happiness level gets lower.

2 Literature Review

Smartphone health and well-being self-tracking has been increasingly studied in many research work recently. Methods include manual self-reporting like EMAs [3,10], automated methods of smartphone sensors like GPS, microphone, accelerometer or meta-data [15,16] or using a hybrid approach of both ways [9,18]. For instance, in using manual methods, Jones et al. [8] have examined the adherence to the use of self-tracking apps with patients.

Unlike the first study, Wang et al. [16] have used passive mobile phone sensing data (automated method) to predict the depression of college students. The third kind of research has combined traditional ways with opportunities that are provided by mobile phones. Both approaches have their advantages but mostly it is about awareness (in manual ways) vs efficiency (in automated ways) [12]. Choe et al. suggested the usage of semi-automated self-tracking methods in building self-monitoring systems [5]. One common traditional method to assess one's current mood, Ecological Momentary Assessment (EMA), has been adopted and supplemented by smartphone capabilities (e.g. GPS) in many studies [7]. A study by Kim et al. [9] was conducted to design, build and evaluate a semi-automated tracking app called OmniTrack that allowed users to track their own activity.

People have benefited from construction projects for a long time in almost all aspects of civilization. The level of advancement in the infrastructure of a country is a high indicator of its welfare and development [11]. Despite their huge benefit, their impacts on the environment and nearby communities are huge [6] as they are considered a main source of air pollution [13] in terms of CO2 emission [1]. Further, they cause other direct and indirect impacts on the environment [6] such as noise, dust, solid waste and water waste [4]. These impacts eventually impact people who live around the location of the construction project. For instance, air pollution has been shown to cause many health problems such as pulmonary and cardiovascular disease, cancer and diabetes [14]. Further, noise can cause serious hearing issues if exposed to a long period of time [2]. In addition to their impacts on people's health, they also impact their lives. For example, a study by Xue et al. has shown how a subway construction project had impacted people's transportation, travel, and daily lives such as traffic jams, inconvenience to nearby residents, and impact shopping stores around [17].

3 Methodology

PUREmotion design is based on a systematic usability study that the authors conducted. This included a series of focus groups with the members of the four communities to finalize the app requirements and come up with an initial design followed by three rounds of usability testing with diverse groups of people from the four communities. The app design was revised after each round of usability testing. The app is comprised of a sequence of screens that include a set of survey questions that are presented to the user as he/she progresses in using the app (See Fig. 1). In addition, PUREmotion collects the current location of the user continuously. There are survey eight questions:

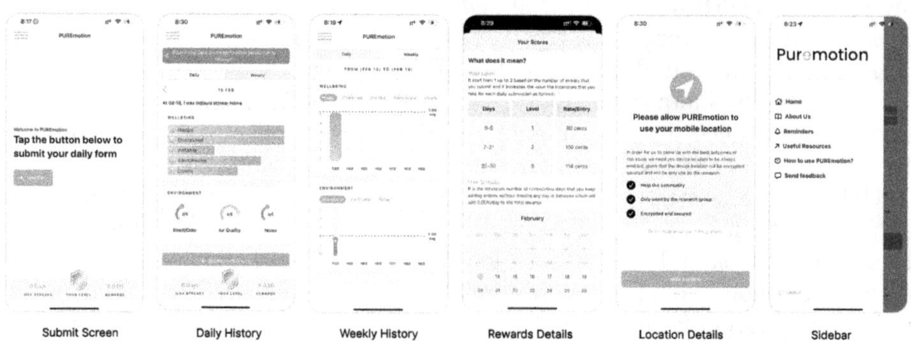

Fig. 1. Screens of other features of PUREmotion app

1. Rate your current feelings for each of the following. Five emotional feelings (Happy, Distressed, Irritable, Alert/Awake, Lonely) are assessed here.
2. Where are you right now? (Indoors, Outdoors)
3. How many hours did you spend outside (not in a building or vehicle) in the last day (24-h)?
4. How satisfied are you with the following qualities of your environment right now? There environmental qualities that are typically affected by construction (Smell/Odor, Air Quality, Noise) are assessed here.
5. Select all the qualities of the air with which you are not satisfied (None, Smell/Odor, Temperature, Visibility, Smokiness, Mustiness, Allergies, Dustiness)
6. How did you move around the community? (Car, Bus, Train, Bicycle, Motorcycle, Walk)
7. Considering the recent impacts of construction, how satisfied are you with your ability to move around the community?
8. Please write any additional comments below.

Questions 1, 4, and 7 allow the user to choose from a 5-point Likert scale: "1: Not at all" to "5: Completely" in Question 1 to understand the current emotional state of the user; "1: Very dissatisfied" to "5: Very satisfied" in Question 4 to assess user's satisfaction with the perceived environmental quality; and "1: Very dissatisfied" to "5: Very satisfied" in Question 7 to assess user's satisfaction with their transportation experiences. Question 2 presents the user with multiple places to choose from after they select either indoors or outdoors. It also allows them to select "Other" if they do not find the place that they are in. Our goal in asking people about their location, perceived environmental quality, and transportation experience is to eventually correlate the impacts of the construction project activities to their feelings reported in the first question. Questions 4 and 5 are designed to assess user's perception of the current environmental quality around them and their level of satisfaction. Questions 6 and 7 are designed to assess how construction activities have affected users' daily commutes.

When users submit an entry, they are able to see the entry on the main screen of the app in two forms, a daily form, and a weekly summary. The weekly summary averages the feelings and perceived environmental quality for each week. Users are able to navigate to any day or any week to see their entries. The first cohort of the research contained 89 participants, and the second cohort increased to 99 participants.

4 Results

4.1 General Statistical Analysis

We have collected a total of 4,565 data points from a total of 188 participants during the two cohorts. After cleaning the datapoints by removing entries that contain null values, we were left with 4350 entries, which is 95% of the original datapoints. Almost half of the users of Cohort 2 were new users, while the other half were returning from Cohort 1. The majority of the participants have submitted 31 to 40 entries in both cohorts, which is the required number for full compensation. In Cohort 2 we raised the amount of compensation from $24 to $40 but there was no significant change in the level of user engagement in the study (Total number of entries).

The daily form starts with asking users about their current feelings (Happiness, Distress, Irritability, Alert/Awake, and Loneliness). To reduce the number of personal differences among participants and make data easier to comprehend and correlated, we combined 1 and 2 as −1 (low), 3 as 0 neutral, and 4 and 5 as 1 (high). We observed that Happy and Alert/Awake feelings were higher, on average, than Irritable, Distressed and Lonely, and among the two cohorts, the average levels of feelings did not have a significant difference.

Table 1 shows the percentages of each place of the top 10 places the users were in when they submitted their entry. As we can see, most of the time people submitted their entries from their homes. We can also see that people submitted their entries from an indoor place almost 90% of the time. Further, there is an increase in the percentage of entries submitted from outdoors in Cohort 2. We attribute this increase to the fact that Cohort 1 was conducted during winter months, while Cohort 2 was conducted during summer months when people tend to spend more time outdoors. This is further confirmed by our next question.

Table 1. Entry submission place (%)

Place	Cohort 1	Cohort 2
Home Indoors	71.66	65.79
Work Indoors	8.23	10.06
Home Outdoors	3.24	9.33
Work Outdoors	1.39	2.23
Car	1.90	1.29
Walk/Bike Indoors	1.39	1.46
Busy Road/Traffic	1.54	1.29
Walk/bike Outdoors	0.51	1.33
Restaurant	1.39	0.60
Store/Retail	1.03	0.64

Table 2. Air Quality Complaints

Issue	Cohort 1 # (%)	Cohort 2 # (%)
None	850 (28.9)	660 (14.64)
Smell/Odor	558 (27.31)	760 (19.41)
Dustiness	401 (19.63)	803 (20.51)
Allergies	330 (16.15)	836 (21.35)
Temperature	309 (15.12)	554 (14.15)
Mustiness	209 (10.23)	350 (8.94)
Smokiness	135 (6.61)	319 (8.15)
Visibility	101 (4.94)	294 (7.51)
Total	2043	3916

Table 3 shows the number of hours people spent outdoors. We can see that the time spent outdoors in Cohort 2 is more than in Cohort 1. Once again, we attribute this to the fact that Cohort 1 was deployed in the winter time while Cohort 2 was deployed in summer time. Table 4 shows the levels of satisfaction with air quality, noise and odor in Cohorts 1 and Cohort 2. To be consistent, we have mapped the scale of these questions from (1 to 5) to (−1 to 1) as we did with the feelings question. While users are generally neutral in both cohorts, we notice that the level of satisfaction decreased slightly in Cohort 2. We hypothesize that the reason for this decrease in satisfaction is because users spent more time outside in Cohort 2, thus increasing their exposure to construction impacts.

Table 3. # hours spent outdoors (%)

# of Hours	Cohort 1	Cohort 2
0–1 h	63.94	40.27
2–3 h	26.24	39.08
4–6 h	5.73	13.13
More than 6 h	4.09	7.52

Table 4. Satisfaction with environmental qualities (Scale: −1 to 1)

Cohort	Air Quality (%)	Noise (%)	Smell/ Order (%)
1	0.301	0.337	0.336
2	0.161	0.135	0.242

Finally, Table 2 shows that the number of people dissatisfied with different air quality features. We can see that the number of people dissatisfied with various air quality features increased significantly in Cohort 2, almost double in many cases. There were about 30 complaints per user in Cohort 1 vs 45 per user in Cohorts 2. Looking at the reports' detail we can see that Smell/Odor was the highest for Cohort 1 while allergies and dustiness in Cohort 2 are equal to smell/odor. Again, we hypothesize that one reason for this increase in Cohort 2 is that users in Cohort 2 spent more time outdoors.

Transportation. The most used transportation method in both cohorts was cars. However, in Cohort 2 people walked and biked way more than in Cohort 1 and the usage of cars decreased a little. Again this can be attributed to the fact that Cohort 1 was deployed in winter while Cohort 2 was deployed in summer. The level of satisfaction for Cohort 1 is a bit higher than cohort 2. In both cohorts, users were least satisfied with driving.

4.2 Pair-Wise Analysis

Table 5. Correlation between the satisfaction of environmental quality and feelings

Feelings	Correlation with							
	Air Quality		Noise		Odor		Transit	
	Cohort 1	Cohort 2	Cohort 1	Cohort 2	Cohort 1	Cohort 2	Cohort 1	Cohort 2
Happy	**0.350**	0.274	**0.288**	0.239	0.330	**0.354**	0.274	**0.305**
Distressed	−0.200	−0.138	−0.214	−0.105	−0.246	−0.204	−0.169	−0.147
Irritable	−0.149	−0.184	−0.157	−0.108	−0.159	−0.210	−0.130	−0.168
Alert/Awake	0.159	0.113	0.163	0.101	0.130	0.178	0.182	0.208
Lonely	−0.240	−0.151	−0.230	−0.141	−0.252	−0.184	−0.233	−0.250

Environmental Quality and Feelings. This analysis investigates the indirect effect of construction activities on people's well-being. In particular, we look at the correlation between satisfaction with perceived environmental quality (Air Quality, Noise, Odor, and Transportation) and levels of different feelings. Table 5 summarizes these correlation values. We can see that there is a strong positive correlation between happiness levels and satisfaction with different environmental quality measures. Further, we notice that there is a strong negative correlation between the levels of negative feelings (Distressed, Irritable, and Lonely) and satisfaction with different environmental quality measures. Among different feelings, the correlation of Happiness with satisfaction with different environmental quality measures is the strongest. The analysis also shows that odor is the most correlated environmental quality and noise is the least.

These correlations provide evidence that people's wellbeing is positively correlated with their satisfaction with environmental quality. Since construction activities result in poor environmental quality, this analysis provides (indirect) evidence that construction activities are contributing to lower wellbeing of users.

Table 6. Correlation between the average feelings and the distance from construction project's Center (C), Highway (HW) and Nearest Point (NP)

Feelings	Correlation of feelings with distance from C, HW and NP								
	Cohort 1			Cohort 2			Long Term		
	C	HW	NP	C	HW	NP	C	HW	NP
Happy	0.041	0.132	0.148	0.008	0.001	0.069	0.11	0.13	**0.24**
Distressed	0.054	−0.01	0.006	0.002	0.077	−0.043	0.08	−0.05	−0.02
Irritable	0.13	0.029	0.055	0.018	0.052	−0.036	0.16	−0.01	0.01
Alert	0.073	0.123	0.131	−0.122	0.031	−0.088	−0.01	0.05	0.1
Lonely	0.023	−0.017	−0.013	0.065	0.118	0.052	0.14	0.08	0.09

Location vs Feelings. We analyze the distance from the construction project activities and the feelings levels. We look at the entries submitted from a location

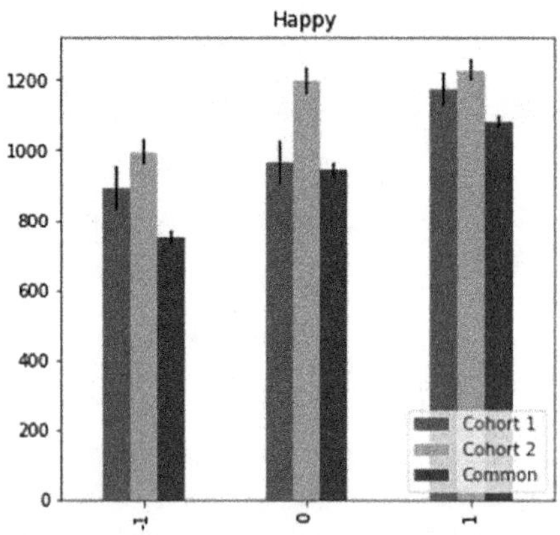

Fig. 2. Happiness vs average distance (Color figure online)

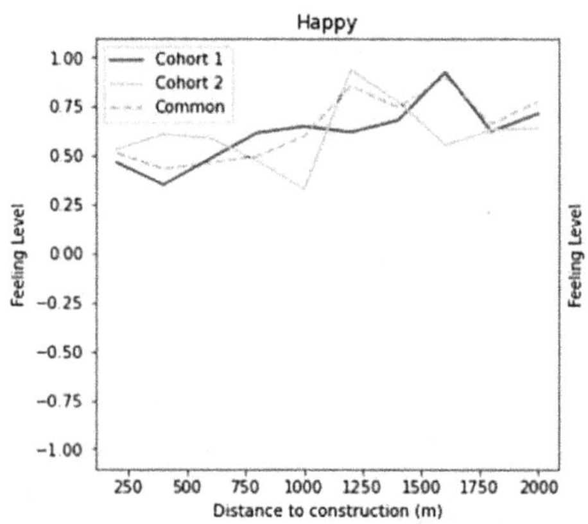

Fig. 3. Happiness vs distance of area

point that is within 3000 m of the center of construction project activities. This allows us to discount any outliers where some of the participants submitted sometimes from very far places. 81% of the data points collected are from the area within 3000 m of the construction project activities.

To understand the correlation between the construction project activities and the well-being of the people in the surroundings, we have looked at the distance between the location from where the entries were submitted and the location of the construction activities. Since the construction activities were spread over a larger area along a highway and specific activities took place at different locations with in this area on different days, we first consider three different construction activity location options: (1) distance from the center of the construction (C), (2) distance from the construction highway (HW), and (3) distance from the nearest construction activity point (NP) - median GPS location of the construction activity locations.

We performed three different types of analyses: 1) PEARSON correlation between the distance from construction and the GPS location from where an entry was submitted; 2) distribution of each feeling over different distances; and 3) average area feelings, where an area consists of all locations within (i) 200 m from the construction activities, (ii) 200–400 m of construction activities, (iii) 400–600 m of construction activities, and so on.

Table 6 shows a summary of the PEARSON correlation analysis. We notice that the only clear positive correlation was between the feeling of Happy over a long term using the Nearest Point method. We believe that this is because happiness is the easiest and most accurate feeling to express, compared to the other feelings. Based on this result, we focused on the feeling of happiness and nearest point distance method for all upcoming analysis.

Figure 2 shows the relationship between the level of happiness (Y-axis) and distance (X-axis). We clearly notice that the level of happiness increases as distance increases, especially over the long term (green). We have conducted an analysis of the variance test and it showed that the means are different (p-value: Cohort 1 = 5.78e−09, Cohort 2: 0.002, Common: 8.189e−11). The difference among other feelings was not significantly different. Figure 3 shows that level of happiness (Y-axis) increases as the distance of the area being considered increases. However, we did not find a clear relationship for happiness and distance beyond 2000 m. Both of these analysis (Figs. 2 and 3) provide a direct evidence that people's wellbeing gets lower as they get closer to the construction activities.

5 Discussion

We now answer our research question: To what extent does the data collected from the PUREmotion app help us understand the impact of C70 construction on the health and wellbeing of the four economic justice communities?

We first investigated the truthfulness of the survey data and found three indicators that tell us that the data is genuine and not random. First, we asked participants in the first question to rate the level of five different feelings. Among these are feelings of happiness, distress, and irritability. We analyzed the correlation between the feeling of happiness and feelings of distress or irritability for each entry submitted. We found this correlation to be strongly negative. This would be expected and the strong negative correlation shows that the entries submitted by the users are genuine. Second, the participants were required to submit 30 entries (one per day) during the study period to receive full compensation. However, we found that more than 50% of the participants submitted more than 30 entries. This shows how motivated a large number of participants were to help us in our study goals, and thus we assume they provided truthful information. Finally, the two cohorts were conducted in the winter and spring seasons. In the second cohort, people reported more walking and biking than driving which is consistent with the expected commute medium for the season.

Our data analysis provides strong evidence that the C70 project has contributed to lower wellbeing of the people living in the vicinity. We found that the level of happiness reported by users decreased and levels of distress, irritability and loneliness reported by users increased as they get closer to the construction activity locations. We found that the average levels of positive feelings (happiness) decreased and average levels of negative feelings (distress, irritability and loneliness) increased as we moved to areas that are closer to the construction activity locations. Finally, we found a strong positive correlation between levels of happiness and satisfaction with environmental quality measures such as Air Quality, Noise, Odor, and Transportation, and negative correlation between the levels of distress, irritability and loneliness and satisfaction with these environmental quality measures, providing an indirect evidence that construction activities contributed to lower wellbeing of the people.

While our results are based on the impact of C70, the insights we gained are relevant for other large infrastructure projects. This is because the impacts of such projects in terms of poor environmental quality and road closures are similar. Further, the communities that live nearby typically tend to be environmental justice communities similar to the ones in the C70 project.

References

1. Ahmed Ali, K., Ahmad, M.I., Yusup, Y.: Issues, impacts, and mitigations of carbon dioxide emissions in the building sector. Sustainability **12**(18), 7427 (2020)
2. Bolaji, B.O., Olanipekun, M.U., Adekunle, A.A., Adeleke, A.E.: An analysis of noise and its environmental burden on the example of Nigerian manufacturing companies. J. Clean. Prod. **172**, 1800–1806 (2018)
3. Chan, L., Swain, V.D., Kelley, C., de Barbaro, K., Abowd, G.D., Wilcox, L.: Students' experiences with ecological momentary assessment tools to report on emotional well-being. Proc. ACM Interact. Mob. Wearable Ubiquit. Technol. **2**(1), 3:1–3:20 (2018)
4. Chen, H.c., Chung, W., Xu, J., Wang, G., Qin, Y., Chau, M.: Crime data mining: a general framework and some examples. Computer **37**, 50–56 (2004)

5. Choe, E.K., et al.: Semi-automated tracking: a balanced approach for self-monitoring applications. IEEE Perv. Comput. **16**(1), 74–84 (2017). ISBN 1536-1268
6. H, L.: Systematic evaluation and assessment of building environmental performance (SEABEP) (2013). Accessed 30 Mar 2015 T15:49+02:00
7. Harari, G.M., Lane, N.D., Wang, R., Crosier, B.S., Campbell, A.T., Gosling, S.D.: Using smartphones to collect behavioral data in psychological science: opportunities, practical considerations, and challenges. Pers. Psychol. Sci. **11**(6), 838–854 (2016). ISBN 1745-6916
8. Jones, S.L., Hue, W., Kelly, R.M., Barnett, R., Henderson, V., Sengupta, R.: Determinants of longitudinal adherence in smartphone-based self-tracking for chronic health conditions: evidence from axial spondyloarthritis. Proc. ACM Interact. Mobile Wearable Ubiquit. Technol. **5**(1), 16:1–16:24 (2021)
9. Kim, Y.H., Jeon, J.H., Lee, B., Choe, E.K., Seo, J.: OmniTrack: a flexible self-tracking approach leveraging semi-automated tracking. Proc. ACM Interact. Mob. Wearable Ubiquit. Technolog. **1**(3), 67:1–67:28 (2017)
10. King, Z.D., et al.: Micro-Stress EMA: a passive sensing framework for detecting in-the-wild stress in pregnant mothers. Proc. ACM Interact. Mob. Wearable Ubiquit. Technol. **3**(3), 91:1–91:22 (2019)
11. Knaap, T., Oosterhaven, J.: Measuring the welfare effects of infrastructure: a simple spatial equilibrium evaluation of Dutch railway proposals. Res. Transp. Econ. **31**(1), 19–28 (2011)
12. Li, I., Dey, A.K., Forlizzi, J.: Using context to reveal factors that affect physical activity. ACM Trans. Comput.-Hum. Interact. (TOCHI) **19**(1), 1–21 (2012). ISBN 1073-0516
13. Shen, L.Y., Lu, W.S., Yao, H., Wu, D.H.: A computer-based scoring method for measuring the environmental performance of construction activities. Autom. Constr. **14**(3), 297–309 (2005)
14. To, T., et al.: Health risk of air pollution on people living with major chronic diseases: a Canadian population-based study. BMJ Open **5**(9), e009075 (2015)
15. Tong, C., Craner, M., Vegreville, M., Lane, N.D.: Tracking fatigue and health state in multiple sclerosis patients using connnected wellness devices. Proc. ACM Interact. Mobile Wearable Ubiquit. Technol. **3**(3), 106:1–106:19 (2019)
16. Wang, R., et al.: Tracking depression dynamics in college students using mobile phone and wearable sensing. Proc. ACM Interact. Mob. Wearable Ubiquitous Technol. **2**(1), 43:1–43:26 (2018)
17. Xue, X., Zhang, R., Zhang, X., Yang, R.J., Li, H.: Environmental and social challenges for urban subway construction: an empirical study in China. Int. J. Project Manage. **33**(3), 576–588 (2015)
18. Zhang, X., Li, W., Chen, X., Lu, S.: MoodExplorer: towards compound emotion detection via smartphone sensing. Proc. ACM Interact. Mob. Wearable Ubiquitous Technol. **1**(4), 176:1–176:30 (2018)

Opinion and Self-confidence in Influence Networks: A Coupled Dynamics Model

Emmanuel Kravitzch[1,3(✉)], Vineeth S. Varma[2], Antoine O. Berthet[3], and Yezekael Hayel[1]

[1] Laboratoire Informatique d'Avignon (LIA), Avignon Université, Avignon, France
emmanuel.kravitzch@alumni.univ-avignon.fr

[2] Département Contrôle Identification Diagnostic (CID), CRAN, Université de Lorraine, CNRS, Nancy, France

[3] Laboratoire des Signaux et Systèmes (L2S), Université Paris-Saclay, CNRS, CentraleSupélec, Orsay, France

Abstract. In this paper, we develop a novel model of opinion dynamics that takes into account the variability of self-confidence. This agent-based model aims at combining both opinion and self-confidence, two notions often assimilated in many modeling works, although well distinct in practice. We analyze our model under two types of opinion space, namely a standard interval and a circular structure under which the system presents some intriguing specificity. We give some insights concerning the global equilibria of the co-evolving dynamics, and also highlight some salient features of the global system's trajectory. Finally, we conduct several numerical simulations to support the analytical results.

Keywords: opinion dynamics · bounded-confidence · influence networks

1 Introduction

1.1 Research Context and Motivation

In many opinion dynamics models, the variable of opinion is taken as a simple scalar value. But since in general, opinion is a complex and fuzzy system of beliefs and reasoning, it can hardly be projected into a simple interval. Then, one shall ask whether the scalar value presented as an "opinion" in the literature is finally nothing else but a confidence degree (with respect to an underlying social norm or behaviour not explicitly defined). To this title, some authors [1,2] propose models where agents hold an opinion on each topic, and the different opinions of the same agent can be structured and influence each other. Hence, the two concepts of *opinion* and *confidence* are confusedly assimilated. The current work aims at developing a framework where the distinction is clearly done: each agent is endowed with an opinion value (which can be in the interval or in a cyclic structure), *plus* a scalar value comprised between 0 and 1, measuring how confident the agent is in his own opinion.

L. M. Aiello et al. (Eds.): ASONAM 2024, LNCS 15213, pp. 58–68, 2025.
https://doi.org/10.1007/978-3-031-78548-1_6

To define the opinion dynamics (OD) structure, we rely on the well-established *bounded-confidence* (BC) interaction, also known under the name of *Hegselman-Krause model* (HK model for short). This socially-inspired mechanism aims at encapsulating both *homophily* and *selective exposure*, two natural psycho-social trends one may follow to reduce cognitive dissonance. Since the pioneering works of Rainer and Krause [3], the BC model has been the scope of intensive study (see e.g. [4,5] and references therein), even until very recently [6], and by researchers coming from various horizons. The interested reader can easily refer to the recent survey of C. Bernardo and her coauthors [7].

Moreover, we also analyse the coupling of opinion and self-confidence in the case where the opinion space is *toric*. This modelling has the main advantages to preserve a total symmetry between all the opinions, and to prevent the *border-effect*.

1.2 Related Literature

Circular opinion has already been investigated in the past. In [8,9], the authors opt for a statistical-physics approach and combine attractive and repelling mechanism. In [10], the problem of convergence time of the HK model *on the circle* is tackled via rigorous and involved mathematical methods. But none of these already existing works encompass in their frame a self-confidence dynamics. This concept of self-confidence is indeed very close to *social power* or *influence*, notions which have been already investigated: authors of [11] indeed consider a dynamical social power furnished to each agent, combined with stubbornness. Nevertheless, they rather focus on network evolution and convergence. Authors of [12] pursue in the same vein but here consider a multiplicity of topics. Some authors indeed mention self-confidence and make a clear distinction with the opinion variables, but the self-confidence is taken constant along time [13]. In his notes of 2016 [14], C. Wang considers dynamical self-confidence, but is designed as an autonomous process, which does not depend of the opinion variables. Finally, to the best of our knowledge, the explicit combination of opinion and self-confidence with an emphasis on the interaction of the two variables still has yet to be explored.

1.3 Paper Scope and Organisation

The goal of this study is to investigate the impact of the self-confidence on the opinion when the former is driven by a dynamics induced by a *network effect*, while the latter is driven by a BC dynamics. In the case we are interested in, we assume that two agents with a low self-confidence are close one to each other, whatever their respective opinion. At the extreme case, two agents with zero self-confidence coincide, whatever their opinion. It is indeed this assumption which acknowledges the very nature of *self-confidence*, and then gives rise to the (semi-)ball structure. If we do not make this assumption, then we simply obtain as phase space a rectangle instead of a semi-ball, and a cylinder's border instead of a ball. In Sect. 2, we formally introduce the multi-agent model coupling opinion with self-confidence. Then, the following Sect. 3 is devoted to its analysis, including an additional variant: the opinion space taken as a *toric* structure (namely the circle $\mathbb{C}$). Finally, the last Sect. 4 before concluding consists of a substantial amount of numerical experiments illustrating the analytical investigations and pave the way for future research.

Notations. For any $K \geq 1$, we write $[K]$ to shorten $\{1, 2, ..., K\}$. We write $\mathbb{1}_U$ for the indicator map of event U, and for any positive scalar $h > 0$ we write $\Delta_h x(t) := x(t+h) - x(t)$ the difference operator. We denote the unitary circle as $\mathbb{C} := \{(x, y) \in \mathbb{R}^2 : ||(x, y)||^2 = 1\}$, the ball $\mathbb{B} := \{(x, y) : ||(x, y)||^2 \leq 1\}$, and the semi-ball $\mathbb{B}_+ := \mathbb{B} \cap (\mathbb{R} \times \mathbb{R}_+)$. Finally, $|P|$ stands for the cardinality of any set P, $\perp\!\!\!\perp$ stands for the symbol of independence between two random variables, and $\lambda(.)$ is the Lebesgue measure.

2 The Model

We define a population of agents of size $K \in \mathbb{N}^*$, where each agent is indexed by an integer $k \in [K]$. The time step is fixed and small: $h \ll 1$. The time axis is taken discrete along the set $\mathbb{N}h$, and at each time point $t \in \mathbb{N}h$, for each agent $k \in [K]$ is assigned an opinion value $\theta_k(t)$. In the first part of the analysis Sect. 3.1, we take the simple interval $[0, \pi]$. Subsequently, we extend the model to the circle $\mathbb{C} := [0, 2\pi]/_{0 \sim 2\pi}$. In this latter case, the curbless structure of $\mathbb{C}$ gives birth to interesting changes in the model's behaviour. In addition to his opinion $\theta_k(t)$, agent k is also endowed with a scalar value, standing for his *self-confidence* degree in his own opinion. We note it $\rho_k(t) \in [0, 1]$. Hence, one can represent the global state of agent k in polar coordinates: $u_k(t) := (\theta_k(t), \rho_k(t)) \in \mathbb{B}_+$ in the case of the semi-ball (resp. $\mathbb{B}$ for the full circle case studied in Subsect. 3.2). As mentioned in the last subsection of the introduction Sect. 1.3, it is precisely the (semi-)ball structure which makes the zero state the only equidistant point of the arc of maximal confidence. Mathematically, one can see the semi-ball (respectively the ball) as the simple rectangle $[0, \pi] \times [0, 1]$ (respectively $\mathbb{C} \times [0, 1]$) where all the points of the horizontal line (respectively circle-shaped section) $\{(\theta, \rho) : \rho = 0\}$ has merged to form one single point.

Since the self-confidence (written as a radius) and the opinion (written as an angle) co-evolve, the confidence profile $(\rho_j(t))_j$ appears in the evolution equation of the opinion $\theta_k(t)$ for agent k and especially his own confidence $\rho_k(t)$. First, define the *radius map* R and the directed *influence network* represented by the matrix-valued map ϕ, two structural elements in the dynamics:

$$R : \rho \in [0, 1] \mapsto R(\rho) := c_0 \pi (1 - \rho), \ c_0 \in [0, 1], \text{ and}$$

$$\phi : (\mathbb{B}_+)^K \mapsto \{0, 1\}^{K^2} \text{ as } \phi_{kj}(u) := \mathbb{1}_{\{|\theta_k - \theta_j| < R(\rho_k)\}},$$

$\forall u \in (\mathbb{B}_+)^K$. The scalar c_0 is a parameter defined beforehand. Hence, if $\phi_{kj} = 1$, then k is influenced by j. Notice that the map R decays according to the confidence variable, and delimits the zone of influence. Let us also define the *neighbour set*. For a given configuration $u \in (\mathbb{B}_+)^K$, denote by $\mathcal{N}_k(u)$ the set of neighbours of agent $k \in [K]$: $\mathcal{N}_k(u) := \left\{ j \in [K] : |\theta_j - \theta_k| < R(\rho_k) \right\}$. When it is clear from the context, we write $\mathcal{N}_k(t)$ instead of $\mathcal{N}_k(u(t))$ for brevity, and respectively $\phi_{kj}(t)$ instead of $\phi_{kj}(u(t))$. Thus, we are now able to state the global (θ, ρ)-dynamics:

$$\Delta_h \theta_k(t) = \frac{h}{K} \sum_j \rho_j (1 - \rho_k) \phi_{kj}(t)(\theta_j - \theta_k), \tag{1}$$

$$\Delta_h \rho_k(t) = h \left(\frac{|\mathcal{N}_k(t)|}{K} - \rho_k(t) \right). \tag{2}$$

The first equation corresponds to a classical BC dynamics with $\rho_j(1 - \rho_k)$ as weighting coefficients. This refinement seems indeed intuitive: the more agent j is self-confident, the more persuasive is she. On the contrary, the more agent k is self-confident, the less steerable is she. This is why this coefficient grows with ρ_j and decays with ρ_k.

The second equation of the dynamics relies on a standard herding behaviour known as *network effect*: each agents gains in self-confidence as long as his neighbour size increases. His ideological proximity with alter egos tends to reinforce him in his own opinion. By contrast, an isolated agent has low self-confidence, and remains able to be influenced. In accordance with the network-effect formalism [15, 16], the self-confidence may be interpreted as the *utility to be connected* to others.

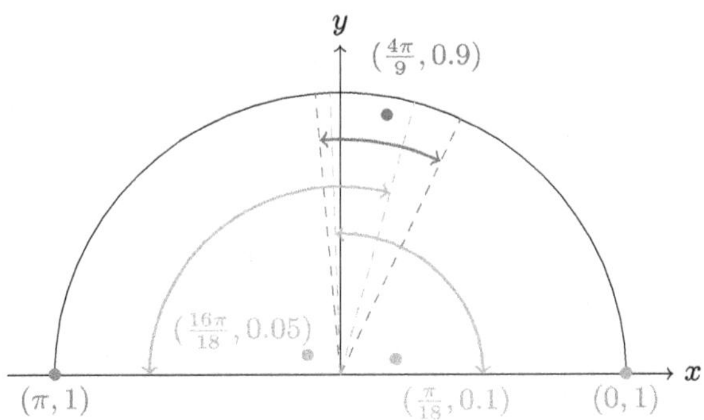

Fig. 1. A simple scheme to explain dynamics (1) with 5 agents and their respective radius of influence.

Figure 1 illustrates a situation with $K = 5$ agents, represented in coloured dots in B_+ whose associated tuples are their respective polar coordinates. The red and brown point respectively represent a left radical and right radical, both with maximal confidence. The blue point correspond to a quasi-centrist agent with high (but not maximal) confidence. On the contrary, the orange and green agents are rather extreme, but with a low self-confidence. This explains why their respective influence zone is larger than the blue one: it is the zone delimited by two dashed segments of the same colour. We thus can see that the blue agent both influences the green and the orange one, but the reciprocal is false. Finally, the maximally-confident agents (the red and brown) cannot be influenced since $R(1) = 0$.

The choice of the initial agents states impacts the system's dynamics. In the rest of the analysis, we consider an *independent and identically distributed* (i.i.d) sequence of tuples: $(\theta_k(0), \rho_k(0))_{k\in[K]} \sim_{\text{i.i.d}} \theta_{\text{in}} \otimes \rho_{\text{in}}$, where the angular and radius distribution are also independent: $\theta_k(0) \perp\!\!\!\perp \rho_k(0)$ for all $k \in [K]$.

Remark 1. Note that if we simply take $(\theta_k, \rho_k)_k \sim_{i.i.d.} \text{Unif}[0,\pi] \otimes \text{Unif}[0,1]$, then a concentration on the center occurs, as illustrated Fig. 2a. In order to construct a sample of K i.i.d agents uniformly distributed on the semi-ball B_+, one shall use the sample $(\theta_k, \sqrt{\rho_k})_k$, where $(\theta_k, \rho_k)_k \sim_{i.i.d} \text{Unif}^{\otimes 2}[0,\pi] \times [0,1]$. For a proof, see the appendix.

3 Analysis

3.1 General Analysis

First, we characterise the rest points of dynamics (1). Relying on the analysis of the classical BC dynamics, which is very similar to the first line of (1), one can assume that clusters emerge in finite time, i.e. there exists a time $T_{\text{clust}} > 0$ such that for any $t > T_{\text{clust}}$, there exists an integer $P \geq 1$ (the actual number of clusters) and a map (assigning to each agent the cluster she belongs to) $\mathfrak{C} : [K] \mapsto [P]$ such that $\mathfrak{C}(k) = \mathfrak{C}(j) \iff |\theta_k(t) - \theta_j(t)| < \min\big((R(\rho_j(t)), R(\rho_k(t))\big)$. In such a clustered configuration, the influence network map ϕ is symmetric and block-wise diagonal, whose blocks are complete graphs, and the size of the blocks correspond to the size of the clusters. We then have two cases according to the number of clusters.

- *one single cluster:* $P = 1$. In that case, we have $|\mathcal{N}_k(T_{\text{clust}})| = K \; \forall k$, implying that

$$\Delta_h \rho_k(t) = h(1 - \rho_k(t)) \text{ for } t \geq T_{\text{clust}} \implies \rho_k(t) \approx (1 - C_k e^{-t}).$$

  However, when $\rho_k(t)$ becomes almost 1, the graph becomes disconnected leading to ρ_k decreasing again. This in turn, makes the graph reconnected and so forth. Thus, we expect a slower convergence speed to exact consensus, i.e., $\lim_{t\to+\infty}(\theta_j - \theta_k)(t) = 0$ for all $k, j \in [K]$ asymptotically.
- *several clusters:* in the case where $P \geq 2$, then we have $\frac{|\mathcal{N}_k(t)|}{K} < 1$, and due to dynamics (1) we have

$$\rho_k(t) \underset{t\longrightarrow+\infty}{\longrightarrow} \rho_k^* := \frac{|\mathcal{N}_k(T_{\text{clust}})|}{K}, \tag{3}$$

  where the convergence occurs exponentially fast from the time instant T_{clust}. Note that in the case where the clusters are of equal size, we have $\rho_k^* = \frac{1}{P}$ for all $k \in [K]$. After clustering and quasi convergence of $\rho_k(t)$, we then get

$$\Delta_h \theta_k(t) \approx \frac{(\rho_j^*)(1-\rho_k^*)}{K} \sum_{j \in \mathfrak{C}(k)} (\theta_j - \theta_k)(t),$$
$$\Longrightarrow \forall j \in \mathfrak{C}(k), \quad (\theta_j - \theta_k)(t) \underset{t \to +\infty}{\longrightarrow} 0.$$

The last convergence is due to the fact that the interaction graph of the variables $(\theta_i, \theta_j)_{i,j \in \mathfrak{C}(k)}$ is complete.

3.2 Circle Case

We now consider the *full circle* (FC) case, that is $\theta_k(t) \in \mathbb{C}$. Choosing the circle $\mathbb{C}$ presents two main modelling advantages: first, there is neither extreme opinions nor central ones, whose connotation is not neutral. More precisely, extreme opinions can be implicitly considered as *negative*, while on the contrary central or *moderate* opinions can be considered as *positive*. A toric modeling of the opinion space is then an interesting way to evacuate this kind of preconceptions. In the FC case, all opinions play a symmetric role, and there is no "natural convergence point" supposed to be in the center. There are indeed many situations where there is no extreme viewpoints, despite an area for disagreement. Second, in many models where the opinion space is an interval, the consensus is reached in the center, which clearly resembles to a modelling artefact of type *edge effect*. Over the circle, this effect does not occur. This topological modification of the opinion space requires to redefine the notion of subtraction and distance, which is done in the next paragraph.

Giving two angles θ_k and θ_j, there are always two paths to reach θ_j from θ_k: the first path d_0 crosses 0, while the other $d_\emptyset$ does not. The latter is of length $d_\emptyset = |\theta_k - \theta_j|$, while the former is of length $d_0 = 2\pi - |\theta_k - \theta_j|$. We can now define the proper subtraction of angles:

$$(\theta_j -_{\mathbb{C}} \theta_k) := \begin{cases} \operatorname{sign}(\theta_k - \theta_j) d_0 & \text{if } d_0 < \pi, \text{ and} \\ (\theta_j - \theta_k) & \text{otherwise.} \end{cases} \tag{4}$$

More simply, we write $d_{\mathbb{C}}(\theta_k, \theta_j)$ for the distance in the circle: $d_{\mathbb{C}}(\theta_k, \theta_j) := \min(d_0, d_\emptyset)$.

In the FC case, all subtraction of angles are made according to the preceding formula (4). Specifically, we then have $\phi_{kj} = \mathbb{1}_{\{d_{\mathbb{C}}(\theta_k, \theta_j(t)) < R(\rho_k)\}}$ and $\mathcal{N}_k = \{j : d_{\mathbb{C}}(\theta_k, \theta_j) < R(\rho_k)\}$.

Although the full-circle case displays a similar behaviour to the semi-circle case, there is indeed one specificity which deserves to be revealed: a *temporal phase separation* between the two variables θ and ρ. Taking $(\theta_{\text{in}}, \rho_{\text{in}}) = \text{Unif}(\mathbb{B})$, there exists a time instant T_{ps} such that

$$\Delta_h \theta_k(t) = 0, \forall t \leq T_{\text{ps}}, \text{ with } T_{\text{ps}} = \inf\{t \geq 0 : \Delta_h \rho_k(t) = 0\},$$

for all $k \in [K]$. To see this, let us compute the opinion dynamics of a single agent from the initial configuration $(\theta_{\text{in}}, \rho_{\text{in}}) \in \text{Unif}(\mathbb{C}) \otimes [0, 1]$. Then, we have the following convergence as the number of agents tends to be infinite:

$$\lim_{K \to +\infty} \left(\theta_j, \rho_j\right)_{j \in [K]} = \frac{d\theta d\rho}{2\pi} \text{ in law,} \tag{5}$$

by Glivencko-Cantelli theorem. Next, for $k \in [K]$ and $(\theta_k(0), \rho_k(0))$ being fixed, we can write

$$\frac{1}{h}\Delta_h \theta_k(t) = \frac{1}{K}\sum_j \rho_j(1-\rho_k)\mathbb{1}_{\{d_{\mathbb{C}} < R(\rho_k)\}}(\theta_j -_{\mathbb{C}} \theta_k)(t) \approx$$

$$(1-\rho_k)\int_{\rho=0}^{\rho=1} \rho \int_{\theta \in \mathbb{C}} \mathbb{1}_{\{d_{\mathbb{C}}(\theta_k, \theta) < R(\rho_k)\}}(\theta -_{\mathbb{C}} \theta_k)\frac{d\theta d\rho}{2\pi},$$

where the second equation relies on the asymptotics (5) mentioned above, with in addition K to be large. Furthermore, we have for any pair (θ_k, ρ_k)

$$\int_{\theta \in \mathbb{C}} \mathbb{1}_{\{d_{\mathbb{C}}(\theta, \theta_k) < R(\rho_k)\}}(\theta -_{\mathbb{C}} \theta_k) d\theta = \int_{\theta_k - R(\rho_k)}^{\theta_k + R(\rho_k)} (\theta - \theta_k) d\theta = 0.$$

We can then conclude that for any $h > 0$ and any initial state $(\theta_k(0), \rho_k(0))$, we have $\frac{1}{h}\Delta_h \theta_k(t) \equiv 0$, provided the entrance law for θ is uniform over the circle. Nevertheless, one has to keep in mind that the uniform density Unif($\mathbb{C}$) is an unstable equilibrium measure, this is why in all numerical experiments presented in the next section, since K is always large but finite, fluctuations arise and the angular symmetry is broken.

As we have shown that the opinion remains constant at least during the first phase of the process, it is then possible to estimate the *pseudo radius equilibrium* (PRE) $\rho^{\star}$, which is clearly different of the *true* limiting value ρ^* defined in equation (3) above. Concretely, $\rho^{\star}$ is the limiting value if we consider a *continuum of agents* written as a density map instead of a large but finite number. This value can be obtained by a repeated use of Eq. (5):

$$\frac{|\mathcal{N}_k(t)|}{K} = \frac{1}{K}\sum_j \phi_{kj} = \int_{\theta \in \mathbb{C}} \mathbb{1}_{\{d_{\mathbb{C}}(\theta, \theta_k) < R(\rho_k(t))\}} \frac{d\theta}{2\pi} = \int_{\theta_k - R(\rho_k)}^{\theta_k + R(\rho_k)} \frac{d\theta}{2\pi} = \frac{R(\rho_k)}{\pi}.$$

Thus, by definition of R, $R(\rho) = c_0\pi(1-\rho)$ for $c_0 > 0$, then the PRE for any agent $k \in [K]$ satisfies the following:

$$\frac{1}{h}\Delta_h \rho_k^{\star} = 0 \iff \frac{\mathcal{N}_k}{K} - \rho_k^{\star} = 0 \iff c_0(1-\rho_k^{\star}) = \rho_k^{\star} \iff \rho_k^{\star} = \frac{c_0}{1+c_0}.$$

This convergence result is finely corroborated by the numerical experiments presented in the next section.

To avoid the phase separation described above, one shall consider an alternative initial distribution for the angular variables:

$$\theta_k(0) \sim_{i.i.d} f_n : x \in [0, 2\pi] \mapsto \frac{1 + \cos(nx)}{2\pi}, \tag{6}$$

for some $n \geq 0$. It is easy to show that f_n is a density map supported on $[0, 2\pi]$. Note that for $n = 0$ and $n = +\infty$, we retrieve the uniform law. We have conducted an experiment with such a non uniform law, and the result is given Fig. 3b.

4 Numerical Simulations

In this section, we present a sample of the results obtained from numerical experiments, conducted with various parameters: the first part Sect. 4.1 deals with the semi-ball case, whereas the second part Sect. 4.1 shows the system's behaviour when the opinion is circular-valued. We systematically take as population size $K = 250$ and step size $h = 0.01$. On the one hand, we took $c_0 = 1$ in the definition of the influence radius map R for the semi-circle case, while on the other hand, we took $c_0 = \frac{1}{2}$ for the FC case. We represent the initial configurations with red stars, and the final ones with blue stars. In the three first cases, it is interesting to notice that all the agents seem to slide across an oval-shaped curve.

4.1 The Semi-circle Case

In the next two figures, the opinion variables evolve in the interval: $\theta_k(t) \in [0, \pi]$ for all $k \in [K]$, and at all time instant $t \geq 0$. The initial angular distribution is the same in the two cases: $\theta_{\text{in}} = \text{Unif}[0, \pi]$, but two types of entrance laws for the self-confidence distribution ρ_{in} are tested: in the first case (Fig. 2a), we took $\rho_k(0) \sim_{i.i.d} \text{Unif}[0, 1]$. We can therefore observe a concentration on the center at initial time. In the second case (Fig. 2b), we use the spatially uniform distribution $(\theta_k, \rho_k) \sim_{i.i.d} \text{Unif}(\mathbb{B}_+)$. In both cases, a quasi-consensus seems to occur around the value $(\theta^*, \rho^*) \approx (\frac{\pi}{2}, 1)$. At equilibrium, all the agents get a maximum confidence because consensus is reached, as depicted in the preceding section. See also that in this case, the convergence is slow due to the exponentially decaying terms $(1 - \rho_k(t))$ in the angular dynamics. Since the entrance law θ_{in} is uniform, the limiting opinion θ^* is unsurprisingly around the center of the interval $[0, \pi]$, whence the interest for the full-circle case.

4.2 The Full Circle Case

The last two figures display the coupled dynamics when the opinion space is circular: $\theta_k(t) \in \mathbb{C}$ for all $k \in [K]$. In both cases, we took $\rho_k(0) = \sqrt{r_k}$, where $(r_k)_k$ is an i.i.d sample of random variables uniform over $[0, 1]$. But in the first case (Fig. 3a), we took $\theta_{\text{in}} = \text{Unif}(\mathbb{C})$. The temporal phase separation is then clearly visible: in the first instants of the process, all agents describe a quasi-exact radial trajectory. Subsequently, as they get closer to the circle of radius $\rho^\star = \frac{1}{3}$, the radial dynamics decays and the 'finite size' effect prevails. Finally, all the agents converge with a self-confidence $\rho^* = \frac{1}{P}$, here $P = 2$.

In the last picture Fig. 3b, we have chosen a non-uniform angular entrance law: $\theta_{\text{in}} \sim f_4$ (see Eq. (6)). The temporal phase separation does not occur, which proves that this phenomenon is very dependent of the initial configuration.

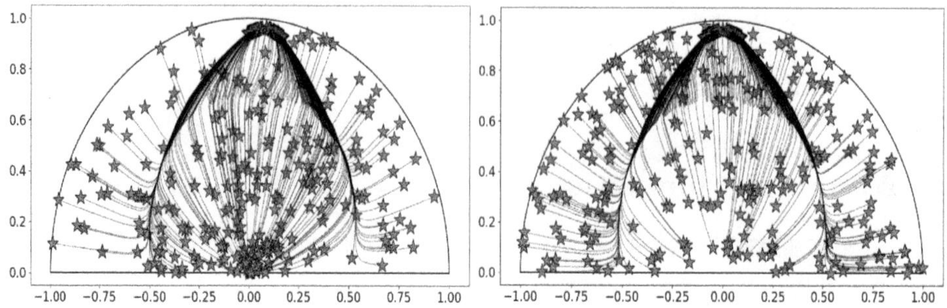

(a) (b)

Fig. 2. Comparison of different entrance laws. On the left (a), we took as *entrance law* uniform in the opinion, that is $\theta_{\text{in}} = \text{Unif}[0, \pi]$, but with concentration on the center by choosing $\rho_{\text{in}} = \text{Unif}[0, 1]$. And on the right (b), we took as *entrance law* the uniform spatial measure: $(\theta_{\text{in}}, \rho_{\text{in}}) = \text{Unif}(\mathbb{B}_+)$.

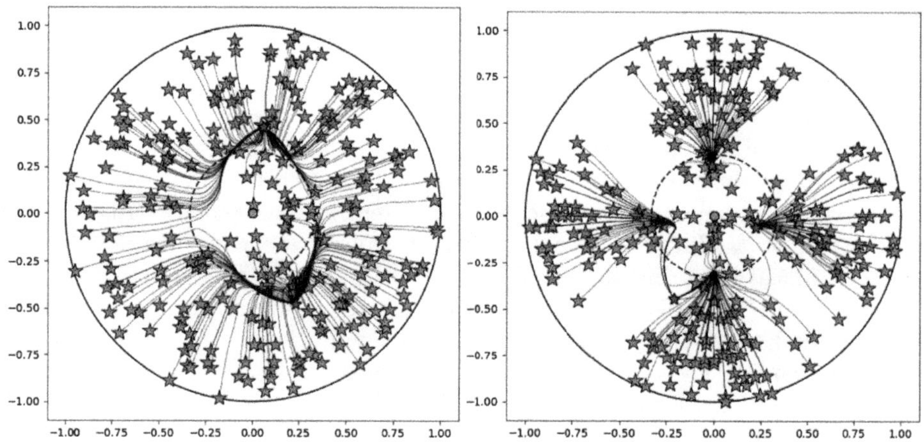

(a) (b)

Fig. 3. Comparison of different entrance laws over the full circle. On the left (a), the entrance law $(\theta_{\text{in}}, \rho_{\text{in}})$ is uniform over the whole ball. *We clearly observe two phases of the process: during the first phase is only the radii evolve. Once the agents' self-confidence get closer to the PRE value, here* $\rho^\star = \frac{1}{3}$*, the radius dynamics decelerates and the angular dynamics expands, namely due to fluctuations, also called 'finite-type' effect.* And on the right (b), we took as entrance law $(\theta_{\text{in}}, \rho_{\text{in}}) = f_4 \otimes \text{Unif}[0, 1]$ (see Eq. (6)). *By contrast with the preceding experiment with uniform angular entrance law, both the angles and the radii evolve from the initial time of the process.*

5 Conclusion

In this paper, we developed and analyzed a tractable multi-agent system intertwining both the opinion and the self-confidence of each agent. We saw that injecting the self-confidence within the classical BC scheme is very intuitive, but the converse is not as straightforward. In the design of the self-confidence dynamics, we opted for a network effect, relating self-confidence and herding behaviour, but this choice deserves to be discussed and refined. More generally, the interaction between the key features of the classical BC model – namely the influence radius map R and the self-confidence dynamics – shall be explored in a more general framework. This constitutes an interesting avenue of research for the future.

Appendix. We show that when we have $(\theta, \rho) \sim \mathrm{Unif}^{\otimes 2}[0, 2\pi] \times [0, 1]$, then the couple $(\theta, \sqrt{\rho})$ is distributed according to a Lebesgue restricted to the ball. Compute the density of the random variable $\sqrt{\rho}$: for any $x \in [0, 1]$,

$$\mathbb{P}(\sqrt{\rho} < x) = \mathbb{P}(\rho < x^2) = x^2 \implies \sqrt{\rho} \sim (2x)_{x \in [0,1]}.$$

Now, for any two angles $\theta_1 < \theta_2$ and any positive value $\rho_0 \in]0, 1[$, define the parcel $\mathcal{P}$ in polar coordinates as $\mathcal{P} := \{(\theta, \rho) \in \mathbb{B} : \theta \in [\theta_1, \theta_2] \text{ and } \rho > \rho_0\}$. It is sufficient to show that $\mathbb{P}\left((\theta, \sqrt{\rho}) \in \mathcal{P}\right) = \frac{\lambda(\mathcal{P})}{\lambda(\mathbb{B})} = \frac{(\theta_2 - \theta_1)(1 - \rho_0^2)}{2\pi}$:

$$\mathbb{P}\left((\theta, \sqrt{\rho}) \in \mathcal{P}\right) = \int_{u=\theta_1}^{u=\theta_2} \int_{x=\rho_0}^{x=1} 2x dx \frac{du}{2\pi} = \frac{(\theta_2 - \theta_1)}{2\pi}(1 - \rho_0^2),$$

which terminates the proof.

Acknowledgments. This work has been fully supported by the Agence Nationale de la Recherche (ANR), through the project NICETWEET (ANR-20-CE48-0009).

References

1. Ahn, H.-S., Van Tran, Q., Trinh, M.H., Ye, M., Liu, J., Moore, K.L.: Opinion dynamics with cross-coupling topics: modeling and analysis. IEEE Trans. Comput. Soc. Syst. **7**(3), 632–647 (2020)
2. Ye, M., Trinh, M.H., Lim, Y.-H., Anderson, B.D.O., Ahn, H.-S.: Continuous-time opinion dynamics on multiple interdependent topics. Automatica **115**, 108884 (2020)
3. Rainer, H., Krause, U.: Opinion dynamics and bounded confidence: models, analysis and simulation. J. Artif. Soc. Soc. Simul. (JASSS) **5**(3) (2002)
4. Blondel, V.D., Hendrickx, J.M., Tsitsiklis, J.N.: On Krause's multi-agent consensus model with state-dependent connectivity. IEEE Trans. Autom. Control **54**(11), 2586–2597 (2009)
5. Kurahashi-Nakamura, T., Mäs, M., Lorenz, J.: Robust clustering in generalized bounded-confidence models. JASSS-J. Artif. Soc. Soc. Simul. **19**(4), 7 (2016)
6. Perrier, R., Schawe, H., Hernández, L.: Phase coexistence in the fully heterogeneous Hegselmann-Krause opinion dynamics model. Sci. Rep. **14**(1), 241 (2024)
7. Bernardo, C., Altafini, C., Proskurnikov, A., Vasca, F.: Bounded confidence opinion dynamics: a survey. Automatica **159**, 111302 (2024)
8. Chen, G., et al.: Deffuant model on a ring with repelling mechanism and circular opinions. Phys. Rev. E **95**(4), 042118 (2017)

9. Gao, S., Huang, C., Han, W., Yang, J.: General consensus with circular opinion under attractive and repulsive mechanisms. Int. J. Mod. Phys. C **31**(05), 2050068 (2020)
10. Hegarty, P., Martinsson, A., Wedin, E.: The Hegselmann-Krause dynamics on the circle converge. J. Differ. Equ. Appl. **22**(11), 1720–1731 (2016)
11. Jia, P., MirTabatabaei, A., Friedkin, N.E., Bullo, F.: Opinion dynamics and the evolution of social power in influence networks. SIAM Rev. **57**(3), 367–397 (2015)
12. Ye, M., Liu, J., Anderson, B.D.O., Yu, C., Başar, T.: Evolution of social power in social networks with dynamic topology. IEEE Trans. Autom. Control **63**(11), 3793–3808 (2018)
13. Ding, Z., Chen, X., Dong, Y., Herrera, F.: Consensus reaching in social network DeGroot model: the roles of the self-confidence and node degree. Inf. Sci. **486**, 62–72 (2019)
14. Wang, C.: An opinion dynamics model with increasing self-confidence. arXiv preprint arXiv:1609.05732 (2016)
15. Economides, N., Flyer, F.: Compatibility and market structure for network goods. NYU Stern School of Business Discussion Paper (98-02) (1997)
16. Björkegren, D.: The adoption of network goods: evidence from the spread of mobile phones in Rwanda. Rev. Econ. Stud. **86**(3), 1033–1060 (2019)

The Emergence of Threads: The Birth of a New Social Network

Peixian Zhang, Yupeng He, Ehsan-Ul Haq, Jiahui He, and Gareth Tyson(✉)

The Hong Kong University of Science and Technology (GZ), Guangzhou, China
gtyson@ust.hk

Abstract. Threads, a new microblogging platform from Meta, was launched in July 2023. In contrast to prior new platforms, Threads was born from an existing parent platform, Instagram, for which all users must already possess an account. This offers a unique opportunity to study platform evolution, to understand how one existing platform can support the "birth" of another. With this in mind, this paper provides an initial exploration of Threads, contrasting it with its parent, Instagram. Our findings reveal that Threads engages more with political and AI-related topics, compared to Instagram which focuses more on lifestyle and fashion topics. Our analysis also shows that user activities align more closely on weekends across both platforms. Engagement analysis suggests that users prefer to post about topics that garner more likes and that topic consistency is maintained when users transition from Instagram to Threads. Our research provides insights into user behaviour and offers a basis for future studies on Threads.

Keywords: Instagram · Threads · Topic analysis

1 Introduction

Meta's new microblogging service, Threads, was launched on July 5th, 2023. It became the fastest-growing 100 million consumer product within five days of its launch. In contrast to prior new social networks, Threads requires an Instagram account, thus enabling rapid social bootstrapping from a user's prior social network. Threads' design allows users to follow their Instagram social connections who already on Threads. The relationship between Threads and Instagram produces a distinct form of attention migration attributed to this unique "parent-child" setup.

The mass sign-up on Threads offers a unique opportunity to characterize the discourse of early adopters on Threads. More importantly, it is possible to track these early adopters on the parent platform and offer a comparative analysis of users' behaviors on the new platform. This is motivated by prior work that has revealed that users tend to exhibit varying norms on different platforms. For example, users share more negative sentiment and work-related content on Twitter than on Instagram [1]. Similarly, Instagram users tend to utilize profile

L. M. Aiello et al. (Eds.): ASONAM 2024, LNCS 15213, pp. 69–78, 2025.
https://doi.org/10.1007/978-3-031-78548-1_7

images with lower smile scores compared to Twitter users [2], and they share more content on weekends than on weekdays [3]. In contrast, users migrating from Twitter to Mastodon did not change topical preferences [4]. We argue that the identification of such communication patterns is important for subsequent research on user modelling and discourse analysis [5].

Given the lack of research on Threads, we present the first analysis of early adopters of Threads, and compare their discourse with the same users on Instagram. This analysis provides a first exploration of Threads communication norms. We explore two research questions:

RQ1: What are topics are discussed on Instagram and Threads, and how do they differ? (Sect. 4).

RQ2: What factors result in a user changing their topic of discussion on Instagram and Threads? Do these factors impact users' selection of topics when they migrate from Instagram to Threads? (Sect. 5).

We find a difference in topical focus, with topics like `threads`, `twitter`, and `trump` being more prevalent on Threads as compared to Instagram. Then, we show that users who consecutively write on similar topics have lower feedback on their posts than those who write on different topics in their consecutive posts.

2 Related Work

We first discuss prior work on multi-social platform comparisons and the platform migration of users.

Multi-social Network Analysis. In order to conduct a comparative analysis of the same users' activities across many platforms, prior studies employ algorithms to establish connections between users' accounts on other platforms [6]. Additionally, they utilize biography websites [2] and examine self-mentions within the material across several platforms [7]. These methods might bring different selection biases among users who utilize biography websites or self-report their personal information. In contrast, the direct relationship between Threads and Instagram obviates the need for intermediary services, simplifying the process of cross-platform analysis. We posit this offers a powerful ground truth dataset. There have been prior studies that focus on analyzing cross-platform behaviours. Studies have found that users exhibit differences in self-description [3] and image usage [2] in profiles, alongside topic preferences [1] and activity time [3]. The engagement for similar content also varies across different platforms [1,8]. Based on these previous findings, we aim to determine whether Threads exhibits comparable topical preferences to its parent, Instagram. To the best of our knowledge, this is the first work to measure user activity on Threads.

Platform Migration. Several works have studied the migration of users to different social media platforms. There are two types of migration: permanent migration and attention migration [10,11]. [12] explore the push and pull factors that trigger migration on Reddit. There has also been prior work studying attention migrations from Twitter to Mastodon [11], and comparing the activities of the same users after migration [4]. A key contribution of our work is the identification of a third type of migration, which we term *parent-child migration.* This is where an existing platform spawns a new service, as is the case with Threads, which was born out of Instagram (*e.g.* only Instagram account owners can access Threads, and Threads posts appear on Instagram timelines as a promotional tool).

To the best of our knowledge, the closest study to our own is [13]. This interesting work explored the characteristics differentiating Twitter users who transition to Threads from those who stay behind. However, in our study, we utilize a larger dataset comprising millions of users on Threads. Furthermore, all of these previous investigations [4, 11, 13] focus on functionally similar platforms. In our research, Threads and Instagram represent distinct platforms that face a special migration [10]. To the best of our knowledge, this represents the first study to amalgamate data from both Instagram and Threads for a comparative analysis of user engagement.

3 Dataset

We now describe our dataset collection, pre-processing, and statistics in the following sections.

Threads. We use the threads-net API[1] to gather Threads data. We use the fact that each Threads account is associated with an auto-generated numeric ID. Starting from August 14th to September 4th (2023), we iterate over all integers from 0 to 12 million to identify any account with IDs in this range. This approach helps us overcome the limitation of searching the Threads platform for specific keywords and provides an extensive search range for early Threads adopters. For each Threads account, we collect the 25 most recent threads posts with their details from that account (25 is the limit imposed by the API). From September 4th to September 13th, we further extend this data by snowballing the identification of additional users, and collecting their posts, based on the reposted threads (*i.e.* users sharing other users' content, similar to retweets on Twitter). In total, there are 1,253,438 Threads accounts with 4,716,626 posts. There are 582,459 (46.47%) users without any threads, 543,522 (43.36%) users with fewer than 25 threads, and 127,457 (10.16%) users with 25 threads. We also find 212 threads posted before the Threads launched (which we exclude from the dataset). A possible explanation could be the presence of test accounts on Threads that Threads developers used to test the application's functions before its launch.

Instagram. Threads and Instagram users share the same username, thus we search for all the usernames from the Threads dataset on Instagram. We use the CrowdTangle API[2] to retrieve all corresponding Instagram accounts and their

[1] https://github.com/dmytrostriletskyi/threads-net.

[2] https://crowdtangle.com.

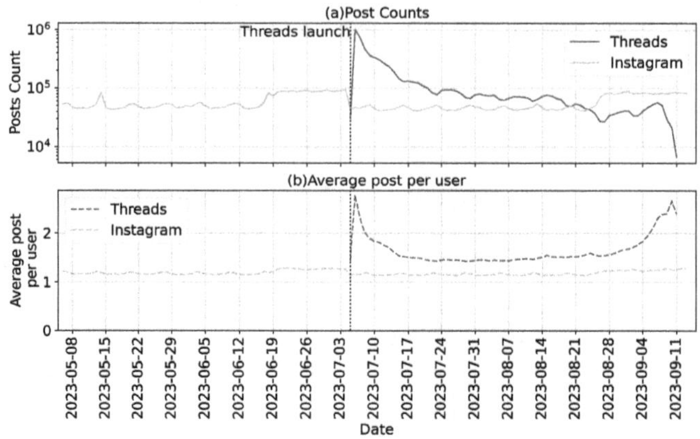

Fig. 1. (a) Daily volume of posts on Instagram and Threads (b) Daily average of posts per users

posts from May 5 to September 13, 2023. In total, we collect 683,168 accounts with 10,862,421 posts on Instagram based on the user list on Threads. Note, we do not collect any Instagram accounts that have their profiles set as private, or any accounts that have changed their usernames. For better comparative analysis with Threads, we split the Instagram data into two parts: ***(i)*** Before Threads launch from May 4th to July 4th with 5,215,434 posts from 575,204 users; and ***(ii)*** After Threads launch from July 5th to Sept 9th with 5,646,987 posts from 587,941 users. This segmentation helps compare users on two platforms after they start posting on Threads. There are 479,877 common users in the two datasets.

Data Pre-processing. We undertake several pre-processing steps. On Instagram. The distribution of posts per user is ($\mu = 9.6, median = 4, min = 1, max = 4,942$), ($\mu = 9, median = 4, min = 1, max = 4,016$) for before and after launch data segment, respectively. Thus, we exclude any users with posts more than the 99th percentile of number of posts per user distribution. This covers 83 accounts for Instagram before Threads launched and 89 after Threads launched. We also exclude the same users on Threads to ensure that the comparison is conducted on the same set of users. Figure 1 shows the (a) total post counts (solid line) and (b) the average post per user (dashed line) on two platforms daily from May 5 to September 13. For the topical analysis, we exclude reposts to avoid repeat counting topics on Threads. This covers 20.83% of Threads' posts. Note, Instagram does not have a repost feature. Hence, we do not apply this to Instagram. On Threads, 2.88% posts do not contain any text (*i.e.* they only contain images), preventing us from performing topical analysis. Thus, we also exclude these. Moreover, in the Instagram dataset, 9.38% of images have embedded text in the images, and these parts of the text have a median length of 18 words. For these posts, we combine the text in the images (CrowdTanlge provides this text) with the users' added caption. We do so because Instagram is an image-centric

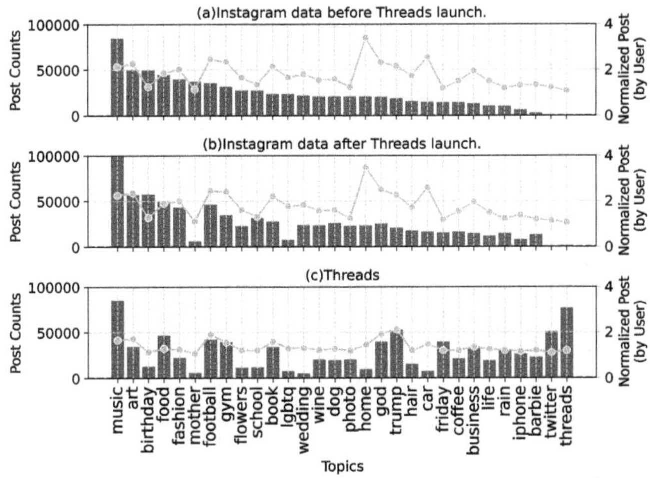

Fig. 2. Posts count for the top-30 topics (left-y) and normalised posts per user (right-y). (a) Instagram before Threads launch. (b) Instagram after Threads launch. (c) Threads. The x-axis is ranked based on the number of posts on Instagram before Threads launch.

platform, and if users have to share a text-based message (such as a quote), they sometimes share it in the image [1]. Finally, we perform the following text pre-processing on the remaining Threads and Instagram posts' text: ***(i)*** We only include English language posts, filtered by langdetect library.[3] 41.47% and 41.82% of total posts are in non-English languages on Threads and Instagram, respectively. ***(ii)*** We remove all hyperlinks. ***(iii)*** We remove any mentions of users (@username). ***(iv)*** We remove all non-alphanumeric characters, including emojis, punctuation, and special symbols (*e.g.* #, @, $). This leaves 3,533,095 posts from 556,730 users on Threads, 2,471,012 posts from 421,426 users on Instagram before Threads launch, and 2,673,157 number of posts from 436,223 users on Instagram after the Threads launch.

4 RQ1: Content-Level Analysis

We now analyse the topics on both platforms.

Topical Extraction. We first look at the most common topics on each platform, and their weekly trends. We use BERTopic [14], which employs sentence transformers and c-TF-IDF to generate topics. We combine the posts from both platforms to create one corpus to ensure consistency and coherence in topic comparisons [4]. Using a threshold of at least 3,500 posts per topic, we identify 206 topics across the combined corpus. These topics cover 50.27% of posts in the data.

[3] https://pypi.org/project/langdetect/.

Topic Volume. We first look at the most common topics within each platform. We see that the prominence of topics varies across the two platforms. Figure 2 presents the absolute number of posts (bars) and the normalized number of posts per user (red line) for the top 30 topics. The x-axis shows the topics ranked in order of their post count in the Instagram dataset before the Threads launch. Normalization is performed by calculating the number of users engaging in a given topic. Figure 2(a) and (b) have similar trends in terms of both the total number of topics and the normalized result. Some topics vary on Instagram before and after the Threads launch in the total number of posts. These topics are related to specific events occurring during that period, such as `mother`, `barbie (film)` and `lgbtq`, which refer to Mother's Day, film-release and LGBT pride, respectively.

Figure 2(c) presents the topics on Threads. When compared with Fig. 2(b), we notice that topics about art (`music` and `art`), food (`food`, `coffee` and `wine`), and sports (`football` and `gym`) are popular on both Instagram and Threads. However, Threads also has distinct topics from Instagram. Topics on social media applications (`Threads` and `Twitter`) are more discussed on Threads. Politics (`trump`) is another topic discussed more on Threads than on Instagram. However, discussion patterns on the two platforms display a different trend. A total of 24,659 (4.43%) users wrote about topic `trump` on Threads. Yet there are only 9,113 (2.09%) users writing about `trump` on Instagram. Interestingly, each user on Instagram is more active though, with 2.26 posts per user about `trump` vs. 1.43 posts per user on Threads. Furthermore, Threads has fewer posts on daily lifestyle topics like fashion (`fashion` and `hair`) and pets (`dog`) as compared to Instagram, where these topics are popular [15]. Moreover, users prefer to share festivals and anniversary topics such as `birthday` and `wedding` on Instagram. Instagram has a higher total number of posts on these topics than Threads. 3,850 users post `wedding` relevant content on Threads, and 12,855 users post wedding on Instagram. 925 (7.20% of 12,855) users post this topic on two platforms. 11,111 users post `birthday` on Threads, and 44,677 users on Instagram. There are 1,914 (4.28% of 44,677) common users who post this topic on two platforms. Finally, we also notice a topic `AI` outside Fig. 2. 10,509 (1.89%) users engaging in discussion on `ai` with an average of 2.12 posts per user on Threads. Only 3,282 (0.75%) users discuss `ai` on Instagram. This indicates a potential interest in AI-relevant topics on Threads that might not be as prominent on Instagram.

Overall, we note that while there are topical similarities on the two platforms, users on Threads post less on fashion-related topics and more on topics related to politics and technology than Instagram.

5 RQ2: Topic-Consistency Exploration

We next analyze the impact of the feedback received by posts in individual users' topical choices. We explore whether users' topical choices change as they move from Instagram to Threads. We first define the term topic consistency and then explore the changes in users' topic consistency within and across the platform.

5.1 Methodology

We define topic consistency as continuing the same topic in two consecutive posts. We further split this into two levels to examine topic consistency at the intra- and inter- platforms scale.

Intra-platform Topic-Consistency. If two consecutive posts by a user on a single platform share the same topic, the preceding post in time is labeled as having intra-platform topic consistency.

Inter-platform Topic-Consistency. Inter-platform topic consistency measures a user's continuation of the same topic from Instagram to Threads. To explore the topic consistency of users when migrating from Instagram to Threads, we inspect posts for each user. We first extract the topics of n recent posts on Instagram before the launch of Threads. We then extract the topics of n posts on Threads. If there is an overlap between these two sets, the user will be regarded as having inter-platform topic consistency. However, it is difficult to select the optimal threshold (n) for the number of posts to calculate this overlap. Hence, we experiment with three thresholds (*i.e.*, $n = 1, 2, 3$) and compare the results of these three alternatives.

Feedback. We also measure feedback on posts, using the number of likes for each posts on both platforms. To calculate the *feedback* of a post, we first calculate the mean value of the likes of all posts published before this post. Taking the mean value of the previous likes and comparing it with the likes received by the current post helps in measuring whether the new post received better or worse engagement. We use the obtained mean value to divide the likes count by the current post, and use Laplace smoothing to avoid a denominator of zero.

5.2 Intra-platform Exploration

We now estimate the feedback measure's impact on topic consistency at the post (intra-platform) and user (inter-platform) levels. We hypothesize that topic-consistent posts tend to have worse feedback since users with unpopular posts tend to focus more on their interests (narrow interests) compared with more popular users [16].

For this, we first divide all the posts into topic-consistent posts and topic-inconsistent posts. From all the posts that have been assigned a valid topic from BERTopic model (same model from RQ1), there are 709,272 (28.25%) topic-consistent posts and 1,801,559 (71.75%) topic-inconsistent posts on Instagram, while 115482 (8.93%) topic-consistent posts and 1178260 (91.07%) topic-inconsistent posts on Threads.

We now analyze the feedback differences on topic-consistent and inconsistent posts. Figure 3(a) shows the distribution of feedback for both Threads and Instagram. We find that the feedback value on Instagram (median = 0.95, std = 10853.97) is higher than that on Threads (median = 0.55, std = 180.74). In general, topic-inconsistent posts (median = 0.88, std = 9637.13) have relatively higher feedback than topic-consistent posts (median = 0.79, std = 7800.72).

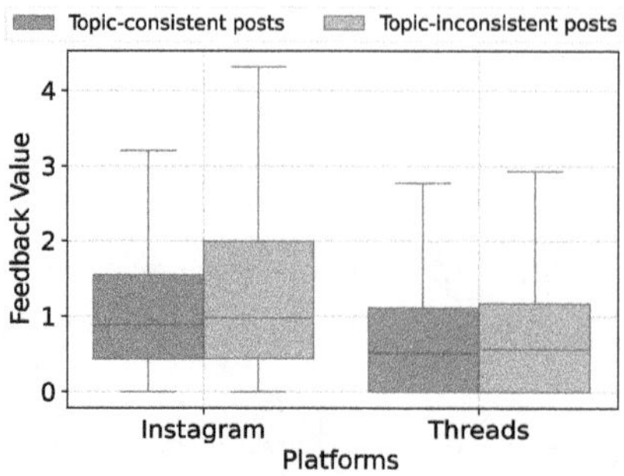

(a)

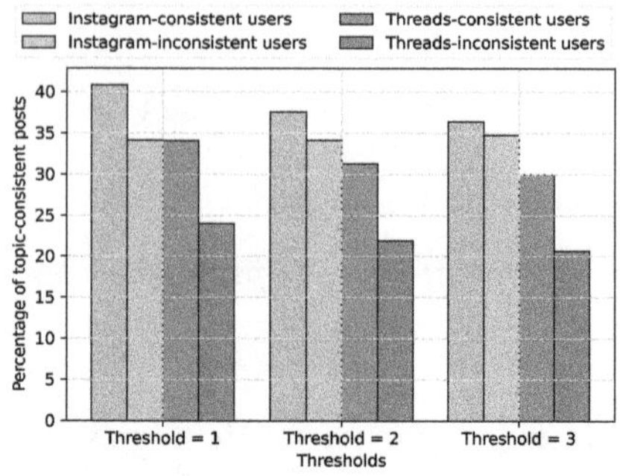

(b)

Fig. 3. (a) Box plot of feedback value for both topic-consistent posts and inconsistent posts on Instagram and Threads. (b) Percentage of topic consistency posts for consistent/inconsistent users on two platforms (with three different thresholds).

Unsurprisingly, these results indicate that topic-inconsistent posts are more likely to have better feedback, *i.e.* users who talk about a topic that receives a high number of likes are more likely to change the topic of their next post. This phenomenon is more common on Instagram than Threads. A potential explanation for this phenomenon is that users who prefer to publish posts on a wide variety of topics tend to be popular with audiences. Naturally, receiving positive feedback reinforces this message for publishers. But users who focus on same topics care little about the feedback.

5.3 Inter-platform Exploration

Next, we utilize the inter-platform consistency definition (users writing similar topics on Instagram and Threads in their consecutive posts) and divide users into consistent and non-consistent groups. We test three different thresholds on number of consecutive posts (from 1 to 3) to define the inter-platform topic consistency. Table 1 shows the number of consistent and inconsistent users with all thresholds.

Table 1. The number of consistent/inconsistent users for different threshold.

	Threshold = 1	Threshold = 2	Threshold = 3
Consistent users	46,051	84,631	102,807
Inconsistent users	131,293	92,713	74,537

We first analyze the percentage of topic-consistent posts by topic-consistent and -inconsistent users. Figure 3(b) shows these percentages for both Instagram and Threads with all thresholds. This figure shows that the percentage

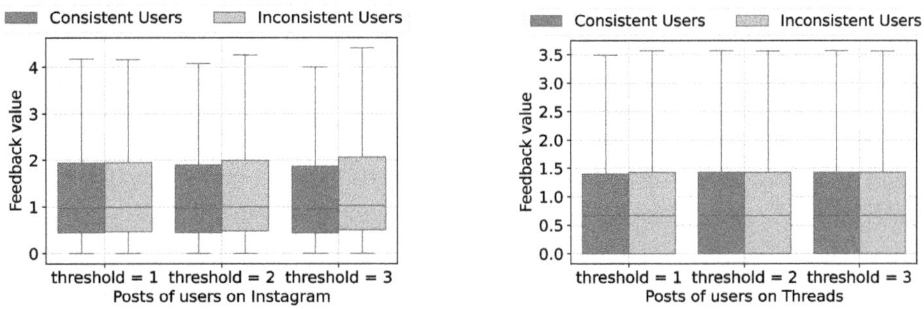

(a) Feedback (Ins)

(b) Feedback (Threads)

Fig. 4. (a) Box plot of feedback for consistent/inconsistent users on Instagram. (b) Box plot of feedback for consistent/inconsistent users on Threads.

of topic-consistent posts (with threshold = 1), on Instagram, from consistent users (40.83%) is higher than inconsistent users (34.14%). The corresponding percentage on Instagram is also higher than Threads. These results indicate that users who maintain topic consistency when they migrate from the "parent" platform to the "child" are also more likely to publish posts on the same topics consecutively. Further, the same group of users is more likely to maintain topic consistency on Instagram but change topics more frequently on Threads.

We finally analyze the differences in feedback on posts from these user groups and show the distribution in Fig. 4. The median feedback count on Threads is less than 1 for both consistent and inconsistent users on all thresholds. On Instagram, the median value is close to 1 for both user types. A K-W test shows that there is no significant difference in feedback received by both user-groups, even though users' feedback on Instagram (median = 0.98, std = 15528.53) is relatively higher than that on Threads (median = 0.67, std = 118.56). These results indicate that there is little correlation between whether users maintain topic-consistency at inter-platforms scale and the feedback they receive. Further, generally users receive better feedback on Instagram compared with Threads.

6 Conclusion

Our study serves as the first characterization of Threads. We focus on the inherited overlaps between Threads (the "child") and Instagram (its "parent"). We have shown that there is a difference in topical focus across the two platforms (RQ1). Instagram has a higher percentage of topics related to lifestyle and fashion, while Threads has more topics related to politics and technology. Finally, we have explored the topic-consistency across platforms and compare the feedback (RQ2). We find consistency is more aligned with the content itself than about the feedback their posts receive (especially on Instagram). We also notice that users who maintain topic consistency when they migrate from Instagram to Threads are more likely to publish posts consecutively on the same topics.

Ethic Statement. We limit our analysis to public posts on both platforms. When gathering data, we witness usernames for both Instagram and Threads users. However, we exclude this from our analysis and make no attempt to de-anonymize users. We further exclude usernames from our content-based analysis.

References

1. Manikonda, L., Meduri, V.V., Kambhampati, S.: Tweeting the mind and instagramming the heart: exploring differentiated content sharing on social media. In: AAAI ICWSM (2016)
2. Zhong, C., Chang, H.-W., Karamshuk, D., Lee, D., Sastry, N.: Wearing many (social) hats: how different are your different social network personae? In: AAAI ICWSM (2017)
3. Lim, B.H., Lu, D., Chen, T., Kan, M.-Y.: # mytweet via instagram: exploring user behaviour across multiple social networks. In: IEEE/ACM ASONAM (2015)
4. He, J., Zia, H.B., Castro, I., Raman, A., Sastry, N., Tyson, G.: Flocking to mastodon: tracking the great twitter migration. In: ACM IMC (2023)
5. Kwak, H., Lee, C., Park, H., Moon, S.: What is Twitter, a social network or a news media? In: ACM WWW (2010)
6. Li, Y., Zhang, Z., Peng, Y., Yin, H., Xu, Q.: Matching user accounts based on user generated content across social networks. Future Gener. Comput. Syst. (2018)
7. Veiga, M.H., Eickhoff, C.: A cross-platform collection of social network profiles. In: ACM SIGIR (2016)
8. Farahbakhsh, R., Cuevas, A., Crespi, N.: Characterization of cross-posting activity for professional users across Facebook, Twitter and Google+. Soc. Netw. Anal. Min. **6**, 1–14 (2016)
9. Kumar, S., Zafarani, R., Liu, H.: Understanding user migration patterns in social media. In: AAAI (2011)
10. Fiesler, C., Dym, B.: Moving across lands: online platform migration in fandom communities. In: ACM on HCI, no. CSCW1 (2020)
11. Jeong, U., Sheth, P., Tahir, A., Alatawi, F., Bernard, H.R., Liu, H.: Exploring platform migration patterns between twitter and mastodon: a user behavior study. arXiv preprint arXiv:2305.09196 (2023)
12. Newell, E., et al.: User migration in online social networks: a case study on reddit during a period of community unrest. In: AAAI ICWSM (2016)
13. Jeong, U., Nirmal, A., Jha, K., Tang, X., Bernard, H.R., Liu, H.: User migration across multiple social media platforms. arXiv preprint arXiv:2309.12613 (2023)
14. Grootendorst, M.: BERTopic: neural topic modeling with a class-based TF-IDF procedure. arXiv preprint arXiv:2203.05794 (2022)
15. Hu, Y., Manikonda, L., Kambhampati, S.: What we Instagram: a first analysis of Instagram photo content and user types. In: AAAI ICWSM (2014)
16. Ferrara, E., Interdonato, R., Tagarelli, A.: Online popularity and topical interests through the lens of Instagram. In: ACM HT (2014)

SocFedGPT: Federated GPT-Based Adaptive Content Filtering System Leveraging User Interactions in Social Networks

Sai Puppala[1], Ismail Hossain[1(✉)], Md Jahangir Alam[2(✉)], and Sajedul Talukder[2(✉)]

[1] Southern Illinois University, Carbondale, IL, USA
saimaniteja.puppala@siu.edu
[2] University of Texas at El Paso, El Paso, TX, USA
{ihossain,malam10}@miners.utep.edu, stalukder@utep.edu

Abstract. Our study presents a multifaceted approach to enhancing user interaction and content relevance in social media platforms through a federated learning framework. We introduce personalized GPT and Context-based Social Media LLM models, utilizing federated learning for privacy and security. Four client entities receive a base GPT-2 model and locally collected social media data, with federated aggregation ensuring up-to-date model maintenance. Subsequent modules focus on categorizing user posts, computing user persona scores, and identifying relevant posts from friends' lists. A quantifying social engagement approach, coupled with matrix factorization techniques, facilitates personalized content suggestions in real time. Additionally, an adaptive feedback loop and readability score algorithm enhance the quality and relevance of content presented to users. Our system offers a comprehensive solution to content filtering and recommendation, fostering a tailored and engaging social media experience while safeguarding user privacy.

Keywords: content filtering · GPT · content diversity · LLM · online social network

1 Introduction

In an era dominated by the pervasive influence of social media, the ability to sift through vast streams of content and deliver personalized, relevant experiences to users has become paramount [2]. Our research addresses this issue by combining advanced machine-learning techniques with privacy-preserving methodologies. We employ federated learning [1], a method that allows collaborative model training across distributed devices while safeguarding sensitive user data [14]. This innovative approach aims to revolutionize content filtering and recommendation systems [3], enhancing user engagement.

L. M. Aiello et al. (Eds.): ASONAM 2024, LNCS 15213, pp. 79–88, 2025.
https://doi.org/10.1007/978-3-031-78548-1_8

Central to our research is the development of context-aware models [4] specifically for social media. Using state-of-the-art language models like GPT [5], we create personalized frameworks that accurately understand and categorize user-generated content. Beyond categorization, we profile user personas to understand their preferences deeply. By analyzing user interactions and content engagement, we aim to provide highly personalized recommendations that resonate on an individual level.

Existing research has explored various methodologies to address content filtering and recommendation in social media ecosystems [3]. Context-aware modeling techniques, using advanced language models like GPT, have shown promise in understanding user-generated content [6]. These models categorize posts and discern user preferences, forming the basis for personalized content delivery. Persona profiling techniques [7] capture diverse user interests and behaviors, enhancing content relevance [8].

Despite these advancements, challenges remain in balancing personalized experiences [9] with user privacy. Federated learning offers a solution by decentralizing model training and keeping data localized. Combining federated learning with context-aware and persona-based modeling, our research aims to advance content filtering and recommendation systems, creating a more personalized, secure, and engaging user experience.

Building upon previous research [10,15], our approach integrates federated learning with context-aware and persona-based modeling techniques to tackle content filtering and recommendation challenges in social media. Using federated learning [11], we emphasize user privacy and data security while enabling collaborative model training across distributed client devices. This decentralized method maintains data integrity and promotes inclusive model development.

Our methodology employs personalized GPT [12] and Context-based Social Media LLM models to categorize content and profile user personas. These models are trained locally on client devices using federated learning, ensuring privacy, and their aggregated results update a global model that captures diverse user interests. By combining federated learning with context-aware modeling and persona profiling, our research aims to enhance content experiences and set new privacy standards in social media analytics. Our experiments demonstrate the approach's effectiveness and scalability, advocating for the coexistence of personalized content and data privacy in social media.

The paper is structured to provide an optimizing content filtering and recommendation systems in social media using federated learning. The methodology section details our technical approach, including model deployment, federated learning techniques, and context-aware and persona-based modeling. Results demonstrate the approach's efficacy and scalability, followed by a discussion of key findings, limitations, and future directions. The conclusion summarizes key insights, emphasizes user privacy, and highlights the practical implications of our work.

2 Methodologies

2.1 Problem Statement

Given the increasing reliance on social media platforms for content consumption and interaction, there is a pressing need for advanced content filtering and recommendation systems that can deliver personalized experiences while ensuring user privacy. Traditional centralized approaches to content filtering may compromise individual data integrity and struggle to adapt to the dynamic nature of user preferences across different social media platforms. To address these challenges, our research aims to develop a decentralized content filtering and recommendation framework utilizing federated learning techniques. By deploying personalized context-aware models tailored for social media content categorization and user persona profiling, our goal is to enable collaborative model training across distributed client devices while preserving user privacy. Through this approach, we seek to optimize content filtering and recommendation systems, offering users personalized and relevant content experiences without compromising their privacy.

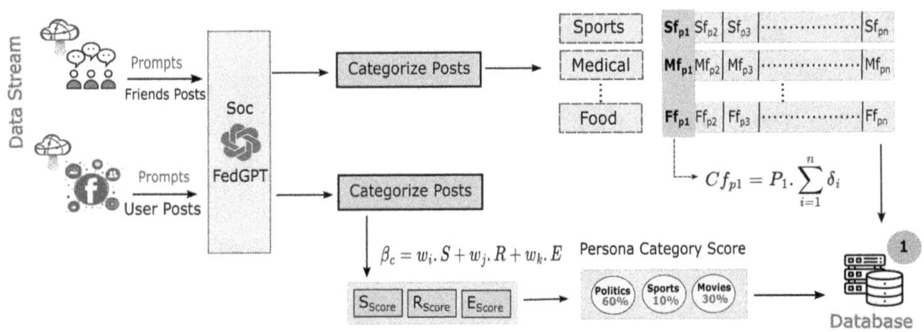

Fig. 1. The system architecture elucidates various components in the Adaptive Content Filtering System approach, with a key focus on GPT, User Persona profile, and Category-based social engagement. Additionally, it provides a concise overview of the distinctive features of the federated learning global server for context-based GPT generation.

2.2 Federated Learning

Federated learning is a decentralized method for training models while maintaining data privacy, as client data remains on their devices. Our approach involves training the GPT model across multiple clients, who share only model updates, not raw data.

Each client starts with common model parameters and trains the GPT model using their local data, resulting in unique model updates. These updates are aggregated to form a global model that incorporates insights from all clients, without exchanging raw data. This iterative process periodically updates the global model to reflect the combined knowledge of all clients. This method preserves data privacy and generates a more personalized GPT model by leveraging diverse data from multiple clients. The equations illustrate how global and local model parameters are updated and aggregated to improve the model iteratively.

$$\begin{aligned} \text{Global Parameter Update:} \quad & \phi^{(t)} \leftarrow \phi^{(t-1)} - \alpha \nabla J(\phi^{(t-1)}, D^{(t)}) \\ \text{Client Update:} \quad & \phi_i^{(t)} \leftarrow \phi^{(t)} - \alpha \nabla J_i(\phi^{(t)}, D_i) \\ \text{Aggregation:} \quad & \phi^{(t+1)} \leftarrow \frac{1}{N} \sum_{i=1}^{N} \phi_i^{(t)} \end{aligned} \tag{1}$$

$\phi^{(t)}$ represents the global model parameters at iteration t. α is the learning rate. $\nabla J(\phi^{(t-1)}, D^{(t)})$ denotes the gradient of the loss function J with respect to the global model parameters using the local dataset $D^{(t)}$. $\phi_i^{(t)}$ represents the local model parameters of client i at iteration t. $\nabla J_i(\phi^{(t)}, D_i)$ denotes the gradient of the loss function J with respect to the local model parameters of client i using its local dataset D_i. The aggregation step computes the updated global model parameters by averaging the local model parameters from all participating clients.

2.3 User Persona Profiler

User interaction data, including likes, shares, and comments, is collected from the LLM platform. Each post is categorized by content (e.g., sports, politics) and sentiment (positive, negative, neutral). Data prepossessing ensures accurate categorization and sentiment labeling by cleaning and standardizing the data. We analyze user engagement within each content category to determine preference distributions and sentiment distributions (positive, negative, neutral). Interactions are weighted to reflect their significance in user preferences and sentiments.

To understand user personas, we compute a composite *Content Score* based on Engagement, Readability, and Sentiment. This score measures content quality and audience resonance, excluding Originality Score due to the difficulty in quantifying uniqueness in large social media datasets.

Engagement Score (E). The Engagement Score encapsulates user interactions with the content, including likes, shares, comments, and views. Each interaction type is assigned a distinct weight based on its perceived value in denoting user engagement. The Engagement Score is normalized to a $[0, 1]$ scale and calculated as follows:

$$\begin{aligned} E = & \frac{w_{\text{likes}} \cdot \text{Likes} + w_{\text{shares}} \cdot \text{Shares}}{MaxE} \\ & + \frac{w_{\text{comments}} \cdot \text{Comments}}{MaxE} \end{aligned} \tag{2}$$

where w_{likes}, w_{shares}, and w_{comments} are the weights assigned to likes, shares, comments, and views, respectively, and $MaxE$ is the maximum possible engagement score for normalization purposes.

Sentiment Score (S). The Sentiment Score reflects the overall sentiment of posts and comment threads on social media. The score is normalized to the $[0, 1]$ range, where 0 represents entirely negative sentiment, 0.5 represents neutral sentiment, and 1 represents entirely positive sentiment.

Category Readability Score (ρ). In the context of assessing user engagement through readability, we developed a Readability Score metric that categorizes posts based on the complexity and clarity of their language. This metric assigns a score of '1' to posts written with simple language, easily accessible to a general audience, enhancing widespread comprehension and interaction. Posts that utilize professional, industry-specific terminology are assigned a higher score of '2', reflecting their targeted appeal to a professional demographic that values technical accuracy or specialized knowledge. Conversely, posts deemed unreadable, either due to poor grammar, lack of coherence, or excessive complexity without proper context, receive a score of '0'.

This scoring approach underscores the importance of clarity and audience-tailored communication in written content. To compute an aggregate Readability Score for each user, we average the scores of all posts they have interacted with, providing a quantifiable measure of the typical readability level of content that the user engages with. This methodology allows for a nuanced analysis of user preferences and content interaction patterns, which is crucial for tailoring content strategy and improving user engagement.

User Persona Score (C). The Persona Score (C_k) (Here (k) represents a specific category) is a weighted sum of the Engagement (E), Readability(R), and Sentiment (S) scores, each contributing to the final score based on predefined weights that reflect their relative importance. The formula for calculating C is as follows:

$$\beta_k = w_i \times E + w_j \times S + w_k \times \rho$$

where w_E, w_R, and w_S are the weights for the Engagement, Readability, and Sentiment scores, respectively, satisfying $w_E + w_S + w_R = 1$.

The comprehensive framework would look similar to below,

$$\beta_{distribution} = \begin{bmatrix} \beta_{\text{news}}/(\beta_{\text{news}} + \beta_{\text{media}} + \beta_{\text{politics}} + \cdots) \\ \beta_{\text{media}}/(\beta_{\text{news}} + \beta_{\text{media}} + \beta_{\text{politics}} + \cdots) \\ \beta_{\text{politics}}/(\beta_{\text{news}} + \beta_{\text{media}} + \beta_{\text{politics}} + \cdots) \\ \vdots \end{bmatrix}$$

This methodology provides a comprehensive framework for evaluating online content, leveraging quantitative metrics to assess user engagement, content accessibility, and emotional resonance. The resultant Content Score serves as a versatile tool to gauge content quality and effectiveness in digital communication for content filtering (Fig. 2).

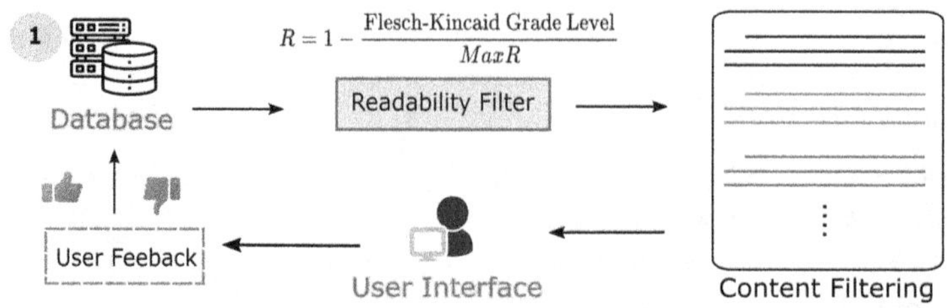

Fig. 2. Continuation for Fig. 1: The workflow architecture of adaptive content filtering starting from database and all the way to feedback collection from user.

2.4 Content Filtering

In our approach to content filtering, we prioritize the personalization of user experience by initially computing a 'user rank' for each individual within a user's social circle. This rank is derived from a carefully formulated scoring system, which quantifies the user's engagement with content shared by each friend. Specifically, the score for each friend is calculated using the following formula:

$$\delta_i = w_l \times L_i + w_c \times C_i + w_{sh} \times Sh_i$$

Here, δ_i, L_i, C_i, and Sh_i represent the overall friend i engagement score, number of likes, comments, and shares, respectively, that the current user has made on the content posted by friend i. The weights w_l, w_c, and w_{sh} correspond to the relative importance of likes, comments, and shares in determining the strength of engagement and affinity towards that friend.

This scoring mechanism allows us to assess the degree of interaction and preference the user exhibits towards each friend's content. By leveraging this data, our system can more accurately filter and prioritize content that aligns with the user's demonstrated interests and interactions within their network. This methodology not only enhances content relevance but also improves user satisfaction by curating a feed that reflects personal connections and preferences.

Upon calculating specific scores for each user derived from their engagement levels, we proceed to tailor the content delivery mechanism. This customization involves filtering categorical posts that align with the interests identified in the user's persona score calculation. For example, if the persona score indicates a preference for sports, the system filters and prioritizes sports-related posts or articles shared by friends within the user's network, particularly those friends who have demonstrated high engagement levels as reflected in our engagement score matrix. This method ensures that the content presented to the user not only aligns with their interests but also emanates from highly interactive sources within their social circle, thereby enhancing user engagement and content relevance. The below equation formulates the importance of the friend post P.

$$P_i = w_C \cdot C + w_L \cdot L + w_S \cdot S + w_T \cdot \frac{1}{T}$$

Here we employ weighted factors where w_C, w_L, w_S, and w_T represent the weights assigned to comments, likes, shares, and latency of the post, respectively. The variables C, L, and S denote the number of comments, likes, and shares that a post has received. The variable T indicates the time elapsed since the post was published. In the expression, multiplication is denoted by $\cdot$ to enhance clarity and visual distinction. The term $\frac{1}{T}$, representing the reciprocal of T, is used to increase the score of more recent posts, thereby emphasizing the timeliness of content in its valuation.

Additional Filtering. Additionally, our system incorporates an enhanced filtering mechanism that enables users to selectively exclude posts with negative scores, utilizing sentiment analysis and predictive scoring of comment thread trends. Furthermore, we apply another layer of filtering based on categories derived from the user's persona profile, allowing for content to be tailored more closely to individual preferences. Posts that do not fit into these specified categories are classified as general posts, providing a broad spectrum of content while still prioritizing personalized user interests. For each post i, we apply a filter based on sentiment and trend predictions:

$$F_i = \begin{cases} 1 & \text{if } S_i > 0 \text{ and } T_i > \tau \\ 0 & \text{otherwise} \end{cases}$$

Here, τ represents a threshold for the trend prediction score that determines if a post is trending positively and thus interesting to the user. F_i represents the filtering status of post i, where 1 indicates the post passes the filter and 0 indicates it does not.

Readability Score (R). The Readability Score assesses the ease with which an audience can comprehend the content. It is determined using the Flesch-Kincaid readability test, which is based on the length of words and sentences. The score is normalized to a $[0, 1]$ scale, with higher scores indicating more accessible content:

$$R = 1 - \frac{\text{Flesch-Kincaid Grade Level}}{MaxR}$$

where $MaxR$ is the maximum grade level used for normalization, ensuring R falls within the $[0, 1]$ range.

Adaptive Feedback Loop. To enhance flexibility in our system, we introduced an additional field for user feedback. For instance, a user who previously expressed interest in Community Services may receive suggestions related to this category based on our historical score calculations. However, recognizing that user preferences can change over time as individuals evolve, we implemented a feature allowing users to indicate their preferences using like or dislike buttons. This user feedback is then utilized to update our database records and serve as filtering criteria for future recommendations, ensuring that users receive more relevant and tailored suggestions.

3 Experiment

3.1 Context GPT

Our system implementation includes different modules, starting with creating a personalized GPT model using federated learning for privacy and security. We connect four client entities to a global server. Each client, equipped with a base GPT-2 model and local social media data collected via web crawlers, trains the model and sends updates back to the server. The global server performs federated aggregation to average these models and redistributes the updated model to clients for further training. This continuous iterative process ensures the model stays current with new data, enhancing adaptability and relevance for our content filtering system.

As depicted in Fig. 1, the Context-based GPT model, tailored for social media content filtering, plays a pivotal role in categorizing user posts. The stream of user posts is directed to the Context GPT model to ascertain their appropriate categories. To enhance the accuracy of category determination, we employ prompt engineering techniques such as zero-shot and few-shot prompt engineering [13]. Once the categories of the streaming posts are identified based on user history, the subsequent step involves determining the user persona scores for each category. The primary objective of user persona profiling is to discern the user's preferences post-filtering. For example, if user history indicates a preference distribution of 30% politics and 70% sports, these insights are stored as the ultimate outcome of the user persona in our database, to be utilized as filtering criteria (Fig. 3).

3.2 User Category Persona

To calculate user persona scores, we analyze user engagement (likes, shares, comments) across different categories from their history. Each category (e.g., politics, sports) is segmented and examined for engagement types. Engagements are assigned weighted scores, with comments, shares, and likes considered. While these weights are currently fixed, they could be adjusted dynamically. Sentiment scores (positive or negative) are also calculated to help filter content. The final user persona score, derived from these analyses, is stored in the database for personalized content suggestions.

3.3 Content Filtering

The next module identifies and ranks relevant posts from a user's friends list. It uses social engagement metrics and a context-based GPT model to categorize posts. After categorization, posts are ranked using content filtering techniques based on user preferences. The aim is to suggest related posts to users and identify similar interests within user groups. Interaction scores help determine the most engaging posts for each category.

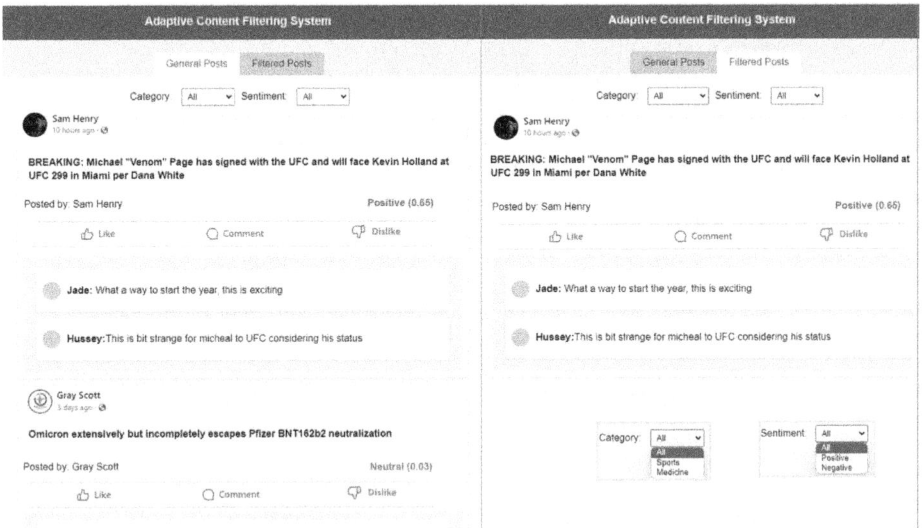

Fig. 3. User interface for Federated GPT-based Adaptive Content Filtering System Leveraging User Interactions in Social Networks

Posts are selected for user suggestions based on user persona and relevance. The system updates in real-time, ensuring constant relevance. Users can filter out negative or positive posts and provide feedback through like/dislike buttons, enhancing personalization. Spam posts are filtered using a readability score algorithm, ensuring high-quality content.

4 Conclusion

Our experiment highlights a multi-module system that enhances user experience and content relevance on social media Using personalized GPT models and federated learning, it ensures privacy and adapts to user preferences. The system improves post categorization accuracy, provides tailored content suggestions, and uses social engagement metrics for relevance. Real-time updates and adaptive feedback enhance interaction, while a readability algorithm maintains quality. The system delivers personalized, high-quality content, with potential future improvements in engagement scores and filtering algorithms.

References

1. Zhang, C., Xie, Y., Bai, H., Yu, B., Li, W., Gao, Y.: A survey on federated learning. Knowl.-Based Syst. **216**, 106775 (2021). Elsevier
2. Liang, T.P., Lai, H.J., Ku, Y.C.: Personalized content recommendation and user satisfaction: theoretical synthesis and empirical findings. J. Manag. Inf. Syst. **23**(3), 45–70 (2006). Taylor & Francis

3. May, A., Chaintreau, A., Korula, N., Lattanzi, S.: Filter & follow: how social media foster content curation. In: The 2014 ACM International Conference on Measurement and Modeling of Computer Systems, pp. 43–55 (2014)
4. Zhu, Y., Nayak, N.M., Roy-Chowdhury, A.K.: Context-aware modeling and recognition of activities in video. In: Proceedings of the IEEE Conference on Computer Vision and Pattern Recognition, pp. 2491–2498 (2013)
5. Floridi, L., Chiriatti, M.: GPT-3: Its nature, scope, limits, and consequences. Minds Mach. **30**, 681–694 (2020). Springer
6. Titus, L.M.: Does ChatGPT have semantic understanding? A problem with the statistics-of-occurrence strategy. Cogn. Syst. Res. **83**, 101174 (2024). Elsevier
7. Li, L., Xiao, J.: Persona profiling: a multi-dimensional model to study learner subgroups in Massive Open Online Courses. Educ. Inf. Technol. **27**(4), 5521–5549 (2022). Springer
8. Chaoji, V., Ranu, S., Rastogi, R., Bhatt, R.: Recommendations to boost content spread in social networks. In: Proceedings of the 21st International Conference on World Wide Web, pp. 529–538 (2012)
9. Neuhofer, B., Buhalis, D., Ladkin, A.: Smart technologies for personalized experiences: a case study in the hospitality domain. Electron. Mark. **25**(3), 243–254 (2015). https://doi.org/10.1007/s12525-015-0182-1
10. Puppala, S., Hossain, I., Talukder, S.: Towards federated learning based contraband detection within airport baggage x-rays. In: 2022 IEEE International Conference on Machine Learning and Applied Network Technologies (ICMLANT), pp. 1–6 (2022)
11. Talukder, S., Puppala, S., Hossain, I.: Federated learning-based contraband detection within airport baggage x-rays. J. Comput. Sci. Coll. **38**, 218–218 (2022)
12. Kim, Y., Park, J.: Towards Personalized AI: Engineering Model Responses for Customized User Interactions in Generative AI Systems (2023)
13. Yong, G., Jeon, K., Gil, D., Lee, G.: Prompt engineering for zero-shot and few-shot defect detection and classification using a visual-language pretrained model. Comput.-Aided Civil Infrastruct. Eng. **38**(11), 1536–1554 (2023). Wiley Online Library
14. Hossain, I., Puppala, S., Talukder, S.: Collaborative differentially private federated learning framework for the prediction of diabetic retinopathy. In: 2023 IEEE 2nd International Conference On AI In Cybersecurity (ICAIC), pp. 1–6 (2023)
15. Talukder, S., Puppala, S., Hossain, I.: A novel hierarchical federated learning with self-regulated decentralized clustering. J. Comput. Sci. Coll. **38**, 222–223 (2022)

Anonymous Dissent in the Digital Age: A YouTube Dislikes Dataset

Sujan Dutta[1], Mallikarjuna Tupakula[2], Sumeet Kumar[2], and Ashiqur R. KhudaBukhsh[1](✉)

[1] Rochester Institute of Technology, Rochester, NY, USA
{sd2516,axkvse}@rit.edu
[2] Indian School of Business, Hyderabad, India
{mallikarjuna_tupakula,sumeet_kumar}@isb.edu

Abstract. On December 13, 2021, YouTube made a major policy change with respect to the visibility of video dislikes. Citing the need to protect the well-being of individual content creators, YouTube stated that dislike information will only be visible to the video owners thus abolishing a decade-long tradition of publicly visible, anonymous outlet for user dissent. This paper makes two key contributions. First, it releases a valuable dataset of dislike information from 8.3 million videos gleaned from 159 popular news and debate outlets to characterize and chronicle the dislike behavior on YouTube. Second, it quantifies and investigates the information gap that this policy change leaves us with.

Keywords: Social Web · User Engagement · YouTube Policy Change

1 Introduction

In December 2021, YouTube introduced a platform-wide policy change that rendered dislike information publicly inaccessible. One month prior to rescinding API access to dislike information YouTube announced this change and gave web science practitioners an advance warning that they were on borrowed time. In our bid to preserve a snapshot of an anonymous outlet for user disapproval, this one-month window allowed us to collect likes-dislikes data for a large number of popular channels. After YouTube's policy change took effect, the platform no longer displays the dislike count. This paper performs a longitudinal analysis of dislike behavior on YouTube on controversial issues and analyzes the possible information gaps YouTube's policy change leaves us with.

Respecting YouTube's rationale to protect small content creators, our analyses preclude small, individual contributors and primarily consider major news networks that are no strangers to audience reactions and TRP battles. Via a substantial dataset of 8.3 million videos that enjoyed overall 2.74 billion likes and endured 333 million dislikes, this paper characterizes and chronicles the like

S. Dutta and M. Tupakula—Joint first authors.

L. M. Aiello et al. (Eds.): ASONAM 2024, LNCS 15213, pp. 89–99, 2025.
https://doi.org/10.1007/978-3-031-78548-1_9

and dislike behavior on YouTube. Central to the theme of inclusive AI, our paper considers (1) news and opinion outlets presented in a wide range of languages (see Table 1); (2) news outlets with minorities as the target audience; and (3) news outlets having a substantial geographical coverage (135 countries).

To summarize, our contributions are the following:

- ***Resource*:** We release a dataset of 8.3 million YouTube videos with like, dislike, video title, and video description information. Beyond the immediate goal of characterizing dislike behavior, our dataset can be valuable to social scientists to understand political events of import in the last decade.
- ***Social*:** We present a detailed characterization of dislike behavior of a global audience. Our analyses reveal that studying dislike behavior may present valuable insights into anonymous dissent. Especially, in countries with questioned democracy. While prior literature investigated potential reasons for YouTube engagement [13] and post-hoc effect of hiding dislikes [11], to our knowledge, ours is the first detailed characterization of dislike behavior at this scale.

Table 1. List of all languages in the dataset.

Language
Albanian, Amharic, Arabic, Azerbaijani, Bengali, Burmese, Chinese, English, Filipino, French, German, Greek, Gujarati, Hausa, Hindi, Hungarian, Indonesian, Italian, Japanese, Kannada, Korean, Malay, Malayalam, Marathi, Odia, Pashto, Persian, Portuguese, Russian, Sinhalese, Spanish, Swahili, Tamil, Thai, Turkish, Ukrainian, Urdu, Vietnamese.

2 Dataset

We collect a dataset[1] that spans 159 YouTube channels with an overall subscriber count of 727,756,097 (4.43 $\pm$ 0.70 million subscribers). Examples include (1) official handles of prominent news networks (e.g., TVC News Nigeria (Africa), BBC News (Europe), CNN (North America), Sky News Australia (Oceania), and Aaj Tak (Asia)); (2) general and parliamentary debates (e.g., Oxford Union, UK Parliament, C-Span); (3) YouTube channels of prominent print media (e.g., New York Times (US), Daily Mail (UK), Bild (Germany), and Times of India (India)).

We consult each country's Wikipedia page that documents major news outlets (cable news channels and print media). For these news outlets, we search and confirm if these outlets have a substantial YouTube presence (we set a threshold of at least 100K subscribers). For these YouTube channels, we scrape the videos using Selenium to obtain the video IDs. We use the publicly available YouTube API to collect the count of likes, dislikes, comments, and views of these videos. In addition, we also collect the video titles and video descriptions (if present). Our dataset is collected between November 13, 2021, and December 12, 2021.

We consider news channels due to the following reasons:

[1] Available at https://github.com/Suji04/YouTube-Dislike-Dataset.

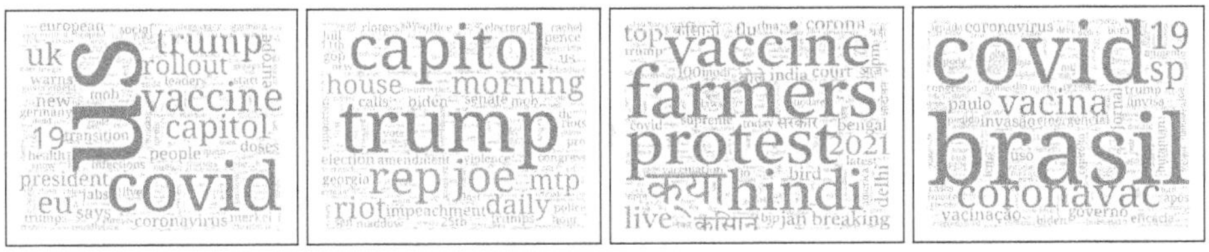

(a) Europe (b) US (c) India (d) South America

Fig. 1. Word cloud visualization of news headlines from prominent news channels around the globe for the week of Jan 6, 2021. EuroNews (Europe); CNN, Fox News, and MSNBC (US); Aaj Tak, Zee News, ABP News and NDTV (India); Band Jornalismo and C5N (South America).

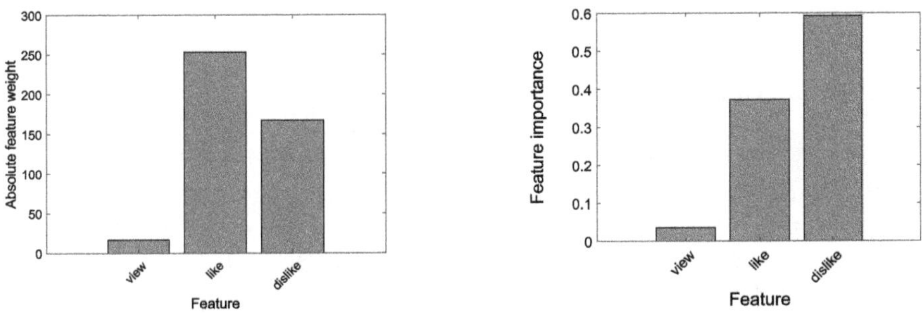

(a) Linear Regression (b) XGBoost

Fig. 2. Relative feature importance of view, like, and dislike in the task of predicting comment count on a given video.

- ***Global participation and audience diversity:*** A broad and diverse audience can be achieved by geographically and linguistically diverse YouTube news channels. Our selected channels span over 130 countries and 38 languages. 64.7% of these channels are operated by official news networks, ensuring a reliable representation of global politics. We take special care in including news outlets from countries beyond the economic superpowers (e.g., TVC News Nigeria) or news outlets catering to a disadvantaged minority group (e.g., Black News Channel) or news outlets with a focus on countries under political instability (e.g., ToloNews from Afghanistan, Syria Al Khbaria for Syria). From Swahili to Pashto – Our linguistic diversity, again, ensures that we reach out to a truly global audience.
- ***Rich topical diversity:*** Within the narrow scope of US politics in 2020–2021, the country witnessed a raging pandemic, a bitterly fought election, a watershed moment in the history of racial justice, and a far from peaceful transfer of power. From climate change to migrant crises, from Superbowl to Wimbledon, and local street crimes to international disputes, the richness of our dataset offers a unique lens to understand public dissent.

- ***News as parallel accounts of world events:*** Finally, this dataset captures multiple parallel accounts of world events as they happen. Figure 1 presents a world-cloud visualization of news titles from news outlets from different geographic regions for the week of Jan 6, 2021. We note that while the US was grappling with an assault on democracy with an unprecedented riot at the Capitol, India was dealing with its own issue of farmers' dissent. At the same time, Brazil was dealing with the COVID-19 vaccine adoption challenges. Even within the same country, two networks can present competing views. For instance, Capitol insurrection is a phrase that seldom appeared on Fox while it was a regular fixture on MSNBC news titles. Together, these outlets paint a comprehensive picture of how the world works, and therefore, our dataset can be of great value to social scientists for a broad range of socio-economic and policy questions.

3 Characterizing Dislikes

Research Question: *How are likes, dislikes, comments, and views correlated?* Table 2 summarizes the pairwise correlations between views, likes, dislikes, and comments. It is intuitive that the more views a video gets, the more likely it would get an audience who are willing to engage through likes, dislikes, and comments. We further note that likes (dislikes) are more correlated with comments than dislikes (likes). This result is also intuitive as not all videos will elicit polarization reactions. On the other hand, a polarized video is likely to receive more comments than a non-polarized one. This intuition is reflected as comments are more correlated with likes and dislikes than views.

Table 2. Pairwise correlation between video views, likes, dislikes, and comments.

	view	*like*	*dislike*	*comment*
view	-	0.66	0.50	0.40
like	0.66	-	0.45	0.58
dislike	0.50	0.45	-	0.50
comment	0.40	0.58	0.50	-

Research Question: *Does dislike present any additional informational value beyond video views and likes?* We design a prediction task where the inputs (features) are views, likes, and dislikes received by a video and the output is the predicted comment count of the video. Figure 2 summarizes the feature weights of individual features upon using two well-known regression algorithms (XGBoost [4] and linear regression [16]). We observe that dislike information is an important feature in predicting comment counts.

3.1 Quantifying Disagreement

To estimate viewership disagreement over a set of videos, we follow a measure introduced by KhudaBukhsh *et al.* [14]. This measure has been applied to quantify the shift in the popularity of fringe conservative news outlets preceding the Capitol riot [15].

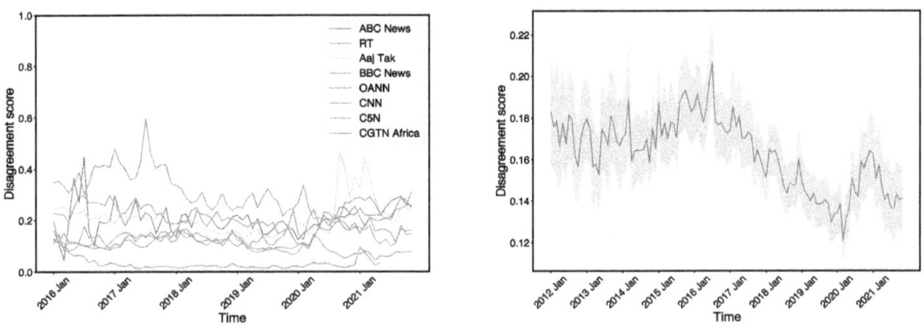

(a) Select channels (b) All channels

Fig. 3. Evolution of disagreement scores over time. Figure 3a considers six channels from six continents: CGTN (Africa); BBC (Europe); Aaj Tak (Asia); ABC News (Oceania); CNN (N. America); and C5N (S. America). OANN and RT are included for contrastive purposes. Figure 3b considers all channels.

For a given video v, let v_{like} and $v_{dislike}$ represent the number of likes and dislikes received by it. The disagreement of v is measured as $\frac{v_{dislike}}{v_{like}+v_{dislike}}$. Let $\mathbb{I}(v^i, \mathcal{F})$ be an indicator function that returns 1 if video v^i satisfies a set of filtering criteria denoted by $\mathcal{F}$. KhudaBukhsh *et al.* [14] mainly fix $\mathcal{F}$ to a specific channel and upload time. For example, $\mathcal{F}$ could be all videos uploaded on CNN in the month of January 2014. The *disagreement score* of the set of videos satisfying the filtering criteria $\mathcal{F}$ is computed as $\frac{\sum_i \mathbb{I}(v^i,\mathcal{F})\ disagreement(v)}{\sum_i \mathbb{I}(v^i,\mathcal{F})}$. A low *disagreement score* indicates that the majority of the viewers liked the content. Whereas, a high value signifies a rising proportion of disapproving viewership. In our work, we vary the filtering conditions $\mathcal{F}$ to characterize like and dislike behavior along different aspects.

Research Question: *Are all news networks uniformly loved (or hated)?* Figure 3a considers a time duration of six years to account for elections, pandemics, protests for justice, and other factors that may disrupt news cycles temporarily with sudden shifts in popularity. We observe that not all news networks enjoy a similar disagreement score profile. For instance, for the most part of the time, CNN content was more disagreed than any other news channel. CNN also experiences a sharp spike starting in 2017, which marks the beginning of the long tussle between CNN and former President Trump, who repeatedly questioned CNN's reporting quality. Elections also created a spike in ABC News Australia at the beginning of 2016. The spike in disagreement scores in India's news channel Aaj Tak coincided with India's second wave COVID-19 crisis, which triggered a nationwide healthcare collapse. In contrast, One America News Network (OANN) has documented evidence of peddling election misinformation and COVID-19 misinformation [8]. We notice a disagreement score profile of being an echo chamber. Finally, RT Russia is a state-sponsored channel with a long history of accusations of being a propaganda news network. However, we see more disagreement on RT Russia than CGTN Africa. A deeper exploration of RT Russia follows in the later sections.

Figure 3 considers our entire dataset and tracks the overall disagreement score. Starting from 2017 to the onset of COVID-19, the overall disagreement score was on a steady decline. Of course, the sharp spike around 2020 February coincides with a global pandemic requiring little explanation, we however wonder what could be the reason for the steady decline that started in 2017. Is it because the hyper-partisan media has polarized and segregated the news audience into different camps so that everyone has found their own cozy echo chamber?

Table 3. Most disagreed trigrams across channels. Aaj Tak uses code-switched Hindi and English in their news titles. We translate the Hindi portions into English. For Russian, we present a translation performed by Google and verified by a native Russian speaker.

Aaj Tak	BBC	CNN	DW	Fox News	RT
abhishek banerjee wife	syria air strikes	assault weapons ban	germany focus europe	pelosi house dems	russian presidential elections
pm modis mann	iran nuclear deal	black lives matter	covid 19 vaccine	speaker pelosi holds	mishustin holds a meeting
opinion poll 2020	brexit trade deal	clinton donald trump	quadriga international talk	jen psaki holds	government members live
monsoon session 2020	leader jeremy corbyn	gop health bill	international talk quadriga	julie roginsky claps	fight covid 19
Gaurav Bhatia बोले [speaks]	labour leader jeremy	religious freedom bill	world max euromaxx	biden delivers remarks	Patriarch kirill congratulated
Bihar opinion poll	pm boris johnson	gop tax bill	chancellor angela merkel	schumer senate dems	head coach of the team
Rajasthan priest burnt	explained 60 seconds	white house trump	hillary clinton wins	covid 19 response	coordinating council fighting
100 Sep 2020	death toll rises	trump travel ban	world stories week	ted cru wins	putin holds a meeting
priest burnt alive	george floyd death	rep adam schiff	stories week reports	trump cruz feud	appeal vladimir putin
bengal jp nadda	aung san suu	sotu john mccain	young global leaders	gov cuomo holds	holds a meeting by members

3.2 Censorship

We now investigate the nature of news that attracts disagreement. We focus on CCTV China, a state-sponsored news outlet. China has documented evidence of social web censorship [1]. In fact, most videos on CCTV China have comments disabled. This implies that the only remaining outlet to express disagreement for this particular channel was through dislikes. Table 4 lists the titles of the top 10 most disagreed videos on CCTV. Research on the popularity of the Chinese Communist Party (CCP) indicates that state-controlled media play an important role in it [12,19]. Nonetheless, we observe that there exists considerable anonymous dissent against CCP that has little documentation.

3.3 Disagreement of Phrases

Research Question: *Are there specific trigger words or phrases that are associated with greater disagreement?* We define disagreement of a word (phrase) as the disagreement score of all videos containing that word (phrase) in their titles. While we note that many such phrases (words) would not necessarily capture the stance of the video, this simple approach may lead to a deeper understanding of trigger words or phrases. On our entire dataset, we find that `climate`

Table 4. Video titles and disagreement scores of the most disagreed videos on CCTV. Translations are obtained using Google Translate and verified by a proficient Chinese speaker.

CCTV	Disagreement score
"Ping Talk" is close to people-Xi Jinping's favorite allusion: "I will be selfless and live up to the people"	0.78
The song "Without the Communist Party, there would be no new China" "Celebrate the 100th anniversary of the founding of the Communist Party of China Theatrical performance "The Great Journey"	0.60
Theme song "Liancheng Family" Singing: Miriam Yeung, Long Shijie, Zheng Qiyuan, etc. [Celebrating the 20th anniversary of Macao's return to the motherland]	0.49
Situational chorus "Roar, the Yellow River" "Celebrating the 100th anniversary of the founding of the Communist Party of China Theatrical performance "The Great Journey"	0.48
The Central Committee of the Communist Party of China and the State Council held a Spring Festival group meeting, Xi Jinping delivered a speech 2020-01-23	0.48
Chinese Ministry of Foreign Affairs: U.S. remarks on Hong Kong appear to be fair, but in fact reveal ulterior motives	0.43
"News Network" races against time and keeps pace with history - General Secretary Xi Jinping's speech at the 2020 Spring Festival group gathering sparked enthusiastic responses 2020-01-24	0.37
VR micro-record "Happiness Coordinates"	0.36
"News Network" and "Seeking Truth" magazines published an important article by General Secretary Xi Jinping" Speech at the Discussion at the Dunhuang Academy" 2020-01-31	0.36
Wang Kai pays tribute to the heroes of the Anti-Japanese War "March of the Volunteers" shows the power of national rejuvenation "Heroes in the Poster"	0.32

Table 5. Predicting disagreement. Each model is trained on a balanced dataset of 16,000 examples with an equal number of *disagreed* and *not-disagreed* examples. The evaluations are conducted on a held-out set of 4,000 examples.

	$\mathcal{D}_{cnn}$	$\mathcal{D}_{fox}$	$\mathcal{D}_{msnbc}$
$\mathcal{M}_{cnn}$	80.9 ± 0.5%	57.3 ± 0.2%	68.1 ± 0.5%
$\mathcal{M}_{fox}$	59.1 ± 0.7%	83.5 ± 0.8 %	56.4 ± 0.4%
$\mathcal{M}_{msnbc}$	63.3 ± 0.6%	55.7 ± 0.2%	81.8 ± 0.4%

`change` (disagreement score: 0.23) is a much more divisive phrase than `weather forecast` (disagreement score: 0.10). Of the words `citizen` (disagreement score: 0.16), `immigrant` (disagreement score: 0.20), and `refugee` (disagreement score: 0.25), `refugee` is more disagreed than `citizen`. The disagreement score gap between refugee and citizen substantially widens when we restrict our focus to EuroNews videos.

Table 3 lists most disagreed trigrams from six prominent news channels: Aaj Tak (India); BBC (UK); CNN (US); DW (Germany); Fox News (US); and RT Russia (Russia). We observe that religious polarization and hate crimes [3], corruption issues surface among the most disagreed trigrams in India [17]. BBC News, being a world news channel with an emphasis on British news, shows both international political controversies (e.g., the Iran nuclear deal or the Syrian war) and internal political crises (e.g., Brexit) among its most disagreed trigrams. CNN and Fox News paint an interesting contrast together. We observe that Fox is primarily fixated on entities (e.g., US President Joe Biden, Former US House Speaker Nancy Pelosi, Senate Majority Leader Chuck Schumer) while CNN disagrees more on issues (e.g., assault weapons ban or black lives matter). We further observe that the most disagreed trigrams in RT Russia indicate anonymous disapproval of Russian President Vladimir Putin.

3.4 Is Disagreement Learnable?

Table 3 indicates that there could be untold stories of anonymous negative reactions to world leaders and dislike information offer vital insights. We next investigate if it is possible to predict controversiality of a news item from video titles and descriptions. Prior literature has considered predicting controversial news using Facebook reactions [2] and early prediction of controversiality on Reddit threads [9].

We cast the problem of predicting disagreement as a binary classification task. We consider the videos in the top and bottom 25 percentile for a given news channel based on the disagreement score. Since disagreement lies on a continuum, we perform this quantization step to segregate the subsets of most-disagreed and least-disagreed videos. All videos featuring in the top 25 percentile and the bottom 25 percentile have the same labels *disagreed* and *not-disagreed*, respectively. Considering the extensive literature on political polarization in hyper-partisan news media [7,10,14], we investigate three YouTube channels: CNN, Fox News, and MSNBC. We use a BERT-based [6] text classifier for this prediction task.

Table 6. Predicting disagreement on BBC Russia and RT Russia.

	$\mathcal{D}_{bbc\text{-}russia}$	$\mathcal{D}_{rt}$
ïż£ $\mathcal{M}_{bbc\text{-}russia}$	$78.9 \pm 2.0\%$	$46.4 \pm 0.8\%$
$\mathcal{M}_{rt}$	$44.0 \pm 0.9\%$	$76.6 \pm 1.6\%$

Table 5 shows that for a given YouTube channel, predicting disagreement is a learnable problem. All models perform quite well on unseen examples from their own channels. However, when we try to predict disagreement on different news channels we observe political alignments coming into play. We observe both $\mathcal{M}_{cnn}$ and $\mathcal{M}_{msnbc}$ predict disagreement on each other's videos better than that on $\mathcal{D}_{fox}$. Interestingly, as a side note, it is worth mentioning that despite being a far-right news channel from a different continent, $\mathcal{M}_{fox}$ predicts disagreement better on ABC News Australia ($62.6 \pm 0.8\%$) than on CNN, and MSNBC. We conduct an additional experiment with RT Russia ($\mathcal{M}_{rt}$) and BBC Russia ($\mathcal{M}_{bbc\text{-}russia}$) (see Table 6). Again, we observe that model predictions on unseen data from one's own channel are reliable. However, both $\mathcal{M}_{rt}$ and $\mathcal{M}_{bbc\text{-}russia}$ perform worse than chance when predicting for each other.

The ability to reliably predict disagreements on a given channel suggests our dataset can predict signals in newer videos after YouTube's policy change. The poor cross-dataset generalization shows the global audience's diversity and conflicting opinions, which our dataset can help explore further.

4 Discussions and Ethics Statements

Despite a growing call for incorporating disagreements in machine learning models [5,18] and moving toward more transparent AI systems in academia, big-tech social web companies are tightening their grip on publicly available information. We see our dataset as an important contribution to capture a decade's worth of disagreements from a global audience. This paper starts a dialogue on web preservation and presents a compelling argument that the whims of platforms can create information gaps that are difficult to bridge. We rest our case with a compelling concluding example.

Figure 4 presents a screenshot of a YouTube news video hosted by CNN in which the former president's both-sidesism drew severe scrutiny. The video now only shows that it has 3K likes. However, what we do not see now is that more than one year back, the video already had 4K dislikes. With the video title and description as input, $\mathcal{M}_{cnn}$ correctly predicts this video as a highly disagreed one. The disagreement score of the word `Charlotseville`, a city in Virginia that seldomly appeared on news without the connection of the car attack, is a staggering 0.35. Nearly 5.9% of the videos in our dataset have more dislikes than likes. In those videos, the disapproval is effectively the majority opinion. Figure 4 presents a compelling example where the dislikes raised a voice against hate which got lost due to YouTube's policy change on dislike visibility.

Fig. 4. A screenshot of a CNN video captured on March 1, 2023. This video was uploaded in the wake of the Charlotesville car attack in which a white supremacist ran over and killed a 35 year old woman protesting in favor of racial justice.

4.1 Ethics Statement

We collected our dataset when dislike information was publicly available. We further note that the internet archive has been tracking more than 25% of these videos. Hence, some of these videos' dislike information is still publicly viewable. Furthermore, respecting YouTube's goal to protect small-content creators, our dataset precludes any individual content contributor and only considers popular news and opinion outlets that are no strangers to public dissent and TRP battles.

References

1. Bamman, D., O'Connor, B., Smith, N.: Censorship and deletion practices in Chinese social media. First Monday (2012)
2. Basile, A., Caselli, T., Nissim, M.: Predicting controversial news using Facebook reactions. In: CLiC-it (2017)
3. Basu, D.: Majoritarian politics and hate crimes against religious minorities: evidence from India, 2009–2018. World Dev. **146**, 105540 (2021)
4. Chen, T., et al.: XGBoost: extreme gradient boosting. R Package Version 0.4-2 **1**(4), 1–4 (2015)
5. Davani, A.M., Díaz, M., Prabhakaran, V.: Dealing with disagreements: looking beyond the majority vote in subjective annotations. TACL **10**, 92–110 (2022)
6. Devlin, J., Chang, M., Lee, K., Toutanova, K.: BERT: pre-training of deep bidirectional transformers for language understanding. In: NAACL-HLT 2019, pp. 4171–4186. Association for Computational Linguistics (2019)

7. Dutta, S., Li, B., Nagin, D.S., KhudaBukhsh, A.R.: A murder and protests, the Capitol Riot, and the Chauvin trial: estimating disparate news media stance. In: IJCAI, pp. 5059–5065 (2022)
8. GarcÃa-Hodges, A.: YouTube suspends OANN for violating its Covid-19 policy (2020). https://www.nbcnews.com/news/all/youtube-suspends-oann-violating-its-covid-19-policy-n1248845, NBC News
9. Hessel, J., Lee, L.: Something's brewing! Early prediction of controversy-causing posts from discussion features. In: Burstein, J., Doran, C., Solorio, T. (eds.) NAACL-HLT 2019, pp. 1648–1659 (2019)
10. Hyun, K.D., Moon, S.J.: Agenda setting in the partisan TV news context: attribute agenda setting and polarized evaluation of presidential candidates among viewers of NBC, CNN, and Fox news. Journal. Mass Commun. Q. **93**(3), 509–529 (2016)
11. Jung, S.G., Salminen, J., Jansen, B.J.: The effect of hiding dislikes on the use of Youtube's like and dislike features. In: ACM WebSci 2022, pp. 202–207 (2022)
12. Kennedy, J.J.: Maintaining popular support for the Chinese Communist Party: the influence of education and the state-controlled media. Polit. Stud. **57**(3), 517–536 (2009)
13. Khan, M.L.: Social media engagement: what motivates user participation and consumption on YouTube? Comput. Hum. Behav. **66**, 236–247 (2017)
14. KhudaBukhsh, A.R., Sarkar, R., Kamlet, M.S., Mitchell, T.: We don't speak the same language: interpreting polarization through machine translation. In: Proceedings of the AAAI Conference on Artificial Intelligence, pp. 14893–14901 (2021)
15. KhudaBukhsh, A.R., Sarkar, R., Kamlet, M.S., Mitchell, T.M.: Fringe news networks: dynamics of us news viewership following the 2020 presidential election. In: 14th ACM Web Science Conference 2022, pp. 269–278 (2022)
16. Mitchell, T.M., et al.: Machine Learning, vol. 1. McGraw-Hill, New York (2007)
17. Parry, J.: The crisis of corruption and the idea of India. Morals Legitim. Between Agency Syst. **12**, 27 (2000)
18. Pavlick, E., Kwiatkowski, T.: Inherent disagreements in human textual inferences. TACL **7**, 677–694 (2019)
19. Stockmann, D., Gallagher, M.E.: Remote control: how the media sustain authoritarian rule in China. Comp. Pol. Stud. **44**(4), 436–467 (2011)

PhD

From Prison Gang to Transnational Criminal Organization. A Social Network Analysis of the Tren de Aragua

Laura Homann and Oscar Alvaro-Montes(✉)

University of Massachusetts Lowell, Lowell, MA 01854, USA
{laura_homann,oscar_alvaro}@uml.edu

Abstract. This research paper explores the evolution of the Venezuelan prison gang Tren de Aragua. This criminal group has emerged as one of the most influential entities in Latin America, exerting control that extends beyond prison walls into broader societal, business, and political spheres. This paper utilizes Social Network Analysis (SNA) to dissect the intricate relationships and networks surrounding the Tren de Aragua. We introduce the concept of "converged governance," which challenges traditional views of the separations between state and non-state actors. This theoretical framework proposes that the clandestine links between these state and non-state entities allow for covert collaborations that maintain internal cohesion while promoting extensive criminal activities. Our study not only provides a detailed map of the Tren de Aragua's network structure through SNA but also emphasizes the prison's role as a critical territory where state and non-state systems intersect. This offers a more systemic insight into the dynamics of criminal governance and its impact on national and regional security. By investigating the conditions under which prison gangs like the Tren de Aragua assume control over state facilities and evolve into transnational criminal organizations, this paper significantly enhances our understanding of systemic challenges by highlighting the complex, often symbiotic relationships between criminal organizations and state actors.

Keywords: Tren de Aragua · Criminal governance · Social Network Analysis

1 Introduction

The growth of prison gangs has significantly impacted correctional systems and public safety. Initially formed for protection around racial and ethnic identities, these gangs have evolved into complex criminal organizations (Maitra 2020). Notable examples include the Mara Salvatrucha and Barrio XVIII among other strong prison gangs in the Latin American region (Cawley 2017; Schmidt 2012, Ruiz 2020, Deneyer Willis 2015, HRW 2022). This evolution of prison gangs into sophisticated criminal organizations presents a dual challenge. Nationally, they contribute to increased violence, instability in correctional facilities, and higher crime rates in broader communities, often recruiting and radicalizing younger populations (Gundur 2020; Oxford Research Encyclopedia of

L. M. Aiello et al. (Eds.): ASONAM 2024, LNCS 15213, pp. 103–116, 2025.
https://doi.org/10.1007/978-3-031-78548-1_10

Criminology 2018). This perpetuates a cycle of crime and strains public resources. Internationally, their transnational networks facilitate organized crime and illegal trafficking, which undermines the rule of law, fosters corruption, and destabilizes governments, particularly in regions with weak institutions, such as in Latin America (Paarlberg 2022). This raises the need for a deeper understanding of how to control their expansion and the hidden interactions between state and non-state actors. Therefore, we study Venezuela's Tren de Aragua prison gang and explore how the power dynamics inside the state contribute to the problem. Specifically, our research aims to answer the question of how prison gangs evolve into sophisticated organizations.

To answer our research question, we will use Social Network Analysis (SNA). SNA will allow us to study the interactions and networks within and around Tren de Aragua, including its influence in society, business, and politics. This method highlights the roles of non-criminal figures in its expansion and the ties between organized crime and political actors (Otero 2023). Our contribution is twofold. Existing theories often argue that criminal or hybrid governance fills power vacuums in weak states either by collaborating or pressuring the government and society (Varese 2011; Felbab-Brown 2018) or as a consequence of internal power relations of the States (Durán-Martínez 2018). However, these theories often view hybrid governance as the state ceding power to criminals (Schlichte & Schneckener 2015). In this regard, we detected two gaps, first the hybrid governance theory places the state in a position of weakness, but we argue that criminal governance can also exist where the state is present and strong. Additionally, the theory does not explain how state and criminal organizations interact without merging their power structures.

Therefore, at the theoretical level, we expand criminal governance theories by introducing the concept of *converged governance*. Converged governance addresses these gaps by showing how power relations between state and criminal actors expand even in strong states. It explains how they create covert channels of communication to interact without merging and explores the specific mechanisms of their interaction. This concept builds on previous theories (e.g., Denyer Willis 2015) by examining how non-state actors form covert networks with state actors, allowing collaboration while maintaining internal cohesion. At the methodological level, we present the first attempt to map the network structure of Tren de Aragua using SNA. This offers a new perspective on the dynamics of criminal systems like Tren de Aragua and their relationships with political and economic systems. Our analysis emphasizes how prisons serve as territories where state and non-state systems intersect.

2 The Origins and Growth of Tren de Aragua

The Tren de Aragua originated as a prison gang in Venezuela's Tocorón prison in the early 2000s and gained prominence between 2013 and 2015 through extortion and job-selling activities in a railway project (Martin 2023). Since then, they have expanded into organized crime, engaging in extortion, kidnapping, homicide, drug and arms trafficking, human trafficking, and contraband (InSight Crime 2020, 2023). Their unprecedented territorial expansion poses a significant threat to South American and regional security (U.S. Department of the Treasury 2023). Tren de Aragua's evolution highlights the complexities of combating organized crime in politically and economically unstable regions, necessitating coordinated international efforts to dismantle their operations.

3 Criminal Governance and State Presence

Criminal governance refers to the ability of criminal organizations to control territories and activities, especially where state influence is weak or absent. This concept is often linked to state absence theories (Rotberg 2004; Call 2011; UNODC 2024), where non-state actors, including criminal groups, fill the void left by the state's inability to provide services or protection (Abello-Colak & Guarneros-Meza 2014). Naím's 'State Retreat theory' (2012) suggests that criminal governance reflects not a total absence of the state, but a significant weakening of its influence. This blending of criminal and lawful governance can severely impact the rule of law in affected areas (Felbab-Brown et al. 2018). These theories help explain the rise of criminal governance in areas where state authority retreats. Other scholars argue that criminal organizations operate as parallel authorities (Lessing 2016; 2021). This coexistence, also called hybrid regimes, does not imply direct collaboration but reflects a tactical understanding, where each maintains its own sphere of influence (Jaffe 2013; Pimenta et al. 2021; Plancarte Escobar 2021).

In summary, theories of state absence, retreat, or hybrid regimes explain the conditions for criminal governance. By doing so, they answer the question of *why* criminal organizations gain political influence. However, they do not explain *how* state-criminal interactions work. To address this, we introduce converged governance. Based on Luhmann's systems theory (2012), we argue that while political, economic, and criminal systems operate independently, they interact within the broader societal context. The state and non-state actors avoid direct collaboration but use covert networks to achieve their goals. The covert interactions between state and non-state actors highlight a complex dynamic where both maintain their distinct identities while engaging for mutual benefit. This leads to our first hypothesis:

> **H1:** *Criminal organizations and the state converge by utilizing covert connections to interact without diluting their borders.*

Following this hypothesis, we aim to delve into the interaction strategies between state and non-state actors, specifically examining the Venezuelan government and Tren de Aragua. It posits that these entities deliberately avoid direct interactions to safeguard their exposure, maintain legitimacy, and protect internal secrets.

4 Criminal Governance, Legitimacy, and Corruption in SNA

To examine these state-criminal relationships, we apply SNA to Tren de Aragua. SNA applications in organized crime are often categorized into "micro" and "macro" approaches (Calderoni 2014). Micro approaches, such as Natarajan's (2000) and Morselli's (2005, 2009) studies, show how a small group of highly connected individuals drives most criminal activities. Macro-level studies by Malm and Bichler (2009, 2011) reveal the importance of connections, such as kinship and business partnerships, in sustaining organized crime. While existing research shows SNA's value in understanding criminal organizations, Calderoni (2014) notes the need for more empirical studies, particularly outside law enforcement.

By focusing on Tren de Aragua's evolution, this paper contributes to literature by showing the multiplexity of the actors involved. According to Smith & Papachristos (2016), multiplexity refers to the intersecting layers of criminal, personal, and legitimate ties within a network. These connections are essential for the survival of criminal organizations, which often use legitimate facades to protect their operations and gain social or political support (Morselli & Giguere 2006; Lauchs et al. 2011). Here, corruption often acts as a bridge between the criminal 'underworld' and the legitimate 'upper world' (UNODC 2024; Blumberg 1974). However, by uncovering a diverse set of social and political connections that rely on trust and reciprocity (Cartier-Bresson 1997; Costa 2017), SNA reveals that these relationships are more than mere financial exchanges. This interlacing of legitimate and criminal actors is at the heart of our converged governance theory and leads to our second hypothesis:

> ***H2.*** *The convergence of Tren de Aragua or similar criminal organizations with the political sector is influenced by the strength of their internal ties and the gang's multiplexity, which determines the scope and nature of their governmental connections.*

This hypothesis suggests that the internal network dynamics of a criminal organization like Tren de Aragua are critical in shaping its interactions with governmental entities. Specifically, it posits that the stronger and more numerous the internal connections within the organization, the more effectively it can engage or converge with the government. These internal ties foster cohesion and coordination, which are essential for the organization to effectively navigate and interact with complex governmental structures.

5 Power Dynamics and Secrecy in Criminal Networks

Understanding the power dynamics between criminal organizations and the state is crucial to answering our research questions. Durán-Martínez (2018) argues that power relations between the state and criminal organizations significantly affect the frequency and visibility of violence in societies. The presence of direct or indirect communication channels between these actors influences the type of violence experienced. Factors such as leadership, hierarchy, communication, and external influences play essential roles in the operations and longevity of criminal entities. The structure of hierarchy within these organizations impacts internal power dynamics, with a clear spectrum of roles from leaders to subordinates (UNODC 2023). Leadership styles vary, with some leaders being directly involved in criminal activities, while others exert influence more subtly. According to Porter and Alison (2006), leadership in many criminal networks is often characterized by indirect strategies rather than explicit commands, with leaders tactically guiding operations and setting agendas. With this in mind, it is important to understand the strength of ties, whether weak or strong, within a network. Weak ties offer operational flexibility and broader network reach, while strong ties foster solidarity and cooperation.

Calderoni's (2014) research on drug-trafficking groups shows that most members are marginal players involved in specific tasks. They often represent weak ties that support a horizontal, flexible structure. This flexibility allows criminal activities to be driven by opportunities rather than formal organizational arrangements. This also extends to non-criminal actors, such as business professionals and political figures, whose involvement helps maintain the network's secrecy and extend its influence into legitimate spheres.

These often-weak ties are vital for the network's adaptability and survival (Smith & Papachristos 2016; Granovetter 1973). But as important as the weak ties are, as important are the strong ones. According to Krebs (2002) and Perliger (2010), influential figures or hubs play central roles in coordination, recruitment, and information dissemination. Centrality measures help identify these hubs: closeness measures an actor's overall connectedness, betweenness shows how often an actor connects others and controls information flow, and eigenvector centrality assesses an actor's influence based on the quality of their connections. This understanding of power dynamics and influential hubs informs the basis for our Hypothesis 3.

> **H3:** *The level of sophistication of a criminal network is determined by its ability to establish ties with members of other societal sectors. These diverse connections enhance the network's operational capabilities and social legitimacy, which, in turn, enables more extensive and resilient criminal activities.*

In this study, we assume that Tren de Aragua's network likely presents ties to cultural, economic, and social systems in the Aragua environment. This hypothesis reflects the adaptability and complexity of the Tren de Aragua as a criminal organization that interacts with various aspects of its environment.

6 Methodology – SNA Applied to the Tren de Aragua

Our study employs SNA to explore the complexities of Tren de Aragua's organizational network and its connections with various entities. We first outline our data collection process and then proceed with the analysis. Using 'apify,' we systematically scraped data from investigative journalism outlets such as Insight Crime, Al Jazeera, BBC News, Transparencia Venezuela, BioBio Chile, and DW, covering both Spanish and English sources. This diverse selection is essential due to the transnational nature of Tren de Aragua and the varying regional focus of these publications. We also referred to official communications from the governments of Venezuela, the US, and Colombia. After data scraping, we manually searched for individual names and details about their roles and activities within the organization, examining their relationships with each other and external entities, like political. Each name and related information were systematically cataloged in a matrix.

We established links between individuals and organizations based on two criteria. First, we define a solid link when the source explicitly mentions a direct connection between two actors. Second, we identify an environmental link when there is an indirect connection. This refers to individuals or groups without explicit ties to Tren de Aragua but who hold strategic positions in institutions that have facilitated the gang's expansion. These connections are considered circumstantial or environmental. In such cases, we hypothesize that if two individuals are associated with the same organization (e.g., the 'Aragua Prison system') and operate in interconnected areas, the likelihood of indirect interactions or collaboration increases. Even without direct evidence of a link, their interdependent positions within related institutions suggest a high probability of indirect interactions. This method of combining direct and indirect links offers a more comprehensive view of the network's structure and dynamics. This process highlights the critical methodological choices involved in mapping organized crime networks. One approach is to rely on the group's own definitions of its boundaries. However, when direct access to members is limited, a nominalist approach is necessary, where researchers define boundaries based on theoretical interests and data availability (Perliger 2010). This is the approach we adopted. The network we analyzed is a snapshot from approximately 2015 to 2018.

7 Data Analysis

Figure 1 shows the network of Tren de Aragua. We compiled a list of 25 members, ranging from leaders, prisoners, and other gang members to politicians and relevant community members. The red dots show the core group of the network, and blue, black, and purple represent subgroups and peripheral figures. The central positioning and numerous connections of Tren de Aragua's core group suggest they are the key players in the network, potentially involved in decision-making processes, coordination, and dissemination of information. To better understand the structure of the network, we calculated the degree and normalized degree (Table 1).

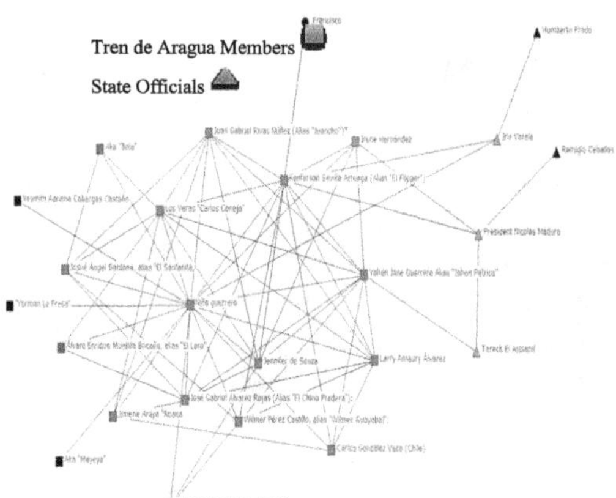

Fig. 1. The Tren de Aragua Network

Niño Guerrero has the highest degree of 19, indicating a high level of direct connections in the network. The degree normalized gives a proportionate measure of connections relative to the network size. Niño Guerrero's normalized degree is 0.79, which is significant in a network of this size. Followed by Niño Guerrero, Kenferson Servilla Artega shows a degree of 13, and Yohan José Guerrero shows a degree of 11. Although there are communication channels with peripheral actors, these are distributed among different core members, a key feature of criminal organizations that protects their leadership by creating layers of members around the core (Lynn 2005). This structure serves as a buffer to insulate leadership from external threats. So, while these nodes can be considered the central nodes that hold significant influence within the network, interesting are the amounts of peripheral figures.

Those actors have lower connectivity degrees, such as Iris Varela (Degree = 3), who was the then-prison minister (InSight Crime 2020), or Yorman la Fresa (Degree = 1), who may take over specific or limited roles to ensure functionality of the network. These connections indicate that the number of connections a node has does not always correlate with its influence or control over the network. This becomes even more clear when we look at Jennifer de Souza and Larry Amaury Álvarez. Both are part of the secondary echelon and hold strategic positions. Jennifer de Souza, allegedly the spouse of Kenferson Sevilla Artega ("El Flipper"), administers Barrio JK, a non-profit entity purportedly providing social services but suspected of being a front for Tren de Aragua's criminal operations (Insight Crime 2021). Understanding the role of such entities is crucial to understanding how Tren de Aragua was able to grow so quickly.

7.1 Is the Tren de Aragua a Scale-Free Network?

In the next step, we examined if the Tren de Aragua is a scale-free network. In a scale-free network, a small number of nodes have a very high degree, while the majority of nodes have a low degree. One way to test for this is by applying the 80/20 rule: if the top 20% of nodes account for at least 80% of the total ties (edges), the network can be considered scale-free.

Table 1. Degree and Normalized Degree Values of first 5 actors

	Name	Degree	Nom Deg
1	Niño Guerrero	19.000	0.792
2	Kenferson Sevilla Arteaga	13.000	0.542
3	Jennifer de Souza	8.000	0.333
4	Larry Amaury Álvarez	8.000	0.333
5	Yohan José Guerrero	11.000	0.458

Since 45.52% is less than 80%, the network does not meet the 80/20 rule in order to be considered a scale-free network. This means that even though we have a few nodes with relatively high degrees (e.g., Niño Guerrero & Kenferson Sevilla Artega) and many nodes with low degrees (e.g., Yesmith Adriana Cabargas, Yorman La Fresa, or Tareck El Aissami), the high-degree nodes do not stand out as clear hubs in the network. Another way of testing this is examining if the degree distribution follows a power-law. If the network is scale-free, we should see a relatively straight line when plotting the data points on a log-log scale. In the Tren de Aragua network, we used the poweRlaw package in R

to fit the degree distribution to a power-law model, which uses normalized values for the fitting process. The plot (Fig. 2) shows the degree distribution with a fitted power-law line. While the network displays some scale-free characteristics, such as a few high-degree nodes, the data points do not perfectly align with the power-law line, especially at higher degrees. This suggests that the degree distribution does not strictly follow a power-law, and the high-degree nodes are not significant hubs, which confirms that the Tren de Aragua network is not a scale-free network.

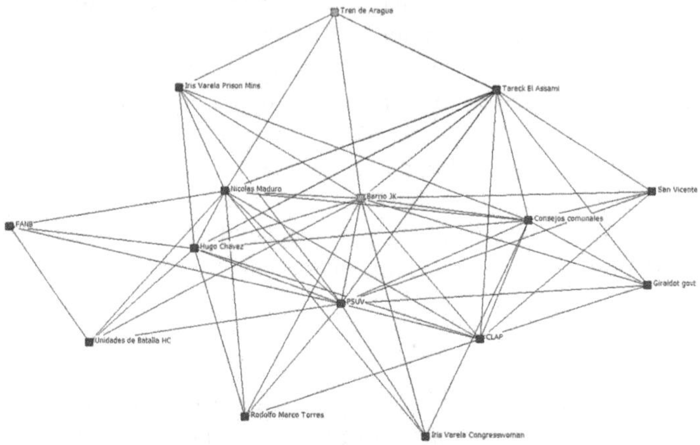

Fig. 2. Tren de Aragua – Politician Map

Centrality measures provide further insights into the actors' roles within the network. The analysis reveals a clear hierarchy, with Niño Guerrero at the top. His closeness centrality score of 0.03 indicates that he is relatively close to all other actors in the network and that, on average, he can communicate with all other nodes in the network more quickly than others. This suggests that he plays a key role in operational efficiency and potentially disseminating information or exerting influence rapidly across the network. His high betweenness centrality of 167.15 further emphasizes his control over information flow as it positions him as a gatekeeper who can facilitate or restrict communication between different parts of the network. This control is crucial in a criminal organization, where coordinating complex operations and managing information is key to maintaining power. Additionally, his eigenvector centrality of 1.00 confirms that he is connected to other highly influential members, showing that his connections are not only numerous but also strategically powerful. This makes Niño Guerrero an invaluable and well-protected figure within the organization's core power structure.

Kenferson Sevilla Arteaga (El Flipper), with a closeness centrality of 0.03, is also central to the network. However, his betweenness centrality of 75.14, though significant, is less than half of Guerrero's, which indicates that while he has influence, he is not as critical for the flow of information. El Flipper controlled the San Vicente neighborhood in Maracay and managed the "Fundación Somos El Barrio JK," which received material support from the government of Aragua (Insight Crime 2004). His eigenvector centrality of 0.84 suggests influential connections but does not match Guerrero's reach. Jennifer de

Souza and Larry Amaury Álvarez, with closeness centralities of 0.02, are less connected, indicating more specialized roles. Their betweenness centralities (4.08 for de Souza and 1.66 for Álvarez) suggest they have some influence on the flow of information but are not vital gatekeepers. Their eigenvector centralities (0.65 for de Souza and 0.68 for Álvarez) indicate connections to influential members, but their reach is narrower compared to Guerrero and El Flipper. This aligns with Insight Crime's (2020) findings, which describe Larry as one of Guerrero's trusted lieutenants and Jennifer de Souza as El Flipper's wife. Both are linked to the core group but likely have more specialized, focused roles within the organization. Peripheral figures like Jimena Araya "Rosita" and Aka "Mayea" have lower centrality scores across all measures, indicating more isolated roles within the network. "Rosita," allegedly a romantic partner of Niño Guerrero (Carabobo 2013), and Aka "Mayea," trusted by Guerrero and operating in Colombia (Trujillo 2023), are not central to the network's operations or communication channels but are likely to fulfill specialized roles for operational success. Overall, the centrality of key actors supports our second hypothesis as it shows that the strength and multiplexity of internal ties empower Tren de Aragua's convergence with governmental systems. The connection between Tren de Aragua and the government, mostly via the foundation Barrio JK, provides an environmental link to the government without directly exposing any internal members. We will explore this concept further in the next section.

7.2 Tren de Aragua's Political Network

Throughout our research, we uncovered several state authority figures directly or environmentally linked to Tren de Aragua. The connections between some Venezuelan government officials and the gang are facilitated through government-supported social organizations. Organizations like the Consejos Comunales and the Comité Local de Abastecimiento y Producción (CLAP), promoted by Nicolás Maduro's administration, have been instrumental in these interactions. Women associated with Tren de Aragua often play significant roles in these organizations. Transparencia Venezuela (2022) reported that CLAP collaborated with Barrio JK to deliver public services, provide food support, and manage public resources. Another key organization is the Unidades de Batalla Hugo Chávez, linked to the United Socialist Party of Venezuela (PSUV), which has also played a strategic role in Tren de Aragua's expansion (Insight Crime 2023). These connections align with our first hypothesis, as Barrio JK, CLAP, and Unidades de Batalla serve as intermediaries, facilitating interaction between Tren de Aragua and the Venezuelan government while maintaining their independence. Also, these "civic organizations" confirm our third hypothesis as they highlight the sophistication of Tren de Aragua's network. Beyond traditional networking with businesses and government corruption, Tren de Aragua uses civic spaces to interact while preserving its own, separate identity. This also helps legitimize its presence.

Several government officials have also been mentioned in connection to Tren de Aragua's growth. Tareck El Aissami, Governor of Aragua in 2012, reportedly oversaw the gang's expansion, facilitated by "peace zones" and the weakening of local police forces (Insight Crime 2023). Although no direct link between El Aissami and Tren de Aragua has been confirmed, environmental connections are likely. Another significant figure is Iris Varela, Minister of Penitentiary Services in 2017, who was photographed with members of Barrio JK and collaborated with them on the "Ecosocialist project Peace House" (Unit 2023). These relationships underscore the complex interplay between governance and criminality as they allow resource exchange between state actors and criminal groups while preserving the gang's internal identity and secrecy.

7.3 Discussion

Studying the network structure of Tren de Aragua reveals valuable insights into the intersection of criminal organizations and government. In specific, it highlights the complexity of their converging relationships. Our first hypothesis suggests that criminal organizations and the state converge by using covert connections while maintaining distinct borders. This is strongly supported by the analysis. Covert connections are evident in the interactions between Tren de Aragua and non-criminal entities like Barrio JK, CLAPs, and Unidades de Batalla, which are linked to both criminal and political figures (see Figs. 1 & 2). Though proving the full extent of these connections is challenging due to missing data, the use of intermediaries like Barrio JK shows a calculated strategy to facilitate connections with political and social actors.

This supports the notion that covert channels are key to the convergence between Tren de Aragua and the government. Our second hypothesis, which posits that the strength and multiplexity of internal and external ties enable the organization to engage with the government, also finds strong support. Tren de Aragua's network shows a clear hierarchy, with central figures like Niño Guerrero holding significant influence and control. This internal hierarchy empowers the organization to engage with governmental systems for mutual benefit. The high centrality of key members demonstrates that the strength of internal ties is crucial for facilitating these connections. The organization's internal cohesion and coordination are key factors in its ability to converge with governmental entities and reinforce the idea that strong, multiplex internal relationships drive its capacity to interact with political structures.

Lastly, the third hypothesis suggests that the social network will extend beyond government and criminal organizations to include ties with other systems. The analysis provides intriguing evidence of this broader web of relationships within the Tren de Aragua network. While the primary focus has been on connections between criminal organizations and government entities, the analysis reveals additional ties to social and potentially economic systems. For instance, Fig. 1 shows a link to Francisco the villager, which may indicate connections to the local community (Risquez 2023; Insight Crime 2023). In combination with the previously mentioned involved civic organizations, these ties illustrate the multifaceted nature of Tren de Aragua's interactions with broader societal systems. However, there are limitations to the analysis, particularly in mapping criminal organizations due to restricted data access. The covert nature of these ties and security concerns pose significant challenges to comprehensive data collection. Moreover, relying on open-source information may not offer a complete understanding of the organization's operations. Despite these challenges, the analysis sheds valuable light on the convergence of criminal entities with the government and the complex dynamics within these networks. It also offers a nuanced view of how criminal organizations engage with multiple societal systems.

8 Conclusion and Future Research

We have thoroughly explored the evolution of Venezuela's Tren de Aragua and the concept of criminal governance. Central to our investigation was the question: how do prison gangs evolve into sophisticated criminal organizations? This question guided our analysis and helped uncover the multifaceted relationships and networks that enable prison gangs to exert influence and expand beyond prison walls. Using SNA, we provided insights into the convergence of criminal organizations with socio-political systems and revealed the complex internal and external relationships within Tren de Aragua. The network of Tren de Aragua shows a clear hierarchy, with figures like Niño Guerrero holding significant influence and control. This hierarchy empowered the organization to engage with governmental systems, which eventually allowed the takeover of the Tocorón prison. While much of our analysis focuses on criminal and political ties, it also hints at broader connections to social and possibly economic systems. These multifaceted connections highlight the complexity of the organization's interactions with society at large. In specific, it illustrates how covert interactions between the "upper world" and the "underworld" are orchestrated to maintain influence and control across multiple domains.

Current approaches to managing prison gangs and their expansion are diverse and specialized, such as enhanced surveillance and stricter prison control measures to improve inter-agency cooperation (NIJ 2020; NCJRS 2001). While these strategies have shown some effectiveness in places like the U.S., they are likely to face significant challenges in Venezuela, where issues like corruption are more prevalent. This is particularly evident with the U.S. Department of the Treasury's designation of Tren de Aragua as a transnational criminal organization (2023), which emphasizes the gang's expansive criminal network and the need for coordinated international efforts. Addressing the influence of prison gangs like Tren de Aragua in Venezuela, therefore, requires overcoming several challenges, including corruption within the prison system, the adaptability of criminal organizations, and the socio-political conditions that enable them to flourish. Our research suggests a multi-faceted strategy to combat the growth of prison gangs like Tren de Aragua. First, gaining more detailed insights into the operations of socio-political organizations, such as Barrio JK, CLAP, and Unidades de Batalla, is crucial to understanding the evolution of criminal governance. Additionally, inter-institutional collaboration is essential for tackling their expansion, with a focus on enhanced data sharing and improved cooperation between national and international law enforcement agencies. This coordination is vital for targeting key figures within the network to disrupt communication and operational coherence. Moreover, targeted interventions, including educational and vocational programs within prisons, could help reduce gang recruitment. Lastly, systemic reforms that address the root causes of corruption and improve prison management are critical to preventing future gang growth.

The application of SNA has been instrumental in revealing these complex dynamics. Our research also uncovered connections between Tren de Aragua and other criminal organizations, both locally and internationally. This opens a promising path for future research into creating a global network map of criminal organizations to expand our understanding of the global reach and influence of these groups.

Appendix: Databases can be accessed through the next link:https://docs.google.com/spreadsheets/d/1CczMrm5AwjJSl1_Hc0UTuApSxxG_FLOiRwKik9evv0/edit?usp=sharing

Acknowledgments. We thank University of Massachusetts Lowell, in particular Angelica Duran-Martinez, Director of the Program in Global Studies, and Professor Arie Perliger, Ph.D., of the School of Criminology and Justice Studies, for their invaluable guidance and support..

Disclosure of Interests. The authors have no competing interests to declare that are relevant to the content of this article.

References

Abello-Colak, A., Guarneros-Meza, V.: The role of criminal actors in local governance. Urban Studies **51**(15), 3268–3289 (2014)

Blumberg, A.S.: UPPERWORLD CRIME (From Current Perspectives on Criminal Behavior - Original Essays on Criminology) (1974)

Call, C.T.: Beyond the 'failed state': Toward conceptual alternatives. Eur. J. Int. Rel. **17**(2), 303–326 (2011)

Cartier-Bresson, J.: Corruption networks, transaction security and illegal social exchange. Polit. Stud. **45**(3), 463–476 (1997). https://doi.org/10.1111/1467-9248.00091

Cawley, M.: Mexico cartel-US gang ties deepening as criminal landscape fragments. InSight Crime (2017)

Calderoni, F.: Social network analysis of organized criminal groups. Global Crime **15**(3–4), 224–250 (2014)

Costa, (Eliminate) (2017)

Denyer Willis, G.: The killing consensus: Police, organized crime, and the regulation of life and death in Brazil. University of California Press (2015)

Duran-Martinez, A.: The politics of drug violence: criminals, cops and politicians in Colombia and Mexico. Oxford University Press (2018)

Escobar, R.P.: Quality of democracy and Game theory: explaining the consolidation of the mexican hybrid regime. Politikon IAPSS J. Polit. Sci. **48**, 47–69 (2021)

Felbab-Brown, V.: A state-building response to organized crime, illicit economies, hybrid threats, and hybrid governance. In: Comolli, V. (eds.) Organized Crime and Illicit Trade. Palgrave Macmillan, Cham (2018). https://doi.org/10.1007/978-3-319-72968-8_6

Felbab-Brown, V.: A state-building response to organized crime, illicit economies, hybrid threats, and hybrid governance. In: Comolli, V. (eds) Organized Crime and Illicit Trade. Palgrave Macmillan (2018). https://link.springer.com/book/10.1007/978-3-319-72968-8

Gundur, R.: Prison Gangs. In: Oxford Research Encyclopedia of Criminology. Oxford University Press (2020)

Granovetter, M.: The strength of weak ties. Am. J. Sociol. **78**(6), 1360–1380 (1973)

Human Rights Watch. Ecuador: Gang control of prisons enabled massacres (2022). https://www.hrw.org/video-photos/interactive/2022/07/28/ecuador-gang-control-prisons-enabled-massacres

Insight Crime: GameChangers 2020: Tren de Aragua y la exportación del crimen organizado de Venezuela (2020). https://insightcrime.org/es/noticias/gamechangers-2020-tren-de-aragua/

Insight Crime: Tren de Aragua: De megabanda carcelaria a empresa criminal transnacional (2023). https://insightcrime.org/wp-content/uploads/2023/08/Tren-de- Aragua-De-megabanda-carcelaria-a-empresa-criminal-transnacional-InSight-Crime-Oct-2023-4.pdf
Jaffe, R.: The hybrid state: crime and citizenship in urban Jamaica. Am. ethnologist **40**(4), 734–748 (2013)
Krebs, V.E.: Mapping networks of terrorist cells. Connections **24**(3), 43–52 (2002)
Lauchs, M., Keast, R., Yousefpour, N.: Corrupt police networks: uncovering hidden relationship patterns, functions and roles. Policing Soc. **21**(1), 110–127 (2011). https://doi.org/10.1080/10439463.2010.540656
Lessing, B. (2016). Inside Out: The Challenge of Prison-Based Criminal Organizations. The Brookings Institution
Lessing, B.: Conceptualizing criminal governance. Perspect. Polit. **19**(3), 854–873 (2021)
Lynn, M.: Organizational Buffering: Managing Boundaries and Cores. Digital Commons @ ACU, College of Business Administration, Management Sciences. Abilene Christian University (2005)
Luhmann, N.: Theory of society, vol. 1 (R. Barrett, Trans.). Stanford University Press (2012)
Maitra, D.R.: 'If You're down with a gang inside, you can lead a nice life': prison gangs in the age of austerity. Youth Justice **20**(1–2), 128–145 (2020)
Malm, A., Bichler, G., Nash, R.: Co-offending between criminal enterprise groups. Glob. Crime **12**(2), 112–128 (2011). https://doi.org/10.1080/17440572.2011.567832
Martin, S.: What is Tren de Aragua? The Venezuelan criminal gang that has expanded throughout South America Voz (2023). https://voz.us/what-is-tren-de-aragua-the-venezuelan-criminal-gang-that-as-expanded-throughout-south-america/?lang=en
Morselli, C.: Contacts, Opportunities, and Criminal Enterprise. Maldives: University of Toronto Press (2005)
Morselli, C.: Inside criminal networks. In: Studies of Organized Crime. Siegel, D. Springer (2009)
Morselli, C., Giguère, C.: Legitimate strengths in criminal networks. Crime Law Soc. Change **45**(3), 185–200 (2006). https://doi.org/10.1007/s10611-006-9034-4
Natarajan, M.: Understanding the structure of a drug trafficking organization: a conversational analysis. In: Mangai, N., Mike, H. (eds.) Illegal drug markets: from research to prevention policy. Crime prevention studies, vol. 11, pp. 273–298. Criminal Justice Press/Willow Tree Press, Monsey (2000)
National Institute of Justice. Experts identify priority needs for addressing correctional agency security threats. National Institute of Justice (n.d.). https://nij.ojp.gov/topics/articles/experts-identify-priority-needs-addressing-correctional-agency-security-threats
Otero, E.: Venezuela regaining prison control after seizing Tocoron jail (2023). Reuters.: https://www.reuters.com/world/americas/venezuela-completes-first-phase-retaking-control-prisons-run-by-prisoners-2023-09-23/
Oxford Research Encyclopedia of Criminology. Comparative approaches to the study of prison gangs. Oxford University Press (2018)
Paarlberg, M.A.: Transnational gangs and criminal remittances: a conceptual framework. CMS **10**(24) (2022). https://doi.org/10.1186/s40878-022-00297-x
Perliger, A., Pedahzur, A.: Social network analysis in the study of terrorism and political violence. Southern Illinois University Carbondale, OpenSIUC, Political Networks Paper Archive (2010)
Pimenta et al.: (Eliminate) (2021)
Porter, L.E., Alison, L.J.: Leadership and hierarchies in criminal groups: scaling degrees of leader behaviour in group robbery. Leg. Criminol. Psychol. **11**(2), 245–265 (2006)
Rotberg, R.I.: When states fail: causes and consequences. Princeton University Press (2004)
Ruiz, P.: Mara Salvatrucha (MS-13) and Barrio 18: Gangs, Terrorists, or Political Manipulation? Small Wars J. (2020). https://smallwarsjounal.com/jrnl/art/mara-salvatrucha-ms-13-and-barrio-18-gangs-terrorists-or-political-manipulation

Schmidt, M.: U.S. Gang Alignment with Mexican Drug Trafficking Organizations. Combating Terrorism Center at West Point **5**(3) (2012). https://ctc.westpoint.edu/u-s-gang-alignment-with-mexican-drug-trafficking-organizations/

Schlichte, K., Schneckener, U.: Armed groups and the politics of legitimacy. Civ. Wars **17**(4), 409–424 (2015). https://doi.org/10.1080/13698249.2015.1115573

Smith, C.M., Papachristos, A.V.: Trust thy crooked neighbor: multiplexity in Chicago organized crime networks. Am. Sociol. Rev. **81**(4), 644–667 (2016). https://www.ojp.gov/ncjrs/virtual-library/abstracts/trust-thy-crooked-neighbor-multiplexity-chicago-organized-crime

Trujillo, M.A.: Metropolitan Police in Cucuta Launch Special Operation to Intercept 'El Nino' Guerrero, Leader of 'El Tren de Aragua'. BNN (2023)

United Nations Office on Drugs and Crime. Organized Crime/Cybercrime Module 13 Key Issues: Conceptualizing Organized Crime and Defining the Actors Involved (n.d.). https://www.unodc.org/e4j/en/cybercrime/module-13/key-issues/conceptualizing-organized-crime-and-defining-the-actors-involved.html

Unit, V.I.: Hybrid Governance and the Criminal Fiefdoms of Tren deAragua. InSight Crime (2023). http://insightcrime.org/investigations/hybrid-governance-criminal-fiefdoms-tren-de-aragua/

U.S. Department of the Treasury. Treasury sanctions Hamas operatives and financial facilitators. U.S. Department of the Treasury (2023). https://home.treasury.gov/news/press-releases/jy2459

UNODC World Drug Report 2024. United Nations Office on Drugs and Crime (2024)

Varese, F.: Mafia movements: how organized crime conquers new territories. Princeton University Press(2011)

Office of Justice Programs. An overview of the challenge of prison gangs. Office of Justice Programs (n.d.). https://www.ojp.gov/ncjrs/virtual-library/abstracts/overview-challenge-prison-gangs

References for the Creation of the SNA Available in the Next Link:

https://studentuml-my.sharepoint.com/:x:/r/personal/oscar_alvaro_uml_edu/_layouts/15/Doc.aspx?sourcedoc=%7B140830E8-D1A1-4874-947B-A4CB7E1EE3C8%7D&file=Tren_SNA.xlsx&action=default&mobileredirect=true

Methodology for Identifying Social Groups Within a Transactional Graph

Maxence Morin[1,2(✉)], Baptiste Hemery[2], Fabrice Jeanne[2], and Estelle Pawlowski-Cherrier[1]

[1] Normandie Univ, UNICAEN, ENSICAEN, CNRS, GREYC, 14000 Caen, France
maxence.morin@unicaen.fr, fabrice.jeanne@orange.com
[2] Orange Innovation, 14000 Caen, France
estelle.pawlowski@ensicaen.fr, baptiste.hemery@orange.com

Abstract. Social network analysis is pivotal for organizations aiming to leverage the vast amounts of data generated from user interactions on social media and other digital platforms. These interactions often reveal complex social structures, such as tightly-knit groups based on common interests, which are crucial for enhancing service personalization or fraud detection. Traditional methods like community detection and graph matching, while useful, often fall short of accurately identifying specific groups of users. This paper introduces a novel framework specifically designed to identify groups of users within transactional graphs by focusing on the contextual and structural nuances that define these groups.

Keywords: Social Network Analysis · Transactional Graphs · Community Detection · Graph Matching · SubGraph of Interest

1 Introduction

Social network analysis is subject to increasing attention. Relational data are obtained from the interactions of users of a service. They are of great value. Whether it comes from social media, telecommunications services, or financial transactions, companies in these sectors have understood the importance of extracting information from relational data. They exploit information to enrich and personalize services, improve trust by combating fraud and improper use, and so on. A common method to analyze relational data is to transform data into transactional graphs. Nodes represent users, and edges represent transactions. This abstraction allows us to solve some issues in social network analysis thanks to graph theory.

In these graphs, we search for groups of users that present interesting characteristics. Users exchange in a manner that creates communities. These can be identified using community detection algorithms. However, in this paper we are not interested in community detection. We are looking for other specific types of groups in transactional graphs.

L. M. Aiello et al. (Eds.): ASONAM 2024, LNCS 15213, pp. 117–130, 2025.
https://doi.org/10.1007/978-3-031-78548-1_11

We define a new concept that we call Group of Interest (GoI). A GoI is a social group composed of users who share social relations. The difference between communities and GoI is explained in the following paragraphs. As with communities, GoI can overlap. For example, a user can be a member of a family and a member of a group of fraudsters at the same time. We define the term "social truth" to encompass all social groups present in transactional graphs. In the real world, we cannot know the entirety of the social truth. Interviewing people costs money. Thus, we only know a part of the social truth that we call *samples*.

We refer to all the exchanges observed in a service as the "transactional truth". Transactions are traces of underlying social relationships. However, transactional truth is bound by a period of time. This restriction makes transactional truth incomplete, and it may reflect a part or the entirety of the social truth. If we extend the period of time, every social relationship may induce a transaction. In this paper, we aim to uncover the social truth by using only the transactional truth.

One could naturally think of traditional community detection methods to solve this specific problem, but these methods are not sufficient for two main reasons. The first reason is that these algorithms identify communities, which are groups of nodes that are highly interconnected relative to the rest of the graph. The transactions induced by a GoI do not create a highly connected subgraph. For example, in the subgraph shown in Fig. 1a, each node $A, B, C,$ or D does not engage in transactions with everyone. We are far from having a densely connected subgraph. Nonetheless, the users represented by these nodes may know each other and all share a relationship. The underlying relationships may be dense, as illustrated in Fig. 1b. The second reason is that community detection methods cover the entire graph and attempt to identify communities of any type, whereas we are interested in only specific subgraphs.

(a) Observed exchanges in a transactional service

(b) Underlying social relationships between the users. Different types of social relationships are represented by different line styles.

Fig. 1. Differences between the observed transactional truth and the underlying and inaccessible social truth.

The second natural solution one might also think of could belong to the graph matching research domain. Graph matching methods usually require a query to find topologically equivalent subgraphs. But the query does not consider any context surrounding it. We consider that the context is important to increase the performance of the model. Nevertheless, community detection methods as well as graph matching methods can represent a part of our solution to social truth identification. We give some details in Sect. 3.1 to adapt these methods in a new and complete framework.

In this paper, we introduce the notion of a SubGraph of Interest (SGI). Furthermore, we propose a new framework that encompasses the social truth identification problem and formalizes several methods for solving it. We also propose a method to evaluate the framework's performance. The experimental evaluation of the framework will be presented in a further paper.

The rest of this paper is organized as follows. Section 2 is devoted to the problem statement. The framework is presented in Sect. 3. Section 4 describes how to evaluate the framework. Conclusions and further work are discussed in Sect. 5.

2 Problem Statement

2.1 Notations

Let $G = (V_G, E_G)$ be a multigraph comprised of a set of nodes V_G and edges E_G. Each vertex $v \in V_G$ represents a user. Each edge $(u, v) \in E_G$ denotes a relationship such as a financial transaction, a communication, or a friendship. Let $S = (V_S, E_S)$ be a subgraph, which implies $S \subseteq G$ if $V_S \subseteq V_G$ and $E_S \subseteq E_G$.

2.2 Problem Formalization

We define a new concept that we call SubGraph of Interest (SGI). An SGI is a subgraph that represents the transactions made by a GoI. Let $\mathcal{S}_{type}$ be an SGI that represents transactions of a certain type of GoI. Similar to GoIs, SGIs can overlap. Thus, a node can belong to several SGIs. Let $\mathbb{S}_{type}$ be the set containing all the $\mathcal{S}_{type}$ of a unique type. The set $\mathbb{S}_{type}$ is representing one GoI type present in the social truth. The type can be families, groups of colleagues, fraudsters, and so on. For the sake of clarity, we use $\mathbb{S}$ and $\mathcal{S}$ instead of $\mathbb{S}_{type}$ and $\mathcal{S}_{type}$ as we are only interested in a unique type of GoI at a time. Recovering the set $\mathbb{S}$ for each type means, in the end, recovering the entire social truth.

Let $\mathbb{S}_n \subset \mathbb{S}$ be the set of known $\mathcal{S}$ in $\mathbb{S}$ that represents our *samples*. The set $\mathbb{S}_n$ has a fixed size n. These n subgraphs are used as training samples. As mentioned in the introduction, we have to deal with a set $\mathbb{S}_n$ containing very few samples.

The aim of this paper is to recover the entire set $\mathbb{S}$ using only the training sample set $\mathbb{S}_n$ for one given type. The contribution of this paper is twofold:

1. Propose a framework to tackle the previously mentioned problem of social truth identification.
2. Evaluate this framework.

3 Framework Definition

Given a multigraph G and a set of samples $\mathbb{S}_n$, we want the framework to output a set $\hat{\mathbb{S}}$. $\hat{S}$ stands for the elements of $\hat{\mathbb{S}}$. We want this set $\hat{\mathbb{S}}$ to be as close as possible to the set $\mathbb{S}$.

We illustrate the framework as follows:

$$(G, \mathbb{S}_n) \rightarrow \boxed{method} \rightarrow \hat{\mathbb{S}} \text{ such that } \hat{\mathbb{S}} \simeq \mathbb{S} \quad (1)$$

The relation $\hat{\mathbb{S}} \simeq \mathbb{S}$ means that we want the set $\hat{\mathbb{S}}$ to be as similar as possible to the social truth $\mathbb{S}$ (which is only part of the social truth, i.e. for a specific type). We precise this point in Sect. 4.

Now we define three *method* proposals in framework (1). The first two *methods* share the same basis and are gathered in the first approach, detailed in Sect. 3.1. The last *method* relies on a new pruning approach, detailed in Sect. 3.2.

3.1 First Approach

The *method* of the first approach consists of three steps.

1. The first step is a function named SUBGRAPHS. This function takes a multigraph G as input and must return a set of clusters. The role of this function can be fulfilled by already existing functions. We chose functions from either community detection or graph matching research domains.
2. The second step is the FEATURES function. This function takes a subgraph as input and should return a fixed-size vector that encodes information about the given subgraph. Details about this function are provided in a further paragraph.
3. The last step of the first approach is the selection step. The clusters returned by the SUBGRAPHS function are not necessarily relevant. We need to select those that are similar enough to the available *samples* in $\mathbb{S}_n$.

We present this FIRSTAPPROACH function through the Algorithm 1. We go through the details of each step now.

Subgraphs Function. The role of the SUBGRAPHS function is to create subgraphs from the given multigraph G. It seems natural to think of community detection and graph matching methods as good functions for creating clusters.

On the one hand, community detection algorithms [4] yield subgraphs, also referred to as communities. They identify either overlapping or non-overlapping communities. Because SGI can overlap, we are interested in overlapping community detection.

Among the overlapping community detection algorithms, we find algorithms using edge betweenness [6] or modularity based algorithms [20]. We propose to use an overlapping variant of the Label Propagation algorithm [13,28] for its inherent usage of the neighborhood. It creates communities by considering nodes

Algorithm 1 FIRSTAPPROACH and CHECK functions.

function FIRSTAPPROACH(G, $\mathbb{S}_n$, SUBGRAPHS, Γ)
  $\hat{\mathbb{S}} \leftarrow$ SUBGRAPHS(G)
  $results \leftarrow$ LIST()
  for all $\hat{\mathcal{S}} \in \hat{\mathbb{S}}$ **do**
    $\hat{\mathcal{F}} \leftarrow$FEATURES($\hat{\mathcal{S}}$)
    if CHECK($\hat{\mathcal{F}}, \mathbb{S}_n, \Gamma$) **then**
      $results \leftarrow results + \hat{\mathcal{S}}$
      break
    end if
  end for
  return *results*
end function
function CHECK($\mathcal{F}_\kappa, \Psi, \Gamma$)
  for all $\psi \in \Psi$ **do**
    $\mathcal{F}_\psi \leftarrow$FEATURES($\psi$)
    if $d_c(\mathcal{F}_\kappa, \mathcal{F}_\psi) < \Gamma$ **then**
      return *true*
    end if
  end for
  return *false*
end function

interactions with their neighbors. The algorithm uses a threshold to determine a node's membership in a community. It also analyzes the frequency to set the dominant label to propagate.

Nevertheless, we identified two limitations to community detection as a solution for identifying the social truth. The first limitation stems from the definition of a community itself. Often, a community is referred to as a subgraph that is highly connected among its nodes and that shares only a few edges with the rest of the graph [4].

The second limitation is the design of community detection algorithms. These algorithms create clusters of any type, covering the entire graph and providing subgraphs even if they are not relevant. In a sense, community detection algorithms provide more information than we require, as our problem is confined to a specific type of SGI.

On the other hand, graph matching [26] consists in finding all occurrences of a query subgraph within a graph. An occurrence is a subgraph that has the same topology as the query. The problem that consists of ensuring that a subgraph has the same topology as a query is named the isomorphism problem [7]. This problem is NP-complete in the general case [3].

As well as for community detection based methods, considering the problem of social truth identification, one can point out three limitations to graph matching based methods. The first limitation is that a query cannot represent an SGI in $\mathbb{S}$. A query has a fixed number of nodes and edges. SGIs in $\mathbb{S}$ have high variability in the number of nodes and edges. To overcome this limitation, we can look for Approximate Graph Matching [5,26]. Approximate Graph Matching methods find subgraphs that are similar enough to a query. The similarity can be computed using the graph edit distance. The graph edit distance is the required distortion to obtain the subgraph by inserting, removing, or substituting some nodes or edges. However, this metric is generally NP-hard to compute [27].

The second limitation of this approach in our use case is the lack of context around the query subgraph. In a social network, a group of people is defined by its structure, its personal pieces of information, and the way it connects to the rest of the graph. For example, the query in Fig. 2c matches both types A and B in Fig. 2. The only way to distinguish them is by considering their extended neighbors. For this reason, a query should be sampled from the graph to take the context around it into account. To the best of our knowledge, such an approach does not exist in the literature.

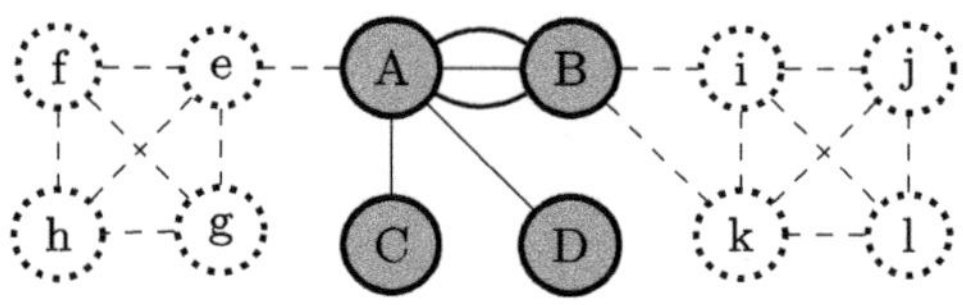

(a) SGI of type A

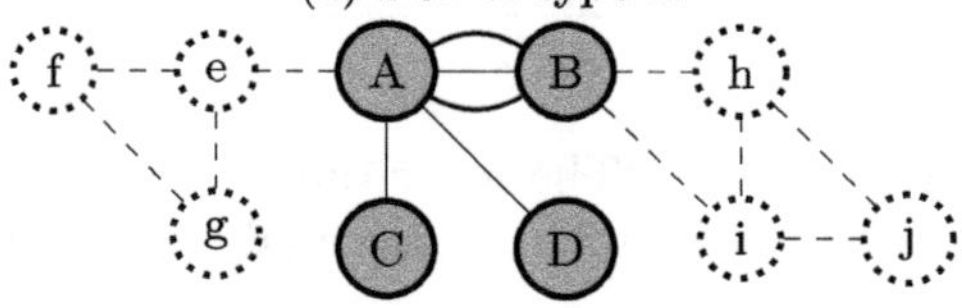

(b) SGI of type B

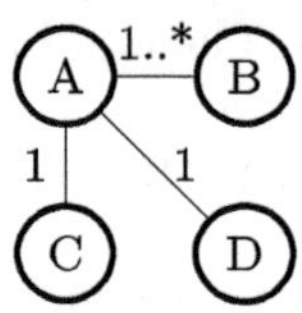

(c) Example of query that matches both types A and B.

Fig. 2. The first two subgraphs represent two different types of SGI. The difference is not in the topology of the subgraphs but in their context. A simple query like in the subgraph 2c is not enough to differentiate them.

The third limitation is that there are n samples in $\mathbb{S}_n$, whereas graph matching requires a unique query. We compensate for this limitation by analyzing subgraphs in $\mathbb{S}_n$ and computing their Maximum Common Subgraph (MCS) [17] to obtain a unique query. Then, we propose to do classic graph matching with this MCS as the query. This approach is the one we chose for the SUBGRAPHS function when considering graph matching.

Features Function. A feature vector enables encoding information about a graph, a subgraph, a node, or an edge. This concept is related to element characterization in graphs. The objective of characterization is to generate a vector of fixed size that encodes information about an element [2,8,12,15,16]. In the FIRSTAPPROACH function presented in Algorithm 1, the FEATURES function takes a subgraph as input. Thus, we are interested in subgraph characterization.

Subgraph characterization requires capturing the local structure, the subgraph's relative position, and nodes' specific information. We achieve this by defining a vector composed of some metrics, as mentioned in reference [12]. For example, we can use the following metrics:

- The number of nodes and edges in the subgraph. These metrics are straightforward and discriminate between subgraphs of different sizes.
- The minimum, maximum, and average of degrees, or clustering coefficients encode the internal structure of the graph.
- The minimum, maximum, and average of the shortest path length within the subgraph encode the diameter of the subgraph and can detect central nodes.
- Transitivity: This is the fraction of all possible triangles present in the subgraph. It encodes how tightly knitted the subgraph is and gives information about the subgraph's internal structure.
- Assortativity is the similarity of connections in the graph with respect to the node degree. This metric informs us how the connected nodes are similar to each other.

This list is non-exhaustive. The selection of one metric over another is related to the dataset and the SGI type we are searching for.

In the state of the art, we find deep learning-based methods to create feature vectors. We find methods such as SubGNN [2], SUGAR [22], or Sub2Vec [1].

Selection Step. The SUBGRAPHS function returns subgraphs that might be irrelevant. For this reason, we have to select only those that could match a subgraph present in $\mathbb{S}$. We propose to resort to the cosine distance $d_c(A, B) = \frac{A \cdot B}{\|A\|\|B\|}$ and define an upper bound Γ for the distance to make sure that our subgraphs are relevant. This guarantees that the property $\hat{\mathbb{S}} \simeq \mathbb{S}$ in framework (1) is fulfilled. This distance is applied to the feature vectors previously defined. Other distances, such as the Euclidean distance or the Manhattan distance, could also be considered.

3.2 Second Approach

In this section, we propose a *method* with a completely different strategy. Instead of predicting where the subgraphs are and whether elements such as nodes or edges belong to an SGI or not, we approach the problem in reverse. We are interested in predicting elements that are not in an SGI. We then prune these elements from the graph. We suppose that the resulting connected components are all SGIs in $\mathbb{S}$. This approach is presented in Algorithm 2.

Algorithm 2 SecondApproach function.

```
function SecondApproach(G, 𝕊_n, Γ_node, Γ_edge)
    V_bad ← V_G
    E_bad ← E_G
    for all u ∈ V_G do
        𝓕_u ← FEATURES(u)
        for all 𝒮 ∈ 𝕊_n do
            if CHECK(𝓕_u, V_𝒮, Γ_node) then
                V_bad ← V_bad − u
                break
            end if
        end for
    end for
    for all e ∈ E_G do
        𝓕_e ← FEATURES(e)
        for all 𝒮 ∈ 𝕊_n do
            if CHECK(𝓕_e, E_𝒮, Γ_edge) then
                E_bad ← E_bad − e
                break
            end if
        end for
    end for
    G_pruned ← ∇(G, V_bad, E_bad)
    return CONNECTED(G)
end function
```

We can relate this approach to the edge betweenness community detection algorithm [6]. The difference is that our approach is not limited to edge pruning. To prune elements, we first need to know which elements do not belong to an SGI in $\mathbb{S}$. This is achieved by comparing the features of elements of the multigraph with elements from samples in $\mathbb{S}_n$. Again, we need feature vectors for nodes and edges in the graph. These vectors can be obtained using two methods.

The first method involves using a simple hand-engineered feature vector, as presented in Sect. 3.1, applied at both the node and edge levels. We design feature vectors using users' and transactions' properties. The second method uses graph embedding techniques such as the Graph Isomorphism Network (GIN) [25] for node features and the Edge2Vec method [23] for edge features.

GIN is based on a Message Passing Neural Network [11]. It creates a feature vector based on the neighborhood of nodes. GIN is as powerful as the Weisfeiler-Lehman graph isomorphism test [24], which ensures that the topology around nodes is correctly encoded.

Edge2vec combines a deep autoencoder [18] to reduce the size of the feature vector and a Skipgram model [14] to ensure the vector correctly encodes the context around edges.

We compare the features of nodes and edges of the multigraph with the training samples in $\mathbb{S}_n$. We propose using again the cosine distance d_c as defined in Sect. 3.1. If a node or an edge differs too much from any of the training samples, we add it to the *bad* set. Let $\mathcal{F}$ represent the features of an element, such as a node or an edge. We consider thresholds Γ_{node} and Γ_{edge} as the maximum distance a node or an edge can be from a node or an edge of the training samples. The sets V_{bad} and E_{bad}, referred to as the *bad* sets, contain the elements to be removed from the multigraph G. We consider all the connected components obtained by the function CONNECTED to be the results $\hat{\mathbb{S}}$.

We define a pruning function ∇ that takes a multigraph G, the bad sets V_{bad} and E_{bad}, and returns a graph from which the nodes and edges from V_{bad} and E_{bad} have been removed. We express the ∇ function as follows:

$$\nabla(G, V_{bad}, E_{bad}) = G_{pruned} \tag{2}$$

We define four specific pruning functions: ∇_{simple}, ∇_{node}, ∇_{edge}, and $\nabla_{majority}$.

The ∇_{simple} pruning function removes all the nodes and edges present in the *bad* sets. We express the ∇_{simple} pruning function as follows:

$$\nabla_{simple}(G, V_{bad}, E_{bad}) = (V_G \setminus V_{bad}, E_G \setminus E_{bad}) \tag{3}$$

When we remove a node, we also remove all edges that were connected to that node.

The ∇_{node} is expressed as follows:

$$\nabla_{node}(G, V_{bad}, E_{bad}) = (V_G \setminus V_{bad}, E_G) \tag{4}$$

The weakness of the ∇_{node} pruning function is the case of two connected subgraphs. For example, in Fig. 3, we cannot distinguish the subgraph $ABCD$ from the subgraph $EFGHI$ by removing nodes only. We need to remove the edge (C, E) to reveal the two connected components.

We formalize the ∇_{edge} pruning function as follows:

$$\nabla_{edge}(G, V_{bad}, E_{bad}) = (V_G, E_G \setminus E_{bad}) \tag{5}$$

The weakness of the ∇_{edge} pruning function is the multigraph context. We need to remove all edges between two nodes to create two disconnected components. For example, in Fig. 3, in the case of multiple edges between nodes C and E represented by a dashed line, all edges between C and E must be removed.

The $\nabla_{majority}$ pruning function removes nodes and edges, but spares nodes if the majority of their edges are not in E_{bad}. We define the function as follows:

$$edgeMajority(v) = \frac{|E(v) \setminus E_{bad}|}{|E(v)|} \tag{6}$$

Here, $E(v)$ represents the set of all edges connected to node v, and $E(v) \setminus E_{bad}$ represents the set of edges connected to v that are not in E_{bad}. The ratio thus gives the proportion of good edges to total edges connected to v. Next, we define the set of all spared nodes as follows:

$$V_{spared} = \{v \in V_G \mid edgeMajority(v) \geq 0.5\} \tag{7}$$

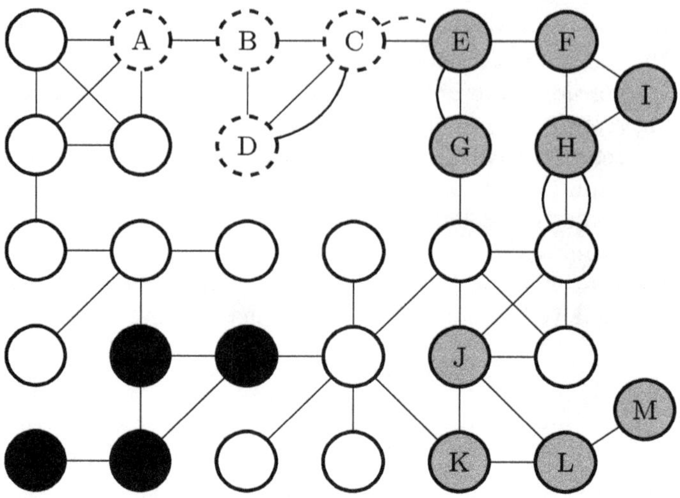

Fig. 3. An example of a multigraph. Dashed nodes are from a subgraph in $\mathbb{S}_n$. Dashed and green nodes compose three subgraphs from $\mathbb{S}$. Black nodes are from a subgraph that shares a similar topology but not the same context.

In this expression, V_G represents the set of all nodes in the multigraph. The condition $edgeMajority(v) \geq 0.5$ ensures that only those nodes for which the majority of their edges are not in E_{bad} are included in V_{spared}. Finally, we can formalize the pruning function as follows:

$$\nabla_{majority}(G, V_{bad}, E_{bad}) = (V_G \setminus (V_{bad} \setminus V_{spared}), E_G \setminus E_{bad}) \quad (8)$$

The expression $(V_{bad} \setminus V_{spared})$ identifies the set of nodes that are bad and not spared. Essentially, it removes the spared nodes from the set of bad nodes, ensuring that nodes with a majority of good connections are not pruned.

This *method* of pruning nodes and edges from the multigraph is a new perspective on the problem of social truth identification. We reversed the problem of the SGIs position's prediction by focusing on the removal of elements from the multigraph to create connected components.

4 How to Evaluate the Framework

In this section, we propose an adaptation of the *precision* and *recall* metrics to evaluate the performance of the *method* in the framework (1). We need to compare the output $\hat{\mathbb{S}}$ of the framework with the social truth $\mathbb{S}$. We describe a *match* function that is used in the adaptation of the *precision* and the *recall*. The *match* function is a boolean function. It guarantees that two subgraphs are similar enough to be matched. It takes $\hat{\mathcal{S}} \in \hat{\mathbb{S}}$ and $\mathcal{S} \in \mathbb{S}$ as inputs and returns *true* if they match, *false* otherwise. The *match* function uses the three boolean functions φ_{extra}, $\varphi_{missing}$, and φ_{size} that we describe now.

1. The φ_{extra} boolean function checks that the ratio of the nodes found in $\hat{\mathcal{S}}$ that are not in $\mathcal{S}$ over the size of $V_{\mathcal{S}}$ is under the threshold Γ_{extra}. We express it as follows:
$$\varphi_{extra}(\hat{\mathcal{S}}, \mathcal{S}) = \frac{|V_{\hat{\mathcal{S}}} \setminus V_{\mathcal{S}}|}{|V_{\mathcal{S}}|} < \Gamma_{extra} \tag{9}$$
2. The $\varphi_{missing}$ boolean function checks that the ratio of nodes in $\mathcal{S}$ that are not present in $\hat{\mathcal{S}}$ over the number of nodes in $V_{\mathcal{S}}$ is under the threshold $\Gamma_{missing}$. We express it as follows:
$$\varphi_{missing}(\hat{\mathcal{S}}, \mathcal{S}) = \frac{|V_{\mathcal{S}} \setminus V_{\hat{\mathcal{S}}}|}{|V_{\mathcal{S}}|} < \Gamma_{missing} \tag{10}$$
3. The φ_{size} boolean function checks that the size difference between $\mathcal{S}$ and $\hat{\mathcal{S}}$ over the size of $V_{\mathcal{S}}$ is under a threshold Γ_{size}. We express it as follows:
$$\varphi_{size}(\hat{\mathcal{S}}, \mathcal{S}) = \frac{abs(|V_{\hat{\mathcal{S}}}| - |V_{\mathcal{S}}|)}{|V_{\mathcal{S}}|} < \Gamma_{size} \tag{11}$$

Every threshold is set according to the context of the problem.

We need to consider the three boolean functions together to declare two subgraphs as matching. We express the *match* function as follows:

$$match(\mathcal{S}, \hat{\mathcal{S}}) = \begin{cases} \varphi_{extra}(\hat{\mathcal{S}}, \mathcal{S}) \ \ \& \\ \varphi_{missing}(\hat{\mathcal{S}}, \mathcal{S}) \ \ \& \\ \varphi_{size}(\hat{\mathcal{S}}, \mathcal{S}) \end{cases} \tag{12}$$

Iverson bracket [10] is a notation that generalizes the Kronecker delta. This function is defined to take the value 1 if the statement is true and the value 0 otherwise. It is generally denoted as follows, where P is the statement:

$$[P] = \begin{cases} 1 \text{ if } P \text{ is true;} \\ 0 \ \text{ otherwise.} \end{cases} \tag{13}$$

We define a subgraph $\mathcal{S}_a$ as relevant if it matches at least one subgraph in a set of subgraphs Ψ. The Ψ variable can take the value $\mathbb{S}_n$, $\mathbb{S}$, or $\hat{\mathbb{S}}$. We use the Iverson bracket to formalize the *relevant* boolean function as follows:

$$relevant(\mathcal{S}_a, \Psi) = \Big(\sum_{\mathcal{S}_b \in \Psi} [match(\mathcal{S}_a, \mathcal{S}_b)] \Big) \geq 1 \tag{14}$$

We use *precision* and *recall* metrics to evaluate the framework. The *precision* metric counts the proportion of relevant SGIs among selected SGIs. The *recall* metric counts the proportion of relevant SGIs selected among all selectable relevant SGIs. We define it as the ratio of $\hat{\mathcal{S}}$ that matches an element in $\mathbb{S}$ out of all the predicted elements in $\hat{\mathbb{S}}$. We formalize *precision* as follows:

$$precision = \frac{\sum_{\hat{\mathcal{S}} \in \hat{\mathbb{S}}} [relevant(\hat{\mathcal{S}}, \mathbb{S})]}{|\hat{\mathbb{S}}|} \tag{15}$$

A higher precision means that the function finds more subgraphs that are close enough to the social truth.

We formalize *recall* as follows:

$$recall = \frac{\sum_{\mathcal{S} \in \mathbb{S}} [relevant(\mathcal{S}, \hat{\mathbb{S}})]}{|\mathbb{S}|} \quad (16)$$

A higher recall means that a greater number of the social truth subgraphs are matched.

In conclusion, the adaptation of precision and recall metrics, integrated with the match function, provides a robust framework for evaluating the performance of the method in the framework (1). We can also calculate an F-measure score [9] such as an F1-score [21].

5 Conclusions and Future Work

In this paper, we introduce a novel framework designed to address the challenge of social truth identification within transactional graphs. This is a critical task for deciphering complex social interactions in large datasets. We present three distinct methods that combine traditional graph analysis techniques with innovative approaches to recover the hidden social truth. By adapting precision and recall metrics, we have established a method to evaluate the performance of the proposed framework. This ensures that the identified SubGraphs of Interest are relevant and representative of the underlying social structures.

For the future, we propose two interesting directions that involve integrating more advanced deep learning techniques into our framework:

1. Enhancement of Community Detection: Incorporating deep learning techniques into community detection algorithms could significantly improve their performance across diverse datasets. Exploring methods such as the Neural Overlapping Community Detection model [19] and other deep learning-based approaches could offer substantial improvements in detecting structures within multigraphs.
2. Deep Learning as a distance function: Applying deep learning to enhance the distance function d used with hand-crafted feature vectors and a pruning approach could increase the framework's robustness and performance. This integration would potentially enable more accurate identification of various types of SGIs, making the framework more effective across different scenarios.

Experiments using the framework are in progress and will be presented in a separate article. We will present the performance and efficiency in terms of time and memory of the framework with large datasets.

Acknowledgments. The authors would like to thank Orange and the ANRT for funding the thesis.

References

1. Adhikari, B., Zhang, Y., Ramakrishnan, N., Prakash, B.A.: Sub2Vec: feature learning for subgraphs. In: Phung, D., Tseng, V.S., Webb, G.I., Ho, B., Ganji, M., Rashidi, L. (eds.) PAKDD 2018. LNCS (LNAI), vol. 10938, pp. 170–182. Springer, Cham (2018). https://doi.org/10.1007/978-3-319-93037-4_14
2. Alsentzer, E., Finlayson, S., Li, M., Zitnik, M.: Subgraph neural networks. In: Advances in Neural Information Processing Systems, vol. 33, pp. 8017–8029 (2020). https://doi.org/10.48550/ARXIV.2006.10538
3. Babai, L.: Graph isomorphism in quasipolynomial time [extended abstract]. In: Proceedings of the Forty-Eighth Annual ACM Symposium on Theory of Computing, pp. 684–697. STOC 2016, Association for Computing Machinery, New York, NY, USA (2016). https://doi.org/10.1145/2897518.2897542
4. Fortunato, S.: Community detection in graphs. Phys. Rep. **486**(3–5), 75–174 (2010). https://doi.org/10.1016/j.physrep.2009.11.002
5. Gallagher, B.: Matching structure and semantics: a survey on graph-based pattern matching. In: AAAI Fall Symposium: Capturing and Using Patterns for Evidence Detection (2006). https://api.semanticscholar.org/CorpusID:12917214
6. Girvan, M., Newman, M.E.J.: Community structure in social and biological networks. Proc. Natl. Acad. Sci. **99**(12), 7821–7826 (2002). https://doi.org/10.1073/pnas.122653799
7. Grohe, M., Neuen, D.: Recent advances on the graph isomorphism problem. In: BCC (2020). https://api.semanticscholar.org/CorpusID:226237505
8. Grover, A., Leskovec, J.: Node2vec: scalable feature learning for networks. In: Proceedings of the 22nd ACM SIGKDD International Conference on Knowledge Discovery and Data Mining. ACM (2016). https://doi.org/10.1145/2939672.2939754
9. Hand, D.J., Christen, P., Kirielle, N.: F*: an interpretable transformation of the F-measure. Mach. Learn. **110**(3), 451–456 (2021). https://doi.org/10.1007/s10994-021-05964-1
10. Iverson, K.E.: A Programming Language. Wiley, USA (1962)
11. Kipf, T.N., Welling, M.: Semi-supervised classification with graph convolutional networks. in: International Conference on Learning Representations. Poster (2017). https://doi.org/10.48550/ARXIV.1609.02907, https://openreview.net/forum?id=SJU4ayYgl
12. Li, G., Semerci, M., Yener, B., Zaki, M.J.: Effective graph classification based on topological and label attributes. Stat. Anal. Data Min. **5**(4), 265–283 (2012). https://doi.org/10.1002/sam.11153
13. Liu, Q., Su, Y., Peng, Q., Chen, K., Lu, Y.: An overlapping community detection algorithm for label propagation based on node influence. In: 2021 3rd International Conference on Advances in Computer Technology, Information Science and Communication (CTISC), pp. 183–187 (2021). https://doi.org/10.1109/CTISC52352.2021.00041
14. Mikolov, T., Chen, K., Corrado, G., Dean, J.: Efficient estimation of word representations in vector space. In: International Conference on Learning Representations. Poster (2013). https://doi.org/10.48550/ARXIV.1301.3781
15. Narayanan, A., Mahinthan, C., Venkatesan, R., Chen, L., Liu, Y., Jaiswal, S.: graph2vec: learning distributed representations of graphs. In: Proceedings of the 13th International Workshop on Mining and Learning with Graphs (MLG). Workshop on Mining and Learning with Graphs, Association for Computing Machinery (2017). https://doi.org/10.48550/ARXIV.1707.05005

16. Perozzi, B., Al-Rfou, R., Skiena, S.: DeepWalk: online learning of social representations. In: Proceedings of the 20th ACM SIGKDD International Conference on Knowledge Discovery and Data Mining, pp. 701–710. Association for Computing Machinery (2014). https://doi.org/10.1145/2623330.2623732
17. Roy, I., Chakrabarti, S., De, A.: Maximum common subgraph guided graph retrieval: late and early interaction networks. In: Proceedings of the 36th International Conference on Neural Information Processing Systems. NIPS 2022, Curran Associates Inc., Red Hook, NY, USA (2024)
18. Salakhutdinov, R., Hinton, G.: Semantic hashing. Int. J. Approximate Reasoning **50**(7), 969–978 (2009). https://doi.org/10.1016/j.ijar.2008.11.006, https://www.sciencedirect.com/science/article/pii/S0888613X08001813, special Section on Graphical Models and Information Retrieval
19. Shchur, O., Günnemann, S.: Overlapping community detection with graph neural networks. In: Proceedings of the 25th ACM SIGKDD International Conference on Knowledge Discovery and Data Mining. Deep Learning on Graphs Workshop (DLG), Association for Computing Machinery (2019). https://doi.org/10.48550/ARXIV.1909.12201
20. Singh, D., Garg, R.: NI-Louvain: a novel algorithm to detect overlapping communities with influence analysis. J. King Saud Univ. Comput. Inf. Sci. **34**(9), 7765–7774 (2022)
21. Sokolova, M., Japkowicz, N., Szpakowicz, S.: Beyond accuracy, f-score and ROC: a family of discriminant measures for performance evaluation. In: Sattar, A., Kang, B. (eds.) AI 2006. LNCS (LNAI), vol. 4304, pp. 1015–1021. Springer, Heidelberg (2006). https://doi.org/10.1007/11941439_114
22. Sun, Q., et al.: SUGAR: subgraph neural network with reinforcement pooling and self-supervised mutual information mechanism. In: Proceedings of the Web Conference 2021, pp. 2081–2091. WWW 2021, Association for Computing Machinery (2021). https://doi.org/10.1145/3442381.3449822
23. Wang, C., Wang, C., Wang, Z., Ye, X., Yu, P.S.: Edge2vec: Edge-based social network embedding. ACM Trans. Knowl. Discov. Data **14**(4) (2020). https://doi.org/10.1145/3391298
24. Weisfeiler, B.Y., Leman, A.A.: The reduction of a graph to canonical form and the algebra which appears therein. Nauchno-Technicheskaya Informatsia, Series **2**(9), 12–16 (1968). https://www.iti.zcu.cz/wl2018/pdf/wl_paper_translation.pdf
25. Xu, K., Hu, W., Leskovec, J., Jegelka, S.: How powerful are graph neural networks? (2019). https://doi.org/10.48550/arxiv.1810.00826
26. Yan, J., Yin, X.C., Lin, W., Deng, C., Zha, H., Yang, X.: A short survey of recent advances in graph matching. In: Proceedings of the 2016 ACM on International Conference on Multimedia Retrieval, pp. 167–174. ICMR 2016, Association for Computing Machinery, New York, NY, USA (2016). https://doi.org/10.1145/2911996.2912035
27. Zeng, Z., Tung, A.K.H., Wang, J., Feng, J., Zhou, L.: Comparing stars: on approximating graph edit distance. Proc. VLDB Endow. **2**(1), 25–36 (2009). https://doi.org/10.14778/1687627.1687631
28. Zhu, X., Ghahramani, Z.: Learning from labeled and unlabeled data with label propagation. Tech. rep., Carnegie Mellon University (2002). https://pages.cs.wisc.edu/~jerryzhu/pub/CMU-CALD-02-107.pdf

Evaluating Deep Graph Network Performance by Augmenting Node Features with Structural Features

Mohamad Abushofa[1(✉)], Amir Atapour-Abarghouei[2], Matthew Forshaw[1], and A. Stephen McGough[1]

[1] School of Computing, Newcastle University, Newcastle upon Tyne, UK
{m.e.a.abushofa2,matthew.forshaw}@ncl.ac.uk,
stephen.mcgough@newcastle.ac.uk

[2] Department of Computer Science, Durham University, Durham, UK
amir.atapour-abarghouei@durham.ac.uk

Abstract. In machine learning, features play a vital role in modeling and understanding the data; their quality and representation essentially determine how accurate the results are. The problem is compounded in the graph-based learning paradigm when one considers how complex and interconnected the data is. However, to achieve more accurate results, augmenting graph data poses specific challenges in the field of graph learning. Feature augmentation is a critical aspect of enhancing data. Moreover, some datasets have limited features and some real datasets do not have features. In this paper, we present our approach termed Feat-Aug which is an extension of our previous work on non-parametric approaches. The aim of this work is to augment node features in graphs on parametric approaches such as Graph Neural Networks (GNNs) with the objective of improving performance in node classification tasks. Our approach combines real features, such as a bag of words in citation networks, which are typically associated with nodes, with structural features extracted at the node level, such as node degree and clustering coefficient. To further enhance these features, we leverage deep learning models to incorporate additional node-level features. The final modified features are the result of the combination of both real and structural features. To evaluate the effectiveness of the approach, we carried out extensive experiments with several real datasets. Moreover, our method consistently outperforms or achieves comparable results to Graph Neural Networks (GNNs) baselines and their variations, such as popular graph neural network models. Crucially, our approach deals with the problem of insufficient real features in certain datasets. This study is a major progression in the field through an effective node classification model. By integrating both real and structural features, our approach holds promise to improve the performance of node classification models.

Keywords: structural features · node features · concatenated features · GNN models

L. M. Aiello et al. (Eds.): ASONAM 2024, LNCS 15213, pp. 131–147, 2025.
https://doi.org/10.1007/978-3-031-78548-1_12

1 Introduction

Graph machine learning tasks, such as drug discovery [1], recommendation systems [2], and traffic prediction [3], have been greatly enhanced by Graph Neural Networks (GNN) and their variants. The essence of GNNs is in the message-passing mechanism, where features from adjacent nodes are incorporated into the features of a given node [4]. Although GNNs have been quite successful in recent years, their model inference performance is improved by the augmentations to nodes and graphs, as a result of modified features. The features augmentation strategies have great success in classic machine learning domains such as computer vision [5,6] and natural language processing [7]. Convolutional neural networks use this method, and the augmentation effects have been particularly fruitful when applied to image data [8]. Unlike graph data, where each node has connections to other nodes, the image pixels do not have any connections. Several data augmentation in image classification tasks independently modify image features, including RGB values and local neighbourhoods in the images [9]. Topology-level [10] and feature-level augmentation [11–13] are the two approaches for node and graph augmentations. At the topology level, a new graph topology is generated through augmentation techniques that modify the adjacency matrix. In contrast, adversarial training improves generalisation by perturbing node features during node-level augmentation techniques [12–14]. Some methods rely on changing the edges by removing or modifying them according to the information within the graph itself [15]. Additionally, some researchers have found great success in data augmentation by using extra topological features to build a dual graph similarity matrix of the graph [16]. Adding more edges between nodes with the same label increases the probability that the central node will benefit from the propagation process, which is another strategy used by researchers [16].

Numerous real-world issues, particularly in achieving high performance in applications, such as social network analysis, protein-protein interaction prediction, and network traffic categorization, involve graph machine learning tasks including graph classification, and node classification. Therefore, to improve the performance we should consider graph augmentation by supplementing the node features. In this paper, we looked at how to combine structural and real features at the node level to enhance the node features within the network.

In graph natural networks, nodes have linked properties like age names in social networks [17] and bags of words in citation networks [18]. On the other hand, the structural features are taken from the graph structure and provide the graph with a unique signature. The node degree, node centrality, cluster coefficient, and average node degree are a few examples of structural node properties. A mix-up technique described by Wang et al. [19] involves altering the data to create new features.

To the best of our knowledge, while some previous data augmentations based on GNNs, such as the baseline Graph Convolutional Networks (GCN) [20], GraphSAGE [1], and Graph Isomorphism Network (GIN) [21], and their variants, may have implicitly used some structural features alongside real features, our method is the first to explicitly and systematically concatenate these features. This explicit approach enhances interpretability and potentially improves performance by clearly leveraging the strengths of both feature types in the context of node classification using GNNs. The implicit methods may lose their potential advantages in graph machine learning applications as they disregard the explicit evaluation of structural features by not systematically concatenating them with the real features. Therefore, they have not assessed the explicit strength of the structural features themselves and how graph deep learning learns from concatenated features. In contrast, we address this gap by explicitly taking into account both structural and actual features.

The proposed method demonstrates its effectiveness in comparison to state-of-the-art methods. Combining real and extracted characteristics can enhance our understanding of the significance of structural features for graph machine-learning tasks.

Therefore, we are taking the structural features out of the graph and concatenating them with the actual ones. Additionally, our method is different from the earlier methods found in GNNs and their variations since previous methods disregard the structural features and concentrate primarily on actual features.

To facilitate the reproducibility of our results, we have made the source code for our method publicly available. This allows other researchers to easily access and build upon our work. Overall, our contributions in this paper can be summarized as follows:

- We propose a novel method for node augmentation that combines aggregated original features and local structural features as input to the graph deep learning model.
- We identify and utilize a set of relevant structural features for node feature concatenation.
- We conduct extensive experiments to evaluate the performance of our approach on node classification.
- Our experimental results show that our method outperforms state-of-the-art approaches on node classification, demonstrating the effectiveness of our approach.
- We make the source code for our method publicly available to enable better reproducibility of the results[1].
- Our approach demonstrates that the improved performance is due to the capability of a graph deep learning model to capture and learn combined features, which benefits data with insufficient features.

2 Literature Review

Since the advent of graph neural networks, there has been a significant shift in how we handle node feature representations. Initially, the methods for embedding lookups, which are crucial for interpreting node features, were limited in

[1] https://github.com/mohamad-1977/Fet-Aug.

their effectiveness [4]. However, GNNs have introduced advanced techniques for managing these node properties. Specifically, GNNs utilize aggregators—sophisticated mechanisms that compile and integrate features from neighboring nodes.

This aggregation process is central to GNNs' ability to effectively capture the complex relationships and interconnectedness inherent in graph data. By leveraging these aggregators, GNNs can construct a more comprehensive and nuanced representation of each node, taking into account not just node attributes but also the influence of its network context [22]. Also, it has succeeded in inductive learning [1,23]. The graph convolution methods, which are based on spatial-based methods, have demonstrated significant success compared to the spectral GNNs [1,2], which are based on convolutions using spectral graph theory [24], they are also expensive to employ as they process the whole adjacency matrix [20,25,26].

Researchers have been seeking ways to make machine-learning models capable of improving the generalisation and efficiency of graph machine-learning tasks such as node classifications, link prediction, and graph classifications. Recently, they achieved significant success in augmentations techniques in many domains such as computer vision [5] and natural language processing [7].

Augmentation in graph machine learning tasks is also a popular topic. The graph augmentations have two approaches: topology-level augmentation and feature-level augmentation.

Works done by Chen et al. [27], and Rong et al. [28], focused on graph topological augmentation, which involves adding or removing edges or nodes and consequently changing the graph structures. To improve the generalization and performance of the models, numerous studies, like work done by Yoo et al. [29], focus on random edge insertion or deletion. The work of Yoo et al. [30] performs node splitting or merging, and You et al. has implemented the subgraph sampling [29]. These methods make variations in graph topology and have effectively leveraged the relationships in graph data and improved performance.

Alternatively, node-level augmentation approaches work on altering node features or edge features in the graph. Works done by Yang et al. [31] and by Feng et al. [12] use the perturbations of the node features based on adversarial training. These methods attempt to increase the diversity of node feature representations, which then increases the performance of the models.

Node level and topology augmentation have been showing improved outcomes in many graph machine learning tasks, for instance, recommendation systems [32], drug discovery [33], and traffic prediction [34]. However, all aforementioned methods neglect the structural features and just focus on node-associated features. Thus, this paper proposes a novel approach by incorporating both real and structural features at the node level and evaluating GNNs baseline models. Our approach advances how structural features enhance deep graph learning.

3 Motivation and Requirements

3.1 Motivation

Graph neural networks have gained considerable interest in recent years for their capacity to capture complex connections among nodes in a graph, making them well-suited for node classification. Nevertheless, to qualify this achievement of GNN, the structure depends on the presence of node features as illustrated in Eq. 1, that are rich and high-dimensional. However, in the most realistic graphs, the quantity and quality of available features are limited and some of the real data has no features [35], which weakens the performance of GNNs. In addition, all previous works based on GNN baselines, GCN [20], GraphSAGE [1], Graph Isomorphism Network [21], and their variants, neglected to integrate structural features by concatenating them with real features, which can give more information about the nodes. There have been lots of attempts to augment graph data as we mentioned in the literature. One of these approaches is the Mixup introduced by Wang et al. [19] that consists of perturbing the data to create new or enhanced features. While this method has also demonstrated some good progress; this and all methods in the litterateurs are hampered by just perturbing the data and do not take the structural features into account, To overcome this drawback we suggest a different way of node augmentation by performing feature augmentation(Feat-Aug) that combines real and structural features, creating more meaningful node embedding. The proposed approach is motivated by the Mixup technique yet varies in that, it includes a combination of actual and extracted features, instead of perturbing the data.

$$h_i^{(l)} \leftarrow \text{AGGREGATE}\left(h_i^{(l-1)}, h_j^{(l-1)} : j \in N(i)\right) \oplus W^{(l)} \tag{1}$$

3.2 Requirements

Incorporation of the suggested method of feature augmentation has to satisfy several requirements.

- A graph must be available for analysis, represented in a format that can be processed by a GNN model (e.g. Networkx)
- Real features must be available for the nodes in the graph. These may include domain-specific knowledge about the nodes, such as their function or role in a particular system.
- It must be possible to compute the structural features for nodes in the graph.
- A GNN model must be available for training and testing. This model should be capable of processing node features and generating node embedding.
- A node classification task must be defined, along with an objective function for measuring the performance of the model.
- A method for sampling and interpolating nodes, such as the mixup technique used in [19], must be implemented to generate the augmented node features.

4 Methodology

In our approach to enhancing node features, we drew inspiration from concepts explored in our prior research, as detailed in [36], which involved modifying node features through non-parametric approaches in graph classification. In contrast to this, our proposal focuses on parametric approaches (graph deep learning), specifically graph neural networks baseline and its variants. By incorporating structural features derived from the node-level, we aim to uncover structural properties inherent in the graph data. Leveraging these structural features could potentially enhance the accuracy of classification models across a diverse range of tests. We further strengthen this method by using deep graph learning models to reinforce structural features with more node-level attributes as shown in Fig. 1. Our approach performance is evaluated by applying the modified features on graph neural network baselines, such as GCN [20] and GraphSAGE [1] and recent graph augmentation based on GCN, Mix-up [19], in which data is perturbed to produce new features. However, in our work, we do not alter anything within the data; our work combines structural and real features at the node level to produce a modified feature. Our approach involves the following these steps:

1. Mining the node features at local-level features.
2. Create the modified features by combining both extracted features and node-associated features.
3. Use the modified attributes as inputs to graph deep learning models.
4. Employ deep learning models to evaluate the performance with the modified features.

Algorithm 1. Feat-Aug

Input: Graph G
Output: Node embedding

// Extract features from nodes (see Algorithm 2)
$FG \leftarrow \text{ExtractedFeatures}(G)$
$C_{mf} \leftarrow \{\text{original features}\} \cup \{\text{structural features}\}$
for each node $v \in G$ **do**
    $h_v^{(0)} \leftarrow x_v^{(0)} \,||\, f_v^{(0)}$
    for $k = 1$ to K **do**
        $\text{msg}_v^{(k)} \leftarrow \text{aggregate}\left(\left\{(h_v^{(k-1)}, h_u^{(k-1)}) \mid u \in \text{neighbors}(v)\right\}\right)$
        $h_v^{(k)} \leftarrow \text{combine}\left(h_v^{(k-1)}, \text{msg}_v^{(k)}\right)$
    end for
    $L \leftarrow \text{loss}(h_v^{(k)}, y)$
    $h_v^{(t)} \leftarrow h_v^{(t-1)} + \text{gradient}(L, h_v^{(t)})$
end for
return node embedding

Algorithm 2. Extracted Features

Require: Graph G

$FG \leftarrow []$ ▷ Initialize feature matrix for G

for each node v in G **do** ▷ Extract features for each node v in G

  $\underline{features} \leftarrow [\delta(v), C(v), K(v), c(N(v)), \overline{d(N(v))}, |E_{\text{ego}}(v)|,$

  $\overline{c(N(v))}, |E_{\text{oego}}(v)|, |N(\text{ego}(v))|]$

  $FG \leftarrow FG \cup \{features\}$

end for

return FG ▷ Return a matrix of node × features

Algorithm 3. Concatenated Features

Input: Graph G, feature matrix X (Real features matrix), additional features $\{F(v)\}$ for each node v

Output: M concatenated features matrix

Initialize M with dimensions $(\text{num\_nodes}, \dim(X) + \dim(\{F(v)\}))$

for each node $v \in G$ **do**

  $concatenated\_features \leftarrow \text{concatenate}(X[v], \{F(v)\})$

  $M[v] \leftarrow concatenated\_features$

end for

Return: M ▷ Return a matrix M with dimensions $(\text{num\_nodes}, \dim(X) + \dim(\{F(v)\}))$

Algorithm 1 presents our approach framework, which consists of three distinct steps: feature extraction (Algorithm 2), concatenating real features with extracted features (Algorithm 3), feature normalization, and evaluating the node classification performance.

Feature Extraction. To encode the local features of the graph (see Algorithm 2), several node-level metrics are extracted. While there are many available metrics for node features, each one has distinct characteristics in terms of its sensitivity to topology and its execution time [36].

Our approach extracts 9 properties from every vertex in a graph. Through experimentation, we have determined that the following 9 feature metrics give the best balance between classification and run-time. Alternatively, if there are other significant attributes of a graph, different metrics could be employed. A value is extracted for each node for each of the 9 node properties given below $v \in V$:

1. **Node degree** δ: The degree of a node in a network refers to the number of connections it has with other nodes.
2. **Clustering coefficient** c: the clustering coefficient is a measure of the degree to which nodes in a graph tend to cluster together.

$$c_u = \frac{|(v_1, v_2) \in E : v_1, v_2 \in N|}{(2^d u)}.$$

The numerator represents the count of edges that connect neighbouring nodes of node u (where we use $N(u) = \{v \in V : (u,v) \in E\}$ to denote the node neighbourhood). The denominator determines the total number of node pairings in nearby of node u. [4]

3. **the average degree of neighbours** k: The average degree of each node nearby is calculated.

$$k_{nn,i} = \frac{1}{|N(i)|} \sum_{j \in N(i)} k_j$$

   where $N(i)$ represents the set of neighbouring nodes of node i, and k_j denotes the degree of node j which is a member of $N(i)$ [36].
4. **Average clustering of neighbourhood**: The average clustering score of each vertex is calculated by taking the mean of the local clustering scores in its neighbourhood.
5. **The average number of node i's two-hops:** it is two-hop away neighbors, denoted as $\overline{dN(i)}$, is computed as $\frac{1}{d_i} \sum_{j \in N(i)} d_j$.
6. **The average clustering coefficient of** $N(I)$**:** The average clustering coefficient of $N(i)$, represented as $\overline{cN(i)}$, is computed as $\frac{1}{d_i} \sum_{j \in N(i)} c_j$. The variable c_j represents the local clustering score calculated in step two [36].
7. **The number of edges in node** i**'s egonet:** The number of edges in node i's egonet is denoted as $|E_{ego(i)}|$, and ego(i) returns the egonet of node i [37]. where the egonet of a node includes the node itself, its immediate neighbors, and all the edges among them.
8. **The number of outgoing edges from** $ego(I)$**:** The number of outgoing edges from $ego(i)$ is denoted as $|E_{o,ego(i)}|$ [37].
9. **The number of neighbors of** $ego(I)$**:** The number of neighbors of $ego(i)$ is denoted as $|N(ego(i))|$ [37].

Table 1. Data statistics for node classification datasets. The letter 's' represents single-label classification. [19]

Dataset	#Nodes	#Edges	#Classes	#Features
Citeseer	3,327	4,732	6 (s)	3,703
Cora	2,708	5,429	7 (s)	1,433
Pubmed	19,717	44,338	3 (s)	500
Amazon (Photo)	7,650	143,662	8 (s)	745
Amazon (Computers)	13,752	287,209	10 (s)	767
Flickr	1,598,960	132,169,734	107 (s)	200

Concatenating Features. As described in Algorithm 3, the proposed approach to feature augmentation involves extracting structural features, defined previously, from the node level and concatenating them with real features associated with the nodes as shown in Eq. 2. Where f_i represents extracted features and h_i represents the real features at level 0.

Features Normalization. A critical step in this preprocessing pipeline involved normalizing the data features. This normalisation technique was applied to mitigate the potential effects of varying scales and magnitudes across different features. By standardising the feature values to a common scale, we aimed to create a more uniform and interpretable representation space for the model. This process not only facilitated comparisons between different features but also prevented certain features from dominating the learning process because of their larger scales. Furthermore, normalization promoted fairer contributions from all features and improved model robustness and generalization by reducing sensitivity to the absolute magnitudes of feature values. Overall, normalisation of the data features was integral to enhancing the stability, reliability, and performance of our model.

4.1 Node Classification Method

Concatenated Features with the GNN Baseline. After the three steps mentioned above, the goal of this process is to use these features to generate node embeddings that capture the relationships between nodes in the graph, and then use these embeddings for node classification, as shown in Algorithm 1. Therefore, we incorporate the concatenated features into works such as GNN baselines GCN [20], GraphSAGE [1], GraphSAINT [23], and recent node augmentation Mixup [19]. In the GNN baseline, the "message passing" process of a GNN layer, as described in Eq. (1) involves updating the representation of node i by aggregating the representations of itself and its neighbors. By stacking L layers, GNNs enable the final-layer prediction of node i to be based on its L-hop neighborhood. This is commonly referred to as the embedding of node i. Thus, in our method, the hidden states for each node in the graph are initialized by setting $h^{(0)}$ to the node attributes $x^{(0)}$, and then structural features are added to the initial features to produce the modified features $h^{(0)}$, as described in the previous sections. Then, for each layer l in the GNN model, from 1 to l - 1, update the hidden states of the nodes using the AGGREGATE function and the hidden states of the neighbors of each node.

Concatenated Features with Mixup. We incorporate the concatenation data into the work of [19], as shown in Algorithm 4, in which the hidden states for each node in the graph are initialized by setting $h(0)$ to the node attributes x_i. For each layer l in the GNN model, from 1 to L - 1, update the hidden states of the nodes using the AGGREGATE function and the hidden states of the neighbors of each node. For each node in the graph, sample another node j from the set of all nodes. Sample a value for λ from a Beta distribution with parameter α. Interpolate the attributes and labels of the two nodes using λ, creating new attributes $\tilde{x_i}$ and $\tilde{y_i}$ for the interpolated node. Initialize the hidden state of the interpolated node $\tilde{h(0)}$ using the interpolated attributes $\tilde{x_i}$. For each layer l in the GNN model, from 1 to L, update the hidden states of the interpolated node using the AGGREGATE function and the hidden states of the neighbors of the interpolated node. Calculate the classification loss L on the interpolated hidden states and labels. Use backpropagation to minimize L and update the model parameters $W(l)$.

$$h^{(l)} = f_i \,||\, h_i{}^{(l)} \tag{2}$$

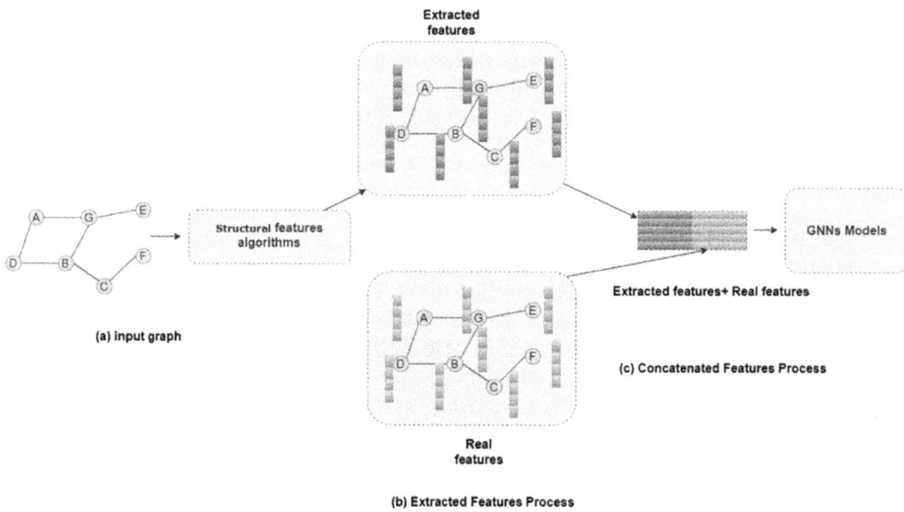

Fig. 1. Architecture Diagram of GNN Models with Concatenated Features, This figure depicts the architecture diagram of GNN models with concatenated features. (a) Input: A labeled graph is provided as input to the extracted features algorithms.(b) Extracted Features Process: The extracted features algorithms process the input graph to extract relevant features. (c) Concatenated Features Process: Once the extracted features process is complete, the features are concatenated to form a feature matrix. Forwarding Concatenated Features to GNN Models: The concatenated feature matrix is then forwarded to the GNN models for further processing and analysis.

5 Experimental Setup and Results

This section evaluates the proposed method using comprehensive experiments. Our proposed approach aims to improve the features of nodes in a graph, enabling it to effectively perform node classification in both transductive and

Algorithm 4. Feat-Aug

Input: Graph G
Output: Node embedding

// Extract features from nodes (see Algorithm 2)
$FG \leftarrow \text{ExtractedFeatures}(G)$
$C_{mf} \leftarrow \{\text{original features}\} \cup \{\text{structural features}\}$
for each node $v \in G$ **do**
  $h_v^{(0)} \leftarrow x_v^{(0)} + f_v^{(0)}$
end for
for $l \leftarrow 1$ **to** $L-1$ **do**
  for each node $i \in G$ **do**
    $h_i^{(l)} \leftarrow \text{AGGREGATE}(h_i^{(l-1)}, \{h_j^{(l-1)}\}, \{W_{ij}^{(l)}\})$
  end for
end for
for each node $i \in G$ **do**
  Sample node j from G
  $\lambda \leftarrow \text{Beta}(\alpha, \alpha)$
  $\widetilde{x}_{ij} \leftarrow \lambda x_i + (1-\lambda)x_j$
  $\widetilde{y}_{ij} \leftarrow \lambda y_i + (1-\lambda)y_j$
  $\widetilde{h}_{ij}^{(0)} \leftarrow \widetilde{x}_{ij}$
  for $l \leftarrow 1$ **to** L **do**
    $\widetilde{h}_{ij,i}^{(l)} \leftarrow \text{AGGREGATE}(\widetilde{h}_{ij,i}^{(l-1)}, \{h_k^{(l-1)}\}, \{W_{ik}^{(l)}\})$
    $\widetilde{h}_{ij,j}^{(l)} \leftarrow \text{AGGREGATE}(\widetilde{h}_{ij,j}^{(l-1)}, \{h_k^{(l-1)}\}, \{W_{jk}^{(l)}\})$
    $\widetilde{h}_{ij}^{(l)} \leftarrow \lambda \widetilde{h}_{ij,i}^{(l)} + (1-\lambda)\widetilde{h}_{ij,j}^{(l)}$
  end for
end for
Calculate classification loss L on $\{\widetilde{h}_{ij}^{(L)}, \widetilde{y}_{ij} \mid i \in G\}$
Back-propagation on $\{W^{(l)} \mid l \in \{1, 2, \ldots, L\}\}$ for minimizing L

inductive settings. The transductive setting means that features at all nodes are available during training, except labels of test nodes, whereas in the inductive setting, none of the feature vectors are available for validation or testing.

5.1 Transductive Setting

A comprehensive evaluation was conducted to assess the effectiveness of our method. Many GNN baselines were compared with our approach, including the Mixup method [19] as well as widely-used GNN models such as GCN [20], GAT [38], and LGCN [39]. We employ the datasets mentioned in Table 1, to evaluate the efficacy of these node classification models. The datasets can be accessed by the public on the Torch Geometric website [40].

We divide the nodes in each graph into three sets: 60%, 20%, and 20% for training, validation, and testing, respectively. We conduct 50 trials of experiments, initializing the weights randomly for each split. Notably, the extracted features have an impact on the GNN baseline methods, as shown in Table 3.

Results. Table 3 shows the results of our proposed approach against the baselines. We use the same experimental setup as in the original Mixup paper [19] to compare our proposed method to this approach. Our proposed method overcomes all methods on the dataset. Our approach achieves, on the Cora dataset, an improvement of 1% over the Mixup [19], and nearly 2% on the CiteSeer and Pumbed dataset; these results show how the structural features impact the graph deep learning models performance.

5.2 Inductive Setting

The datasets utilised for evaluation are Flickr, Amazon(Computer), and Amazon(Photo) [41]. The size of these datasets exceeds the capacity of full-batch implementations of GCN architectures to effectively handle them. Therefore, to compare, we employ the more scalable GraphSAGE [1] and GraphSAINT [23] as the baselines. These methods have similar goals to our proposed work but differ in their approach to augmenting the features of the nodes. The research experiments were conducted for a total of 100 trials, utilising random weight initialization. The findings have been documented in Table 2. GraphSAGE utilises mean, LSTM, and max-pooling as the aggregators, respectively.

Results. We use GraphSAGE-mean and GraphSAINT-GCN as the implementations for our suggested method to investigate whether Feat-AUG can enhance the performance of GNNs in the inductive setting. Concatenating features improves the test F1-micro scores of GraphSAGE-mean by 1.9% on Flickr and yields a 1% improvement over DropEdge + GraphSAGE on the same data. Additionally, it improves the scores by 12% on Amazon (Computers) and 4% on Amazon(Photo) compared to GraphSAGE. Our method + GraphSAGE increases the scores by nearly 10% on Amazon (Computers) and by 5% on Amazon (Photo) compared to DropEdge + GraphSAGE. Furthermore, our method + GraphSAINT achieves nearly similar results on the Flickr dataset compared to GraphSAINT, and on Amazon Computers and Photo, it improves the results by 6% and 2% respectively. Consequently, our proposed method improves them to surpass the performance of the standard approaches.

5.3 Discussion

Our experiments demonstrate that our approach in node classification tasks on different datasets, including Cora, Citeseer, and PubMed for transductive settings, as well as Flickr, Amazon (computer), and Amazon (Photo) for inductive settings, has significant benefits. This indicates that leveraging structural features can greatly enhance the baseline of graph deep learning and its variants.

In addition, previous methods have neglected the use of structural features and focused only on real features associated with graphs. Therefore, by combining the extracted features with real features, the proposed method has the

Table 2. Classification accuracy comparison for inductive setting

Method	Flickr	Amazon (Computer)	Amazon (Photo)
	Accuracy	Accuracy	Accuracy
GraphSAGE-mean [1]	50.1 ± 1.1	80.91 ± 0.83	87.25 ± 0.50
GraphSAGE-LSTM [1]	50.3 ± 1.3	80.25 ± 0.43	87.1 ± 0.30
GraphSAGE-pool [1]	50.0 ± 0.8	80.76 ± 0.23	87.34 ± 0.30
DropEdge [28] + GraphSAGE	50.8 ± 0.9	82.11 ± 0.30	87.64 ± 0.40
GraphSAINT [23]	51.1 ± 0.2	77.4 ± 0.1	87.5 ± 0.3
Ours + GraphSAGE	**52.1** $\pm$ 0.2	**93.2** $\pm$ 0.1	**92.4** $\pm$ 0.2
Ours + GraphSAINT	**51.2** $\pm$ 0.2	**83.4** $\pm$ 0.1	**89.4** $\pm$ 0.2

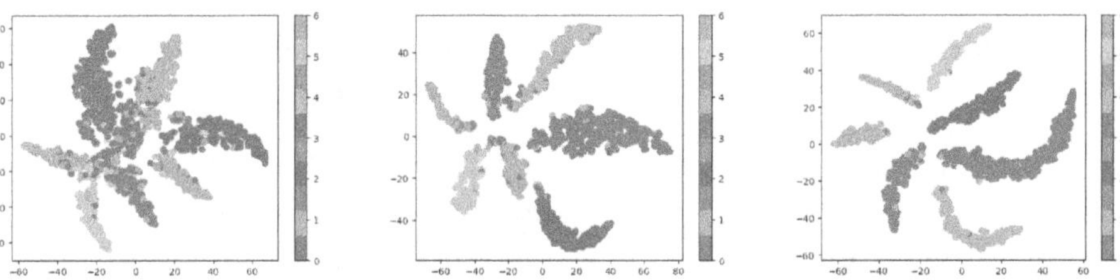

(a) GCN (b) Mixup (c) Ours

Fig. 2. A comparison of Mixup model and Ours Model, showcasing their performance differences in the final representation in the Cora dataset (visualized by t-SNE [42])

potential for further research to capture the benefits of structural features and leverage them in GNN (Graph Neural Networks) variants.

Furthermore, the interpreter of our proposed method for its significant achievements is the capability of a graph deep learning model to capture and learn these combined features. The modified features lead to precise node representations, thereby improving node classification performance.

Moreover, the simplicity of the architectures is the main advantage of our method. With this approach, we can achieve state-of-the-art performance, such as GCN [20], while maintaining low memory and computation costs.

5.4 Model Visualization

Figure 2 presents the last node representations of the GCN, Mixup, and our proposed models on the Cora dataset. From the figures, we notice the clear node representations and discriminative between classes in the cora dataset, indicating that our proposed method is superior to other methods. Figure 3 gives the visualization of the node representations of our proposed method and the Mixup. This figure provides some insight into the training process of both methods and provides a clear comparison, demonstrating how our approach leads to faster convergence.

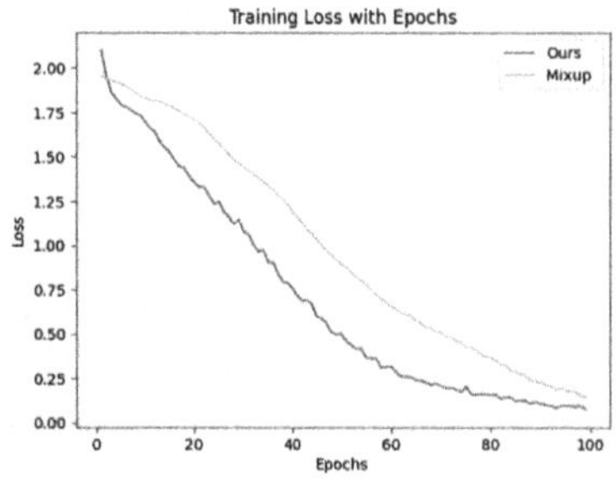

(a) Cora

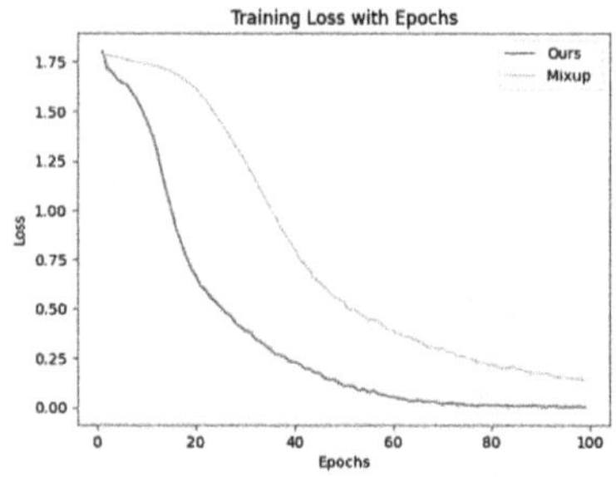

(b) CiteSeer

Fig. 3. Comparison of loss functions between our proposed model and the Mixup model, illustrating the superior compression achieved by our model.

Table 3. Classification accuracy comparison for the Transductive setup. The experimental setup of Mixup is followed to enable a fair comparison.

Method	Citeseer	Cora	Pubmed
	Accuracy	Accuracy	Accuracy
GCN [20]	70.3	81.5	79.0
GAT [38]	72.5 ± 0.7	83.0 ± 0.7	79.0 ± 0.3
LGCN [39]	71.1 ± 0.5	82.2 ± 0.5	79.0 ± 0.2
GMNN [43]	73.6	83.7	81.9
Mixup + GCN [19]	78.7 ± 0.9	90.0 ± 0.7	87.9 ± 0.8
Ours + Mixup + GCN	**80.4** ± 1.4	**91.4** ± 0.1	**88.4** ± 1.1

6 Conclusion

In this paper, we have introduced a novel approach for augmenting node features in graphs to enhance performance in node classification tasks. Our method combines real features, such as a bag of words in citation networks, with structural features, like node degree and clustering coefficient, extracted at the node level. Additionally, we utilize deep learning models to further enrich these structural features, creating a comprehensive node representation. We have conducted extensive experiments on a variety of real-world datasets, including Cora, Citeseer, and Pubmed, to evaluate the effectiveness of our approach. The results demonstrated significant improvements over traditional methods that rely solely on real features, demonstrating the benefits of integrating both real and structural features. This approach not only addresses the issue of limited real features in some datasets but also highlights the potential of combining different types of node features to achieve more accurate and robust node classification.

Our approach represents a step forward in the field, as previous methods for augmenting graph features have largely focused on the use of real features and have neglected the potential benefits of incorporating structural features.

By considering both real and structural features when augmenting the data, we are able to provide a more comprehensive representation of the nodes within the graph, leading to improved performance in node classification tasks. Future work will see an investigation of the potentials of our proposed approach for other graph-based tasks, such as graph generation and link prediction. We also plan to further explore the use of deep learning models to incorporate additional node-level features and evaluate the impact on the performance of the model. We believe that these efforts will help to facilitate further progress in the field and contribute to the development of more effective graph classification models.

References

1. Hamilton, W., Ying, Z., Leskovec, J.: Inductive representation learning on large graphs. In: Advances in Neural Information Processing Systems, vol. 30 (2017)
2. Ying, R., He, R., Chen, K., Eksombatchai, P., Hamilton, W.L., Leskovec, J.: Graph convolutional neural networks for web-scale recommender systems. In: Proceedings of the 24th ACM SIGKDD International Conference on Knowledge Discovery and Data Mining, pp. 974–983 (2018)
3. Guo, S., Lin, Y., Feng, N., Song, C., Wan, H.: Attention based spatial-temporal graph convolutional networks for traffic flow forecasting. In: Proceedings of the AAAI Conference on Artificial Intelligence, vol. 33, no. 01, pp. 922–929 (2019)
4. Hamilton, W.L.: Graph Representation Learning. Morgan & Claypool Publishers, San Rafael (2020)
5. Cubuk, E.D., Zoph, B., Mane, D., Vasudevan, V., Le, Q.V.: Autoaugment: learning augmentation strategies from data. In: Proceedings of the IEEE/CVF Conference on Computer Vision and Pattern Recognition, pp. 113–123 (2019)
6. Atapour-Abarghouei, A., Bonner, S., McGough, A.S.: A King's ransom for encryption: ransomware classification using augmented one-shot learning and Bayesian approximation. In: 2019 IEEE International Conference on Big Data (Big Data), pp. 1601–1606. IEEE (2019)
7. Feng, S.Y., et al.: A survey of data augmentation approaches for NLP. arXiv preprint arXiv:2105.03075 (2021)
8. Xie, Q., Dai, Z., Hovy, E., Luong, T., Le, Q.: Unsupervised data augmentation for consistency training. In: Advances in Neural Information Processing Systems, vol. 33, pp. 6256–6268 (2020)
9. Wang, Y., Wang, W., Liang, Y., Cai, Y., Liu, J., Hooi, B.: NodeAug: semi-supervised node classification with data augmentation. In: Proceedings of the 26th ACM SIGKDD International Conference on Knowledge Discovery and Data Mining, pp. 207–217 (2020)
10. Adjeisah, M., Zhu, X., Xu, H., Ayall, T.A.: Towards data augmentation in graph neural network: an overview and evaluation. Comput. Sci. Rev. **47**, 100527 (2023)
11. Deng, Z., Dong, Y., Zhu, J.: Batch virtual adversarial training for graph convolutional networks. arXiv preprint arXiv:1902.09192 (2019)
12. Feng, F., He, X., Tang, J., Chua, T.-S.: Graph adversarial training: dynamically regularizing based on graph structure. IEEE Trans. Knowl. Data Eng. **33**(6), 2493–2504 (2019)
13. Kong, K., et al.: Robust optimization as data augmentation for large-scale graphs. In: Proceedings of the IEEE/CVF Conference on Computer Vision and Pattern Recognition, pp. 60–69 (2022)

14. Suresh, S., Li, P., Hao, C., Neville, J.: Adversarial graph augmentation to improve graph contrastive learning. In: Advances in Neural Information Processing Systems, vol. 34, pp. 15:920–15:933 (2021)
15. Liu, X., Sun, D., Wei, W.: A graph data augmentation strategy with entropy preservation. arXiv preprint arXiv:2107.06048 (2021)
16. Song, R., Giunchiglia, F., Zhao, K., Xu, H.: Topological regularization for graph neural networks augmentation. arXiv preprint arXiv:2104.02478 (2021)
17. Chunaev, P.: Community detection in node-attributed social networks: a survey. Comput. Sci. Rev. **37**, 100286 (2020)
18. Thekumparampil, K.K., Wang, C., Oh, S., Li, L.-J.: Attention-based graph neural network for semi-supervised learning. In: arXiv preprint arXiv:1803.03735 (2018)
19. Wang, Y., Wang, W., Liang, Y., Cai, Y., Hooi, B.: Mixup for node and graph classification. In: Proceedings of the Web Conference 2021, pp. 3663–3674 (2021)
20. Kipf, T.N., Welling, M.: Semi-supervised classification with graph convolutional networks. arXiv preprint arXiv:1609.02907 (2016)
21. Xu, K., Hu, W., Leskovec, J., Jegelka, S.: How powerful are graph neural networks? arXiv preprint arXiv:1810.00826 (2018)
22. Zhou, J., et al.: Graph neural networks: a review of methods and applications. AI Open **1**, 57–81 (2020)
23. Zeng, H., Zhou, H., Srivastava, A., Kannan, R., Prasanna, V.: Graphsaint: graph sampling based inductive learning method. arXiv preprint arXiv:1907.04931 (2019)
24. Bruna, J., Zaremba, W., Szlam, A., LeCun, Y.: Spectral networks and locally connected networks on graphs. arXiv preprint arXiv:1312.6203 (2013)
25. Defferrard, M., Bresson, X., Vandergheynst, P.: Convolutional neural networks on graphs with fast localized spectral filtering. In: Advances in Neural Information Processing Systems, vol. 29 (2016)
26. Gasteiger, J., Bojchevski, A., Günnemann, S.: Predict then propagate: graph neural networks meet personalized pagerank. In: arXiv preprint arXiv:1810.05997 (2018)
27. Chen, D., Lin, Y., Li, W., Li, P., Zhou, J., Sun, X.: Measuring and relieving the over-smoothing problem for graph neural networks from the topological view. In: Proceedings of the AAAI Conference on Artificial Intelligence, vol. 34, no. 04, pp. 3438–3445 (2020)
28. Rong, Y., Huang, W., Xu, T., Huang, J.: Dropedge: towards deep graph convolutional networks on node classification. arXiv preprint arXiv:1907.10903 (2019)
29. You, Y., Chen, T., Sui, Y., Chen, T., Wang, Z., Shen, Y.: Graph contrastive learning with augmentations. In: Advances in Neural Information Processing Systems, vol. 33, pp. **33**, 5812–5823 (2020)
30. Yoo, J., Shim, S., Kang, U.: Model-agnostic augmentation for accurate graph classification. In: Proceedings of the ACM Web Conference 2022, pp. 1281–1291 (2022)
31. Yang, L., Zhang, L., Yang, W.: Graph adversarial self-supervised learning. IN: Advances in Neural Information Processing Systems, vol. 34, pp. 14:887–14:899 (2021)
32. Jing, M., Zhu, Y., Zang, T., Yu, J., Tang, F.: Graph contrastive learning with adaptive augmentation for recommendation. In: Joint European Conference on Machine Learning and Knowledge Discovery in Databases, pp. 590–605. Springer (2022)
33. Zhong, Z., Mottin, D.: Knowledge-augmented graph machine learning for drug discovery: from precision to interpretability. In: Proceedings of the 29th ACM SIGKDD Conference on Knowledge Discovery and Data Mining, pp. 5841–5842 (2023)

34. Zhu, J., Wang, Q., Tao, C., Deng, H., Zhao, L., Li, H.: AST-GCN: attribute-augmented spatiotemporal graph convolutional network for traffic forecasting. IEEE Access **9**, 35:973–35:983 (2021)
35. Ding, K., Xu, Z., Tong, H., Liu, H.: Data augmentation for deep graph learning: a survey. ACM SIGKDD Explor. Newsl. **24**(2), 61–77 (2022)
36. Abushofa, M.E., Abarghouei, A.A., Forshaw, M., Mcgough, A.S.: FEGR: feature enhanced graph representation method for graph classification. In: Proceedings of the International Conference on Advances in Social Networks Analysis and Mining, pp. 371–378 (2023)
37. Berlingerio, M., Koutra, D., Eliassi-Rad, T., Faloutsos, C.: Netsimile: a scalable approach to size-independent network similarity. arXiv preprint arXiv:1209.2684 (2012)
38. Velickovic, P., Cucurull, G., Casanova, A., Romero, A., Lio, P., Bengio, Y.: Graph attention networks. stat **1050**, 20 (2017)
39. Gao, H., Wang, Z., Ji, S.: Large-scale learnable graph convolutional networks. In: Proceedings of the 24th ACM SIGKDD International Conference on Knowledge Discovery and Data Mining, pp. 1416–1424 (2018)
40. Morris, C., Kriege, N.M., Bause, F., Kersting, K., Mutzel, P., Neumann, M.: Tudataset: a collection of benchmark datasets for learning with graphs. arXiv preprint arXiv:2007.08663 (2020)
41. Xue, Y., Liao, Y., Chen, X., Zhao, J.: Node augmentation methods for graph neural network based object classification. In: 2021 2nd International Conference on Computing and Data Science (CDS), pp. 556–561. IEEE (2021)
42. Van Der Maaten, L.: Accelerating t-SNE using tree-based algorithms. J. Mach. Learn. Res. **15**(1), 3221–3245 (2014)
43. Qu, M., Bengio, Y., Tang, J.: GMNN: graph Markov neural networks. In: International Conference on Machine Learning, pp. 5241–5250. PMLR (2019)

Beyond the Surface: Navigating Complex Systems via ABMs and Hypergraphs

Daniele De Vinco(✉)

Università degli Studi di Salerno, Fisciano, Italy
ddevinco@unisa.it

Abstract. Interactions, organization, and emergent behaviors are fundamental to all complex systems. These characteristics can be observed in every real-world scenario, prompting the question: how do these interactions among elements serve as the foundation of complex systems? Over the years, researchers have developed various methods to answer such a question. One widely used bottom-up approach is agent-based simulation, which deduces a system's properties by examining its components. Alternatively, a system can be modeled via a top-down approach, such as hypergraphs, mathematical structures abstracting higher-order interactions. These two methods provide different perspectives on the same system, each offering unique and valuable insights. In this Ph.D. project, my primary focus is on social systems, aiming to shed light on the dynamics of group behaviors in online social media.

Keywords: Complex systems · Hypergraphs · ABMs · Stance detection · LLM · Social networks

1 Introduction

Many closely interconnected components constitute various natural, social, and artificial structures, giving rise to unintelligible behaviors if we focus on a single component at a time [25]. Such organizations are commonly referred to as complex systems, and as examples, we can name traffic control, weather forecast, or policy-making. Usually, the analysis of complex systems involves creating a model that can replicate the real system. These models are typically the result of the collaborative efforts of multiple experts from various fields. Complex systems are characterized by key features such as non-linearity (when small changes in one part of the system can lead to disproportionately large and unpredictable effects in other parts), self-organization, and adaption [8]. Investigating these phenomena means dealing with the following challenges:

- *Data complexity and access*: Complex systems often require and generate vast amounts of data, making analysis and interpretation challenging. Further, online platforms frequently deny or hide their data behind paywalls, making data acquisition difficult.

L. M. Aiello et al. (Eds.): ASONAM 2024, LNCS 15213, pp. 148–163, 2025.
https://doi.org/10.1007/978-3-031-78548-1_13

- *Modeling complexity*: Developing accurate mathematical or computational models for complex systems can be difficult due to their non-linear and dynamic nature.
- *Computational resources*: Simulating or analyzing complex systems may require significant computational resources regarding time, money, and equipment.
- *Interdisciplinary nature*: Understanding complex systems often requires knowledge and cooperation from multiple domain experts.

During my Ph.D. project, I have been focusing on studying social systems, particularly by analyzing social media platforms. Among these platforms, Reddit stands out as a prominent example of an online social data source. Due to the richness and quality of its data, Reddit has become a key resource for researchers who have extensively studied its data to validate various social science theories. While analyzing user behavior in online forums like Reddit sheds light on the fundamental processes through which groups of individuals develop collective thought, it also has significant practical applications, such as enhancing user experience, increasing engagement, and automatically detecting bullying in online conversations [10].

Researchers can use top-down strategies, bottom-up strategies, or a combination of both to understand the intricacies of human behavior. These approaches are complementary and offer different perspectives on the systems to study.

The bottom-up approach involves examining individual components within the complex system and understanding its behavior starting from its basic elements. Agent-based models (ABMs) represent a robust modeling technique that uses such an approach [14]. Specifically, in ABMs, modelers define agents and environments to replicate specific aspects or properties of the underlying reality. These models provide a detailed understanding of how local interactions give rise to system-level behavior, offering fine-grained control and manipulation of individual components. This approach is particularly useful when the behavior of individual elements within the system is well understood.

The top-down approach involves examining the system's global behavior and properties and then deconstructing and understanding the underlying components and interactions. One of the most expressive mathematical structures capable of capturing high-order interactions is a hypergraph, a generalization of a graph where a (hyper)edge connects an arbitrary number of nodes [9]. Hypergraphs are useful for modeling complex systems as they can represent relationships that involve multiple entities simultaneously, unlike graphs, which only model pairwise interactions. Hypergraphs are particularly valuable when examining social networks, as interactions often occur in groups of different sizes. For instance, users might collaborate in teams, participate in group activities, form communities, or work together towards a common goal. All these activities involve more complex connections than simple pairwise relationships. By using hypergraphs, we can more accurately capture and analyze the dynamics of these high-order interactions, leading to better insights and understanding of the system's behavior and structure [41].

Social interactions on online platforms frequently involve written communications. Textual messages can carry multiple meanings, and the context in which they are located can significantly affect their interpretation and nuance. As a result, understanding the semantics of these interactions is crucial and cannot be ignored when examining online user behavior. As technology advances, recent research has proved how the rapid development of large language models (LLMs) has assisted researchers in performing various tasks across multiple fields [39]. After being pre-trained on extensive text corpora, LLMs exhibited vast context-aware knowledge and exceptional semantic comprehension capabilities [12]. For instance, LLMs have demonstrated their effectiveness in stance detection tasks, showing adaptability across different datasets with various prompting schemes and even outperforming supervised models in terms of performance while using fewer resources [15].

My Ph.D. project fits this broader context and has a dual focus. First, it aims to advance the analysis of complex systems by improving ABM frameworks and addressing challenges such as the scalability of large models. Second, it seeks to deepen our understanding of online social interactions through two main approaches: (i) incorporating group interactions and modeling them via hypergraphs and (ii) leveraging LLMs to account for the semantics of these interactions. This multifaceted approach aims to advance our understanding of various real-world phenomena, particularly in social dynamics, as outlined in the research questions in Sect. 3. By doing so, this project aspires to foster novel insights in a more than-ever interdisciplinary world.

The remainder of this paper is organized as follows. Section 2 describes related works, delineating the gap in the existing literature that motivates my research project. Section 3 outlines the main goals of my research project, the preliminary results achieved, and the challenges I am facing. Finally, Sect. 4 summarizes the work and indicates the next steps of the project.

2 Background and Related Work

This section provides an overview of how complex systems have been explored in the literature through the lenses of ABM and hypergraphs. It also briefly explains LLMs and discusses their capabilities.

2.1 ABMs

Among the techniques used to analyze real-world systems, ABMs are one of the most effective [6]. From modeling an epidemic spread to an economic fluctuation of the market, ABMs have been used to develop an accurate representation of reality [11,16]. The main element of an ABM is an *agent*, a basic autonomous entity that the modeler can define with different characteristics and behaviors. Agents can interact with each other and the environment according to a set of rules defined by the modeler. These interactions result in a model of the world being studied, which the modeler can analyze to learn relevant lessons from its emerging behaviors. Structures, patterns, and behaviors emerge through interactions rather than being purposefully coded within the model. ABMs also help researchers determine how a system's micro-level properties, constraints, and laws affect its macroscopic behavior. An agent-based simulation is composed of a three-component structure:

- *Agents.* Each agent is an individual that forms the population under consideration. Agents may act independently or in relation to one another, executing complex functions through simple interactions.
- *Relationships.* The relationships define how agents interact with each other.
- *Rules.* Rules define the agents' behavior, determining the outcomes of their interactions.

Using ABM simulations makes analyzing complex models easier, decomposing the problem into smaller components. However, their implementation can be difficult for domain experts with little experience in software development [1]. The solution to this type of problem lies in multiple frameworks that make it easy to develop and execute a simulation, eliminating the need to know all the necessary technical details. Specifically, simulations have various aspects in common to implement, and these frameworks provide functionality effectively circumventing the need to reinvent solutions for well-known challenges. Good examples of successful ABM engines are Mason, a fast, modular, discrete event multi-agent simulation toolkit written in Java [29], and Netlogo, designed to be used by researchers and educators through a simple interface and language [36].

ABM simulations can be time-consuming because of their particularly high computational load. In addition, their execution time can significantly affect the final result since some simulations achieve greater accuracy as time progresses, as in weather forecasting models. The distributed computing paradigm alleviates such a problem by reducing the running time of long-running simulations without impacting the final outcomes [13]. While, on the one hand, distributed engines have many advantages, on the other, managing the communication and workload between the different machines requires additional effort. In a distributed environment, agents may be simulated on different machines at each time step, thus meaning that the workload should be dynamically adjusted during the simulation. These operations add communication overhead, which adds up to the synchronization step. Further, the whole computation proceeds at the pace of the slowest computer, which could be a barrier to the system's total performance. Another unresolved issue in this context pertains to the reliability of distributed processes, meaning that one should expect functions to either complete successfully or terminate with errors without losing information, even though this may not always be theoretically possible [21]. Frameworks like Mason already provide algorithms and tools to assist their users in implementing distributed ABMs. This design choice allows them to focus only on the simulation-related details without directly dealing with the technicalities of a distributed computation.

2.2 Hypergraphs

A hypergraph is a generalization of a graph where a hyperedge allows the connection of an arbitrary number of nodes. Formally, a hypergraph is an ordered pair $H = (V, E)$ where V is a set of vertices and E is a set of hyperedges. Each hyperedge is a non-empty subset of vertices; that is, $E \subseteq 2^V \setminus \emptyset$, where 2^V is the power set of V [9]. A simple illustration is shown in Fig. 1.

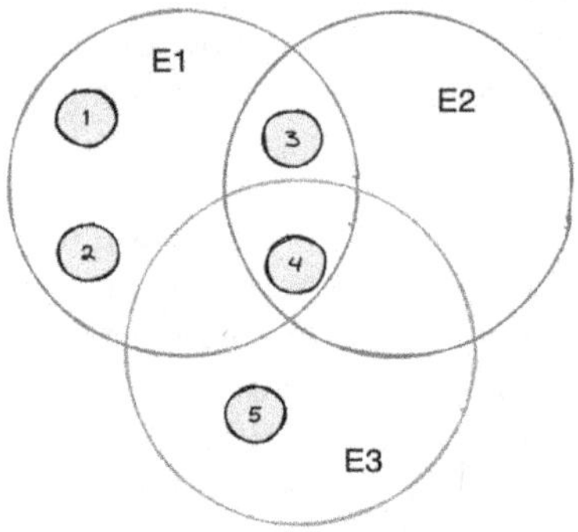

Fig. 1. An example of a hypergraph, with 5 nodes and three hyperedges, E1 = {1,2,3,4}; E2 = {3,4}; E3 = {4,5}.

Hypergraphs model the extremely non-linear interactions amongst a collection of three or more nodes, thus accounting for additional information [30]. For instance, hypergraphs can abstract social systems in which people interact in groups of any size. One example is the co-authorship collaboration network, where a hyperedge represents an article and links together the authors who have collaborated on it. Other examples can be found in biology, ecology, and neuroscience [7]. One of the few downsides of the strong expressiveness of hypergraphs is that studying these structures requires specific algorithms due to their complexity [24,28]. Because of this, hypergraphs have not been utilized as much in past literature as their graph equivalent. The use of hypergraphs has been arising recently as a result of numerous systematic studies showing how converting a hypergraph to a classical graph either results in an information loss that cannot be avoided or generates a significant number of additional nodes and edges that increase the amount of space and time needed for analytic tasks [4].

The literature has often studied an entire graph's topology by observing its substructures, commonly known as motifs. In particular, triangular subgraphs appear to be a powerful signal for identifying high-density communities or local trends within a graph [38]. This concept can be generalized to hypergraphs, where h-motifs describe the connectivity patterns of connected hyperedges [26]. Considering an h-motif involving three hyperedges, it is possible to enumerate 26 instances of these substructures. For example, given a set $E1, E2, E3$ of three connected hyperedges, an h-motif describes its connectivity pattern by the emptiness of the sets shown in Fig. 2. This definition can be generalized to an arbitrary number of hyperedges and expanded by including a temporal layer to control the evolution of these structures along time [27].

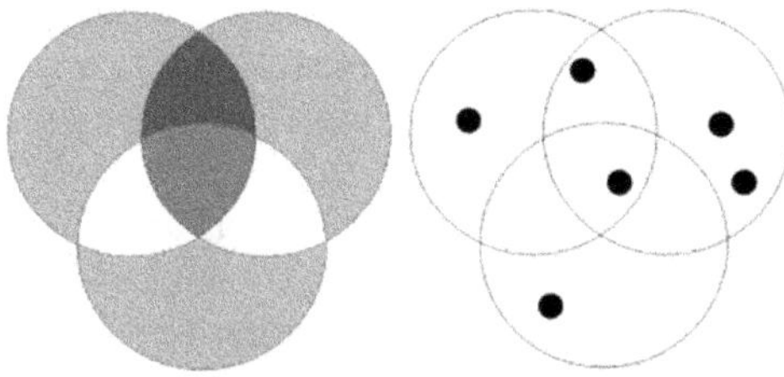

Fig. 2. The image on the left illustrates the intersections between three hyperedges involved in the motif, with colored areas indicating regions of non-empty intersections. The red area represents the complete intersection of three hyperedges; the blue area indicates the intersection of two hyperedges, excluding the third; and the green areas highlight nodes that belong to only one hyperedge. The image on the right depicts a possible instance of this motif. (Color figure online)

H-motifs can be used to build a hypergraph's characteristic profile (CP) to distinguish the structural design principles of real-world hypergraphs [26]. The significance of each h-motif in a hypergraph is measured by comparing the count of its instances against their count in randomized hypergraphs. Specifically, the significance of a h-motif t in a hypergraph H is defined as

$$\Delta_t := \frac{Motif[t] - Motif_{random}[t]}{Motif[t] + Motif_{random}[t] + \epsilon}$$

where ϵ is a parameter fixed to 1 to avoid a possible division by zero when a specific motif does not exist. The CP of a hypergraph H, which summarizes the local structural patterns, is a vector of size 26, where each t-th element is

$$CP_t := \frac{\Delta_t}{\sqrt{\sum_{t=1}^{n=26} \Delta_t^2}}$$

The information conveyed by h-motifs can be leveraged for tasks like hyperedge prediction. According to Lee et al. [26], different motifs produce varying results in this task, possibly because different motifs hold varying levels of importance in different application domains. This finding suggests that further efforts should focus on exploring how selecting the appropriate motifs across various datasets can provide the most valuable insights for the task at hand.

2.3 LLMs

Language models are artificial intelligence models designed to understand, generate, and manipulate human language [20]. These models are used in various applications, such as text generation, translation, and sentiment analysis, to name a few. Language models with hundreds of billions (or more) of parameters, known as large language models (LLMs), are trained on enormous amounts of text data. As an example, we can include GPT [19], LLaMA [37], and Gemma [35]. LLMs generally try to forecast the likelihood of future (or absent) tokens by modeling the generative likelihood of word sequences [40]. These models have been evolving rapidly, from basic language model interfaces to problem-solving agents, demonstrating strong abilities to comprehend natural language and complete challenging tasks via text generation.

Prompting has emerged as the most popular method for using LLMs for various tasks based on the natural language interface. Context learning imbues LLMs with the capacity to perform well on unknown tasks, sometimes surpassing fine-tuned models, by merging task descriptions and demonstration examples into prompts. Advanced prompting approaches, such as the chain-of-thought strategy that incorporates the intermediate reasoning steps into prompts, have been proposed to improve the ability to use complicated reasoning. LLMs enable the development of more intelligent systems (e.g., autonomous AI agents) to tackle various complex tasks in real-world scenarios. With the industry's most recent developments, these tools have many applications, and new ones are found daily [23]. In my PhD project, I leverage LLMs to tackle a stance detection task. Specifically, this task combines hypergraphs for abstracting conversational networks with the analytical power of LLMs to examine their dynamics.

3 Scope

This section details my project's objectives, the research outcomes, and the challenges to tackle across all project topics.

3.1 Project Objectives

The primary objective of my PhD project is to investigate and analyze complex systems by leveraging ABMs, hypergraph-based network models, and LLMs. The specific goals for each of these research areas are outlined below.

ABMs. In the context of ABMs, my goal is to enhance their functionalities to advance the state of the art in ABM engines, particularly regarding their computational limits, as the most widely used ABMs prioritize either performance or usability. Hence, a primary focus of my research is to develop an approach that effectively integrates both features. The main research question my project aims to answer is:

> *How can advances in parallel and distributed computing technologies be harnessed to improve the scalability and efficiency of ABMs for simulating large-scale, real-time, and highly dynamic systems?*

Answering such a question implies validating the following hypotheses:

- H^1_{ABM}**:** Feasibility of developing an efficient and reliable distributed ABM engine.

- H^2_{ABM}: Addressing the potential increase in model development complexity through a comprehensive study of system usability from a user perspective.

Hypergraphs. Another main line of work of my PhD project is analyzing group interactions and the rise of collective behaviors in online social media through the lenses of hypergraphs. In this regard, the main research questions my project aims to answer are:

Are specific structural patterns in hypergraphs dependent on the communities involved, or do they highlight specific user behaviors? Can we predict the formation of these patterns?

Answering this question leaves room for several hypotheses, such as:

- H^1_{HG}: When analyzing data from the same domain (e.g., conversations about general purpose vs. conversations on domain-specific programming languages), users' activity patterns should differ.
- H^2_{HG}: Tracking users across communities and observing their influence can reveal a correlation between structural patterns and specific user behaviors.
- H^3_{HG}: When a community of the same users persists over time, its structural pattern should remain consistent.

LLMs. Regarding LLMs, my project's goal is to assess the feasibility of using these tools to extract meaningful insights from text and enhance our understanding of how user interact and exchange their opinions in online social contexts. Specifically, the main research question my project aims to answer is:

Can LLMs improve our understanding of online social interactions and opinion dynamics?

Answering this question leaves space for different hypotheses, such as:

- H^1_{LLM}: An LLM can serve as a viable ground truth generator in the absence of annotated conversational datasets.
- H^2_{LLM}: Extracting information from text and structures using LLMs can enhance the performance of existing models.
- H^3_{LLM}: For stance identification tasks, incorporating high-order interactions that capture the subtle local context of conversations can improve model performance.

3.2 Results

This section presents an overview of the preliminary results achieved during my PhD research. All findings are discussed in relation to the objectives and hypotheses outlined in Sect. 3.1.

ABMs. As discussed in Sect. 2.1, over the past decade, the demand for more complex, computation-intensive models has led to the development of numerous frameworks and tools for running ABM simulations. To offer a tool integrating reliability, performance, and ease of use, we developed krABMaga [1,3], an ABM simulation engine written in Rust. A high-level representation of the architecture of krABMaga is shown in Fig. 3.

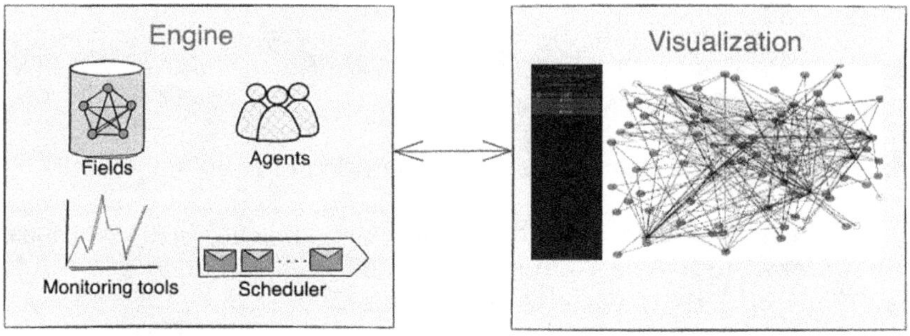

Fig. 3. A high-level illustration of the architecture of krABMaga: the Engine module includes the core functionalities of the framework (on the left), while the Visualization module orchestrates visualization-related features (on the right).

One of the main goals of krABMaga is to make large-scale simulation available without expensive equipment while still offering efficiency and reliability [3]. Leveraging the Rust programming language has been instrumental in achieving this objective thanks to its performance akin to C, enabled by its memory model, and its expressiveness and usability typical of high-level languages like Python [22]. To further support distributed computation, we developed a custom implementation of a K-Dimensional (K-D) Tree data structure to partition the simulation field and distribute the workload evenly across processes. Unlike traditional K-D Trees, which maintain references to each child in the parent node, our modified approach ensures that all nodes retain references to one another throughout the tree. This adaptation simplifies neighbor search and synchronization operations, crucial for KrABMaga simulations.

Figure 4 shows the performance of our framework running the Flocker simulation, commonly known as Boids [33], on a Microsoft Azure cluster with 128 nodes using the K-D Tree. Specifically, Fig. 4a highlights the execution times for simulations with varying numbers of agents (1M, 2M, 5M, and 10M), while Fig. 4b showcases the speedup achieved in each experiment. This second plot reveals an increased speedup as the number of agents grows, attributed to the effective management of agents by the simulation processes. This efficiency becomes more pronounced when the computational workload exceeds the communication overhead. The results closely follow the ideal performance curve, where speedup scales linearly with the number of processors or machines (e.g., 2 processors yield a speedup of 2, 4 processors yield a speedup of 4, etc.). In future work, we aim

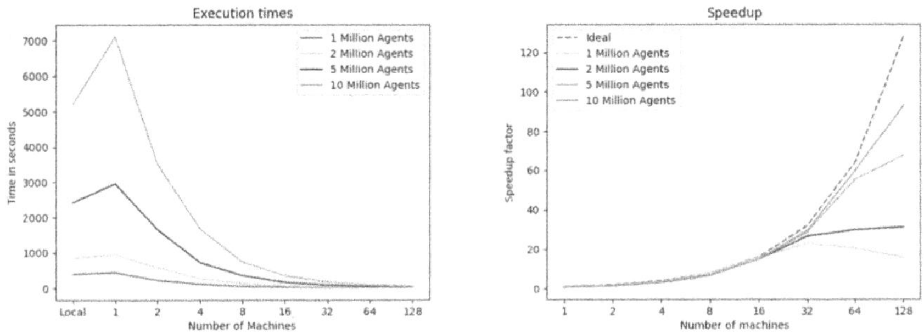

(a) Execution times. (b) Relative speedup.

Fig. 4. Performance of krABMaga in a distributed environment (Microsoft Azure cluster with 128 nodes) during simulations of the Flocker model with a growing number of agents.

to develop a more robust communication layer that is entirely transparent to the modeler to further simplify the simulation process by removing the need to implement complex custom functions.

Currently, we are working on integrating Geographic Information System (GIS) data into our engine. As illustrated in Fig. 5, we implemented a plugin that allows users to upload a file containing a map (e.g., the US territories) and convert it into one of the krABMaga fields, enabling agents to move on top of it. In this manner, krABMaga will empower users to simulate more diverse and realistic scenarios.

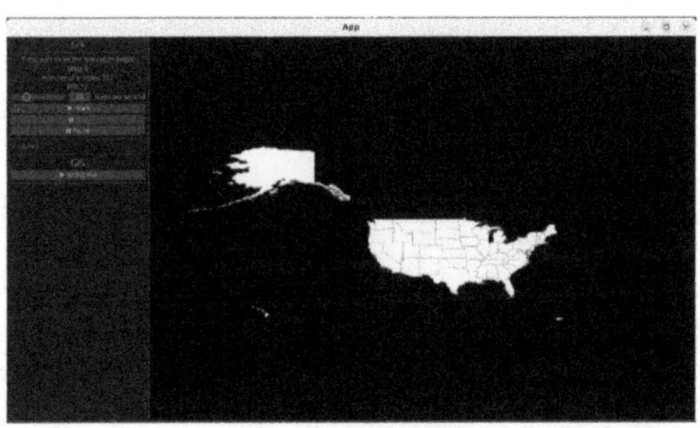

Fig. 5. Graphical user interface of krABMaga, demonstrating the implementation of the GIS plugin.

Hypergraphs. Online social networks generate vast amounts of data. Websites like Facebook, YouTube, or Reddit have millions or billions of users that interact daily. My focus in this context is analyzing group interactions and the structural and semantics patterns of user communities in online social media through the lenses of hypergraphs.

In a first study, we analyzed developers' behavior and the formation of communities around twenty popular programming languages, emphasizing the significant differences in how users engage in different Question & Answer (Q&A) platforms, such as StackOverflow and Reddit [2]. Q&A platforms, and social networks in general, usually consist of discussions in which users reply to each other on the same topic. Hence, conversational networks provide opportunities to identify strongly connected communities and examine whether specific user behavior depends upon the characteristics of the given platform. Looking at the CP of such conversations could give deeper insights into user communication patterns and their similarities across topics, communities, and platforms. Figure 6a shows the CP of 22 different subreddits as an example of this analysis. Preliminary results indicate that conversations on similar topics do not necessarily exhibit similar CPs. This observation is further confirmed in Fig. 6b, showing the CPs of three subreddits related to climate change.

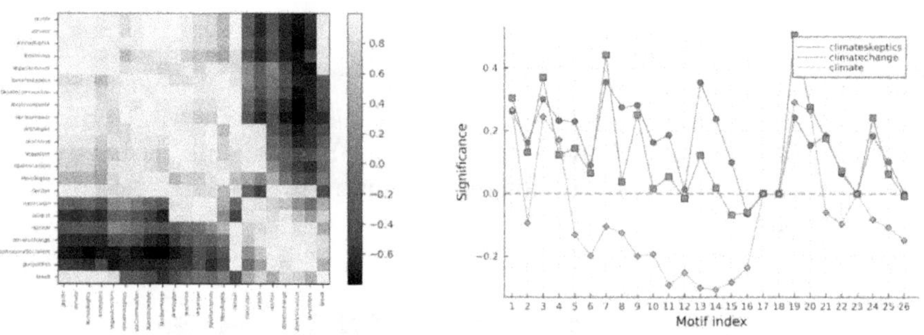

(a) Similarity matrix of CPs of 22 subreddits on different topics.

(b) CPs of 3 climate-related subreddits.

Fig. 6. CP-based similarity across different subreddits.

Unfortunately, access to the API or raw data of many online social platforms is often limited or behind a paywall, posing a barrier for researchers interested in studying online behaviors. As a consequence, having a ready-to-use dataset to carry on specific analyses is not always trivial. To alleviate such a problem, we published an open-source online repository to allow researchers to compare and download data tailored for high-order analysis [5].

LLMs. Another interesting line of inquiry that has emerged recently is the use of LLMs to gain insights into both structural and linguistic attributes of conversational networks. The architecture depicted in Fig. 7 represents our initial attempt to combine hypergraphs (to learn the conversational hypernetwork's structure) and LLMs (to understand the text's semantics) for performing a stance detection task [17]. Specifically, stance detection involves automatically classifying a user's attitude toward a target into one of three categories: *Favor, Against, Neither.*

In our experiments, we focused on the Reddit discussion platform, selecting three subreddits centered around the climate topic (i.e., *climate, climatechange, climateskeptics*), and retrieved all users' conversations over a two-year period. To obtain an initial training set, we used Gemma [35], a recently published LLM, to label a real-world dataset from Reddit, annotating the stance of all the messages, while the results obtained will be validated against a social network dataset annotated by humans for the same task [34].

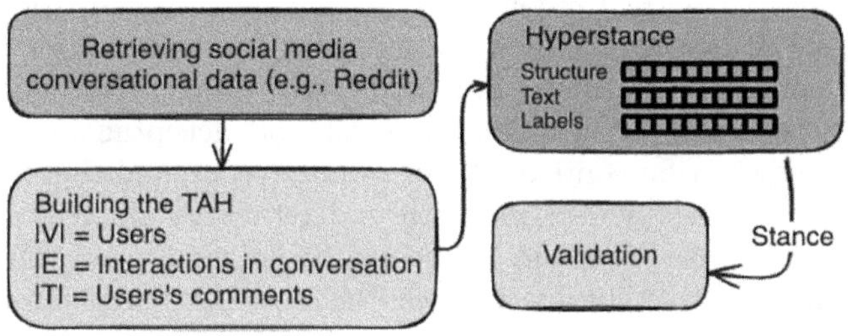

Fig. 7. A proof-of-concept pipeline to investigate a machine learning model combining semantics and structure.

3.3 Challenges

These topics come with many theoretical and practical challenges to consider.

ABMs. Regarding ABM, there are two main problems to deal with. First, ABMs can be computationally intensive, especially when simulating large numbers of agents or complex interactions between agents [31]. This can limit the scale and scope of simulations that can be conducted, particularly on basic computing hardware (e.g., personal laptops). Furthermore, ABMs often require significant parameterization and calibration to represent real-world systems accurately. Identifying appropriate parameter values and validating the model against real-world data can be challenging and may introduce uncertainties in the model [32].

Hypergraphs. Hypergraphs are a growing field, but there is a lack of standardization regarding definitions, models, and algorithms. Due to the arbitrary size of the relationships between nodes, dealing with hypergraphs implies interacting with a more complex network than a traditional graph. One example of this increased complexity is the visual representation of hypergraphs, which is often depicted using Venn diagrams. However, drawing such representations becomes nearly impossible or extremely chaotic as the number of nodes and hyperedges grows. While graphs can be easily represented with points and lines, hypergraphs require circles or other techniques that do not always provide intuitive insights, necessitating the introduction of valid alternatives [18].

Another challenging task on hypergraphs is link prediction. In a graph, link prediction involves predicting the existence of a specific relationship between two nodes. This definition does not translate well to hypergraphs, where a hyperedge may include an arbitrary number of nodes [4]. In this context, predicting the appearance of an h-motif is even more challenging, as it requires understanding not only if a link will connect certain nodes but also if it will form a specific substructure with other existing links.

LLMs. Although LLMs offer numerous advantages, they also present several drawbacks. One of the most notable issues is bias. Training LLMs on massive datasets can inadvertently reinforce preexisting inequities or omit important perspectives due to biases in the data. Moreover, LLMs are often seen as black box models, making it difficult to understand how they reach their conclusions. This lack of interpretability can hinder trust and acceptance of their analyses, especially in scientific contexts where transparency and reproducibility are crucial. Finally, LLMs may struggle with understanding the context of a task, particularly in specialized domains where terminology and conventions vary significantly. Misinterpreting context can lead to inaccurate analyses.

4 Conclusion and Future Work

Studying social systems encompasses a variety of open problems, ranging from examining emergent behaviors to algorithmically analyzing how to prevent the spread of misinformation. My PhD project aims to enhance our understanding of social dynamics in online social media by considering group (or high-order) interactions and to provide tools for easily and efficiently studying human dynamics. Specifically, the current contributions of my PhD research can be summarized as follows:

- **ABMs.** We (i) conducted an in-depth review of existing ABM tools to identify current trends, challenges, and opportunities for innovation; (ii) developed a stable version of krABMaga, an ABM simulation engine, and various simulation models (contributing to H^1_{ABM}); (iii) extended the features and capabilities of krABMaga to include distributed computation and GIS data integration. Additionally, we focused on enhancing the system's usability by creating an easy-to-use GUI (addressing H^2_{ABM}).
- **Hypergraphs.** We (i) developed a community-driven open-source platform to store hypergraph datasets; (ii) examined interactions in online social media groups through the lenses of hypergraphs (contributing to H^1_{HG}); (iii) investigated the application of modern techniques, such as graph neural networks, to analyze high-order interactions (contributing to H^2_{HG}).
- **LLMs.** We (i) analyzed the semantic information within the text linked to each node in a hypergraph using LLMs (contributing to H^1_{LLM}); (ii) developed a model to embed LLM-generated information into a stance detection task (addressing H^2_{LLM}).

In future development, we aim to: (i) continue developing new simulation models, conducting performance benchmarks, and adding features and plugins (e.g., other distributed fields, visualization functionalities) to krABMaga; (ii) thoroughly explore the influence of patterns on opinion change within conversational data; (ii) explore how LLMs can be utilized to enhance our understanding of text in online social contexts.

Acknowledgment. My Ph.D. project is under the supervision of Vittorio Scarano, Full Professor at the University of Salerno.

References

1. Antelmi, A., et al.: Reliable and efficient agent-based modeling and simulation. J. Artif. Soc. Soc. Simul. **27**(2) (2024). https://doi.org/10.18564/jasss.5300
2. Antelmi, A., Cordasco, G., De Vinco, D., Spagnuolo, C.: The age of snippet programming: toward understanding developer communities in stack overflow and reddit. In: Companion Proceedings of the ACM Web Conference 2023. WWW 2023 Companion, New York, NY, USA. Association for Computing Machinery (2023). https://doi.org/10.1145/3543873.3587673
3. Antelmi, A., Cordasco, G., DAmbrosio, G., De Vinco, D., Spagnuolo, C.: Experimenting with agent-based model simulation tools. Appl. Sci. **13**(1) (2023). https://doi.org/10.3390/app13010013
4. Antelmi, A., Cordasco, G., Polato, M., Scarano, V., Spagnuolo, C., Yang, D.: A survey on hypergraph representation learning. ACM Comput. Surv. **56**(1), 1–38 (2023). https://doi.org/10.1145/3605776
5. Antelmi, A., De Vinco, D., Spagnuolo, C.: HypergraphRepository: a community-driven and interactive hypernetwork data collection. In: Dewar, M., et al. (eds.) WAW 2024. LNCS, vol. 14671, pp. 159–173. Springer, Cham (2024). https://doi.org/10.1007/978-3-031-59205-8_11
6. Bankes, S.C.: Agent-based modeling: a revolution? Proc. Natl. Acad. Sci. **99**(suppl_3), 7199–7200 (2002). https://doi.org/10.1073/pnas.072081299
7. Battiston, F., et al.: Networks beyond pairwise interactions: structure and dynamics. Phys. Rep. **874**, 1–92 (2020). https://doi.org/10.1016/j.physrep.2020.05.004
8. Bossomaier, T.R., Green, D.G.: Complex Systems. Cambridge University Press, Cambridge (2000)
9. Bretto, A.: Hypergraph Theory. An Introduction. Mathematical Engineering, vol. 1. Springer, Cham (2013). https://doi.org/10.1007/978-3-319-00080-0
10. Can, U., Alatas, B.: A new direction in social network analysis: online social network analysis problems and applications. Phys. A **535**, 122372 (2019). https://doi.org/10.1016/j.physa.2019.122372
11. Chen, S.H., Chang, C.L., Du, Y.R.: Agent-based economic models and econometrics. Knowl. Eng. Rev. **27**(2), 187–219 (2012). https://doi.org/10.1017/S0269888912000136
12. Chen, Z., et al.: Exploring the potential of large language models (LLMs) in learning on graphs. ACM SIGKDD Explor. Newsl. **25**(2), 42–61 (2024). https://doi.org/10.1145/3655103.3655110
13. Cordasco, G., Scarano, V., Spagnuolo, C.: Distributed mason: a scalable distributed multi-agent simulation environment. Simul. Model. Pract. Theory **89**, 15–34 (2018). https://doi.org/10.1016/j.simpat.2018.09.002

14. Crooks, A.T., Heppenstall, A.J.: Introduction to agent-based modelling. In: Heppenstall, A.J., Crooks, A.T., See, L.M., Batty, M. (eds.) Agent-Based Models of Geographical Systems, pp. 85–105. Springer, Dordrecht (2011). https://doi.org/10.1007/978-90-481-8927-4_5
15. Cruickshank, I.J., Ng, L.H.X.: Use of large language models for stance classification. arXiv preprint arXiv:2309.13734 (2023). https://doi.org/10.48550/arXiv.2309.13734
16. Cuevas, E.: An agent-based model to evaluate the COVID-19 transmission risks in facilities. Comput. Biol. Med. **121**, 103827 (2020). https://doi.org/10.1016/j.compbiomed.2020.103827
17. De Vinco, D., Antelmi, A., Spagnuolo, C., Aiello, L.M.: Deciphering conversational networks: stance detection via hypergraphs and LLMs. In: Companion Publication of the 16th ACM Web Science Conference. Websci Companion 2024. Association for Computing Machinery, New York (2024). https://doi.org/10.1145/3630744.3658418
18. Fischer, M.T., Frings, A., Keim, D.A., Seebacher, D.: Towards a survey on static and dynamic hypergraph visualizations. In: 2021 IEEE Visualization Conference (VIS), pp. 81–85. IEEE (2021). https://doi.org/10.1109/VIS49827.2021.9623305
19. Floridi, L., Chiriatti, M.: GPT-3: its nature, scope, limits, and consequences. Mind. Mach. **30**(4), 681–694 (2020). https://doi.org/10.1007/s11023-020-09548-1
20. Hadi, M.U., et al.: A survey on large language models: applications, challenges, limitations, and practical usage. Authorea Preprints (2023). https://doi.org/10.36227/techrxiv.23589741.v1
21. Hsieh, C.C., Hsieh, Y.C.: Reliability and cost optimization in distributed computing systems. Comput. Oper. Res. **30**(8), 1103–1119 (2003)
22. Jung, R., Jourdan, J.H., Krebbers, R., Dreyer, D.: RustBelt: securing the foundations of the rust programming language. Proc. ACM Program. Lang. **2**(POPL), 1–34 (2017). https://doi.org/10.1145/3158154
23. Junprung, E.: Exploring the intersection of large language models and agent-based modeling via prompt engineering. arXiv preprint arXiv:2308.07411 (2023). https://doi.org/10.48550/arXiv.2308.07411
24. Kallaugher, J., Kapralov, M., Price, E.: The sketching complexity of graph and hypergraph counting. In: 2018 IEEE 59th Annual Symposium on Foundations of Computer Science (FOCS), pp. 556–567. IEEE (2018). https://doi.org/10.1109/FOCS.2018.00059
25. Ladyman, J., Lambert, J., Wiesner, K.: What is a complex system? Eur. J. Philos. Sci. **3**, 33–67 (2013). https://doi.org/10.1007/s13194-012-0056-8
26. Lee, G., Ko, J., Shin, K.: Hypergraph motifs: concepts, algorithms, and discoveries. Proc. VLDB Endow. **13**(11). https://doi.org/10.14778/3407790.3407823
27. Lee, G., Shin, K.: Temporal hypergraph motifs. Knowl. Inf. Syst. **65**(4), 1549–1586 (2023). https://doi.org/10.1007/s10115-023-01837-2
28. Lee, S., Ho, H.J.: Algorithms and complexity for weighted hypergraph embedding in a cycle. In: First International Symposium on Cyber Worlds, 2002. Proceedings, pp. 70–75. IEEE (2002). https://doi.org/10.1109/CW.2002.1180862
29. Luke, S., Cioffi-Revilla, C., Panait, L., Sullivan, K., Balan, G.: Mason: a multiagent simulation environment. SIMULATION **81**(7), 517–527 (2005). https://doi.org/10.1177/0037549705058073
30. Neuhäuser, L., Mellor, A., Lambiotte, R.: Multibody interactions and nonlinear consensus dynamics on networked systems. Phys. Rev. E **101**(3), 032310 (2020). https://doi.org/10.1103/PhysRevE.101.032310

31. Parry, H.R., Bithell, M.: Large scale agent-based modelling: a review and guidelines for model scaling. In: Agent-Based Models of Geographical Systems, pp. 271–308 (2011). https://doi.org/10.1007/978-90-481-8927-4_14
32. Platt, D.: A comparison of economic agent-based model calibration methods. J. Econ. Dyn. Control **113**, 103859 (2020). https://doi.org/10.1016/j.jedc.2020.103859
33. Reynolds, C.W.: Flocks, herds and schools: a distributed behavioral model. In: Proceedings of the 14th Annual Conference on Computer Graphics and Interactive Techniques, pp. 25–34 (1987). https://doi.org/10.1145/37401.37406
34. Sakketou, F., Lahnala, A., Vogel, L., Flek, L.: Investigating user radicalization: a novel dataset for identifying fine-grained temporal shifts in opinion. In: Proceedings of the Thirteenth Language Resources and Evaluation Conference, pp. 3798–3808 (2022). https://doi.org/10.48550/arXiv.2204.10190
35. Team, G., et al.: Gemma: open models based on Gemini research and technology. arXiv preprint arXiv:2403.08295 (2024). https://doi.org/10.48550/arXiv.2403.08295
36. Tisue, S., Wilensky, U.: NetLogo: a simple environment for modeling complexity. In: International Conference on Complex Systems, vol. 21, pp. 16–21. Citeseer (2004)
37. Touvron, H., et al.: Llama: open and efficient foundation language models. arXiv preprint arXiv:2302.13971 (2023). https://doi.org/10.48550/arXiv.2302.13971
38. Watts, D.J., Strogatz, S.H.: Collective dynamics of small-worldnetworks. Nature **393**(6684), 440–442 (1998)
39. Wu, J., Yang, S., Zhan, R., Yuan, Y., Wong, D.F., Chao, L.S.: A survey on LLM-gernerated text detection: Necessity, methods, and future directions. arXiv preprint arXiv:2310.14724 (2023). https://doi.org/10.48550/arXiv.2310.14724
40. Zhao, W.X., et al.: A survey of large language models. arXiv preprint arXiv:2303.18223 (2023). https://doi.org/10.48550/arXiv.2303.18223
41. Zhu, J., Zhu, J., Ghosh, S., Wu, W., Yuan, J.: Social influence maximization in hypergraph in social networks. IEEE Trans. Netw. Sci. Eng. **6**(4), 801–811 (2018). https://doi.org/10.1109/TNSE.2018.2873759

Which Reveals Ideology Better? Comparing Self-Presentation and Public Rhetoric in the Facebook Climate Debate via Embeddings Analysis

Luigi Arminio(✉) and Luca Rossi

IT University of Copenhagen, Copenhagen 2300, Denmark
{luia,lucr}@itu.dk

Abstract. Information about the ideological orientation of social media users is crucial to analyse a large number of social issues. However, the lack of structured data about users' beliefs is a common issue in social science. To address this gap, after a review of the state-of-the-art approaches for automated ideology detection, this research focuses on the use of a text-based methodology for this goal, using word embeddings. Specifically, the study contributes to the existing literature by focusing on ideology detection in the specific case of the polarized climate debate, and by testing the less-utilized OpenAI "ada" model for this task. Moreover, this research compares the accuracy of posts, representing users' public rhetoric, and page descriptions, thought to reveal their self-presentation, to predict users' ideological orientation. Our findings suggest that post-based methods hold the highest accuracy, but clustering-based approaches using page descriptions also yield respectable results while allowing a reduced use of computational resources. Overall, embedding-based methodology is shown to be a valuable tool for analyzing the ideological leanings of users in polarized debates.

Keywords: ideology · detection · embedding · polarization · climate

1 Introduction

When conducting social research, data on individuals' ideological leanings are invaluable for studying a number of social issues, including how political orientation drives macro-level social processes such as collective health behaviour (Geana, Rabb, and Sloman 2021; Hornsey et al. 2021), attitudes towards climate action (Arbuckle 2017), consumption practices (Buechner et al. 2022), political polarization processes (Rogowski and Sutherland 2016) or homophilic behaviour based on political inclinations among individuals in social networks (Del Valle 2022). However, social media studies often lack structured individual data (Chen et al. 2015): this also applies to the political orientation of the users

L. M. Aiello et al. (Eds.): ASONAM 2024, LNCS 15213, pp. 164–178, 2025.
https://doi.org/10.1007/978-3-031-78548-1_14

(Conover et al. 2011). This data scarcity issue has led researchers to employ automated methods for estimating users' political orientation (Barberá 2015; Bond and Messing 2015). Some of these techniques rely on the network patterns that characterize social media interaction and start from homophily assumptions, but their accuracy may vary due to differing levels of homophily across platforms and interactions (Hanteer et al. 2018). Furthermore, the availability of data for network construction is often limited. Alternative ideology detection approaches rely on users' social media posts' textual content, but the most traditional text-based approaches exhibited other significant limitations, such as the weakness in pinpointing words' contextual meanings and the necessity for prelabeled data to train the algorithms (Chinn, Hart, and Soroka 2020). In this sense, in recent years, advanced deep-learning models like word embeddings seemed to offer a promising alternative, although their employment remains relatively limited, although their employment remains relatively limited, primarily confined to a restricted number of political domains and types of textual content, relying on a narrow range of embedding models, neglecting potential advancements. In this context, our research seeks to extend the current understanding of ideology identification from textual data in the social media context through the pursuit of three delineated research goals: firstly, focusing on ideological polarity detection in the context of a specific social cleavage (i.e., the climate discourse) that has been underexplored in the context of embedding-based ideology identification. Second, to test the use of a newer language model that is more recent than the traditionally employed ones in embedding-based research on ideology. Lastly, to compare the efficiency of predicting users' ideology based on their self-presentation (how they describe themselves on Facebook) versus an approach that employs their public rhetoric (their published posts) as textual material for the embedding algorithms.

2 State of the Art

Existing computational social science literature documents various approaches for automated detection of political orientation in social media actors. These methods can be categorized through various lenses. We can classify methods based on their focus on analysis, distinguishing between behaviour-based (focused on how the users act) and identity-based (focused on how the users identify themselves) approaches, or based on the nature of the analysed data, distinguishing between network-based and text-based methods.

Network-based methods include Bayesian ideal point estimation, initially employed with roll call data (Bafumi et al. 2005). In its application in social media studies, it uses MCMC methods to map users on a continuous political spectrum based on a user-to-user adjacency matrix built on follower-followee connections (Barberá 2015). An analogue network-based approach applies singular value decomposition (SVD) on Facebook data, extracting a continuous and unidimensional political spectrum from a user-to-user matrix and labeling users based on their connections to Facebook pages (Bond and Messing 2015).

Within text-based methods, Wordfish stands out as an unsupervised text scaling model that treats ideology as a latent variable (Chinn, Hart, and Soroka 2020; Hjorth et al. 2015) inferred from the semantic choices of the different political actors (Proksch and Slapin 2009). While this method has been repeatedly employed in social media studies (Ceron 2017), it resulted to struggle with shorter texts and stylistic nuances (Ruedin and Morales 2019), which may pose a significant limitation when working with social media data. An alternative approach is Wordscores, a supervised scaling model that identifies political ideology based on a predefined number of dimensions and labelled textual political data (Laver, Benoit, and Garry 2003). However, Wordscores requires extensive labeled datasets (Ruedin and Morales 2019), which limits its flexibility, and it exhibited poorer performance when working with extreme ideologies (Ruedin and Morales 2019).

The constraints of these methods substantiate the importance of exploring alternative advanced approaches, such as the embedding-based methodology. Embedding-based methods, that employ deep learning to convert text (such as the textual content of politically relevant Facebook posts, or the Facebook description of the official page of a political party) into vector representations to effectively capture semantic meaning and contextual relationships within textual documents (Jurafsky and Martin 2008). These methods have been tested both with parliamentary corpora (Rheault and Cochrane 2020) and social media text data (Rao and Spasojevic 2016), but their utilization remains somewhat limited: the study of their performance in the context of automated ideology detection is relegated to a reduced number of studies referring to specific dimensions of the political spectrum to measure political ideology, such as the US *Democratic-Republic* dichotomy (Rao and Spasojevic 2016).

This limited range of application highlights concerns tied to theoretical sociological considerations. By referring to the theory of cleavages, that identifies a range of societal divisions shaping political discourse (Lipset and Rokkan 1967), as the reference theoretical model for political ideology analysis, social polarization can be elucidated as an expression of existing social cleavages (Casañ et al. 2022). In this context, it is noteworthy to acknowledge the presence of several different cleavages shaping the public discourse, which also increased in number in recent years (Alter and Madsen 2021; Ramaciotti Morales 2022; Jahn 1993): the continuous emergence of new cleavages, each exhibiting diverse socio-demographic profiles and behavioral patterns (Lonsdale 2023), which may even be associated with variations in language production (Pelinka 2007; Dassonneville, Fréchet, and Liang 2022), suggests that methodologies that work for text-based ideology detection within one social divide, such as the Republican-Democratic divide, may not necessarily valid when applied to other divides, such as the cleavage on environmental values. More specifically, success in text-based ideological detection achieved within one ideological divide, like the Republican-Democrat dichotomy, cannot be assumed to translate equally across other polarized debates (such as the climate debate) due to the unique lexical items, semantic nuances, and potentially varying communicational tones that characterize each specific debate.

To address the challenge of applying text-based ideological detection within specific social cleavages, this research will focus on a specific politically polarized area of the public debate and conduct an in-depth exploration of how word embedding performs in detecting ideological positions of the actors within this divide. In this study, we focused on the polarized climate change debate, that emerged as a contentious social cleavage in the latest decades (Alter and Madsen 2021). The choice to focus on the social divide around climate change stems from the significant impact of this cleavage in reshaping political landscapes, with shifts in voting behaviour and increased engagement in environmental activism by the younger generations (Alter and Madsen 2021; Ramaciotti Morales 2022; Jahn 1993; Wallis and Loy 2021).

Furthermore, the current literature on embedding-based ideology detection mainly relies on traditional embedding models such as BERT (Baly et al. 2020), Word2Vec (Gurciullo and Mikhaylov 2017), and to some extent, ELMo (Jiang et al. 2019), and GloVe (Kamboj et al. 2020). In this context, we suggest that the inclusion of novel embedding models, such as those recently developed by OpenAI, like Ada (Neelakantan et al. 2022), may provide a relevant contribution to this field.

Concerning other limitations in the current literature on ideological detection, it is worth mentioning the lack of a comprehensive analysis focusing on the choice of the type of textual content to predict users' political orientation. In effect, in social media contexts, users release multiple types of data, that are structurally different. While previous studies employed embedding-based ideological detection in different ways, there remains a gap in understanding which specific kinds of social media text may be more predictive for this task. We believe that in contexts like Facebook, where we can access both textual data concerning the self-presentation of the users, given by their account descriptions, and their public rhetoric, given by the textual content that they produce through the posts, it is useful to determine which performs better in predicting ideology through text embeddings. For these reasons, we decided to compare the embedding-based ideological detection on these two kinds of available textual data.

3 Data and Methods

3.1 Data Collection

To pursue our research goal, we manually compiled a list of politically relevant Facebook pages within the climate debate from multiple countries, distinguishing between those advocating for climate action (pro-actors) and those promoting climate denial or skepticism (counter-actors). Data was collected through the Crowdtangle API from 113 actors and 58 counter-actors, encompassing both their Facebook page descriptions (*i*) and the textual content of their posts (*ii*) over a four-month period (from April 29 to August 29 2023).

3.2 Ideological Scores Generation and Validation

Both textual datasets went through translation into English using the *deep_translator* Python library (Kareem 2023) for linguistic uniformity. Then, the two corpora went through word embedding utilising the *ada* language model (Neelakantan et al. 2022), that allowed us to represent each textual document (post or page description) in a space of 1531 continuous dimensions. After that, we employed t-SNE (Van der Maaten and Hinton 2008) to project the embedded documents into one single dimension in order to transform the textual content of each post in a single scalar value, aimed at politically positioning the users along a continuous scale based on their textual content. In the case of the corpus of posts, it has been necessary to aggregate all the scores from posts of each same page by average, in order to have a single score for each stakeholder. At this point, each stakeholder was labeled through two continuous values: the Content score, representing the average projected embeddings of the published posts by a given stakeholder, and the Self-description score, given by the projected embedding value of the page description of each stakeholder.

After labeling each stakeholder with the two embedding-generated scores, we assessed their accuracy in reflecting their latent ideological position by analyzing the association between each of the two continuous scores and the manually annotated actor position (pro-actor or counter-actor). In light of the continuous nature of the embedding-based scores and the dummy nature of the actor status, we resorted to the Point-Biserial correlation (PB) (Kornbrot 2014). We specifically focused on the absolute value of the PB coefficient, in light of the absence of directional assumptions, to solely assess how effectively the embedding-based scores differentiated between users from distinct classes (pro and counter actors) in a way that mirrors the ground truth's classification, without assuming any predefined relationship between high or low embedding values and a specific class (pro or counter actor). After that, we assessed our data-driven scores' performance in determining actors' ideological polarity within the "pro-actors vs. counter actors" framework by categorizing our word embedding-generated scores.

3.3 Categorization Methods and Performance Evaluation

Initially, we employed and compared two methods to categorize the scores. First, we resorted to mean-based dichotomization, assigning 0 to scores below the mean and 1 to scores above it. Second, we utilized threshold optimization, dividing the continuous score range into 100 equidistant points and then identifying a threshold holding the highest Normalized Mutual Information (NMI) (Forbes 1995) with the dichotomous ground truth (the pro-counter actor status).

The mean-based approach was used as a simple heuristic method for actor classification based on observed score distributions, which tend to be unimodal. The second one, named threshold optimization, maximizes the predictive power of our scores by dichotomizing data based on the ground truth.

After that, a third approach was involved, bypassing dimensionality reduction and instead employing. k-means clustering across all 1531 dimensions of word embeddings. The value of k was predetermined as 2 based on the premise of a polarized ideological environment (pro vs counter actors of climate action) rather than relying on data-driven optimization techniques, aiming to assess the embeddings' performance in accurately capturing the inherently dichotomous and polarized ground truth (Table 1).

Table 1. Summary of Tested Methods for Ideology Detection

Data	Analysis Focus	Categorization Method
Posts	Public Rhetoric	t-SNE & Threshold Optimization
		t-SNE & Mean-Based Dichotomization
Descriptions	Self-Presentation	t-SNE & Threshold Optimization
		t-SNE & Mean-Based Dichotomization
Posts	Public Rhetoric	k-means Clustering (k=2)
Descriptions	Self-Presentation	k-means Clustering (k=2)

Subsequently, we assessed the alignment between the categorizations generated from the three methods and verified actor status using NMI and Matthews correlation coefficient (MCC) (Matthews 1975). We chose to employ both MCC and NMI to provide comprehensive evaluation of our text-driven categorizations: while NMI measures overall agreement between embedding-based categorizations and ground-truth, MCC ensures balanced performance assessment, accommodating class imbalances (Boughorbel, Jarray, and El-Anbari 2017). This was applied to both the data types (descriptions and posts) and all three categorization approaches (optimization-based, mean-based, clustering-based). We only utilized the absolute value of MCC to focus on association magnitude, mirroring our approach with the point-biserial correlation. After that, we selected the optimal model (i.e., the one that showed the highest performance in terms of the above-mentioned metrics across different data types and categorization approaches).

Furthermore, we determined that if any of the content-based classifications (i.e., relying on the released posts) showed meaningful performance, we would have selected the most performative one to measure its sensitivity to time patterns and stylistic variations in content production. The decision to examine the performance sensitivity of the Content-based approach with respect to temporal variations stemmed from the premise that different time scales could influence political entities, leading to language shifts or evolving lexicon due to social events and political demands. To analyze this, we split our dataset, spanning months, into four subsets representing each month. We then computed key performance parameters, like NMI and MCC, for each month separately and for cumulative periods (e.g., first month, first and second month, and so on) to assess the hypothesized temporal variations of the classification quality. Average post length was calculated as the mean word count per post.

As mentioned before, in addition to the temporal sensitivity analysis, we evaluated factors impacting the optimal content-based classification performance. This involved analysing three parameters: posting frequency, average post length, and content diversity in published posts. Posting frequency was measured as the total number of posts per page within the specified time window, capped at 100 posts per page due to the data collection constraints. Content diversity was computed by constructing a corpus for each page, encompassing their published posts within the above-mentioned time frame. We treated each user's set of post embedding scores as a cluster, computing an average embedding vector as the page's centroid. Then, we determined content diversity as the average cosine distance between all posts embeddings on a page and its centroid.

After defining these three parameters, we estimated the relationship between classification accuracy of the actors and each parameter in order to assess their potential influence on the classification performance of the optimal Content-based approach.

4 Results

4.1 Overview of Ideological Categorization Approaches

Both the textual content of the Facebook posts and page descriptions provided valuable insight into categorizing the stances of the main stakeholders involved in the polarized climate debate. In particular, the optimal approach emerged be the one that joins word embeddings and dimensionality reduction for continuous scoring of ideological polarity based on Facebook posts. However, the clustering-based approach's notable performance with page descriptions should be noted: in cases of limited computational resources: this method, by demanding fewer data points than post-based approaches, can facilitate efficient data processing while keeping respectable performance with respect to the optimal approach, while acknowledging a slight compromise in classification accuracy.

Concerning the third method, which employed k-means clustering, it showed more balanced performance with different data types (i.e., posts and descriptions), but none of the applications outperformed the optimal t-SNE-based approach on users' posts. As for the drawbacks of k-means approaches, he absence of a metric (like the Content score) to place actors on a continuous ideological spectrum, which may provide some research benefits. For example, in case of extension of the binary classification model to a tripartite model including a third category between the two extremes (i.e., referring to stances that are not explicitly identifiable as pro-climate or anti-climate action), continuous ideological scoring would facilitate the exploration and the detection of this middle ground. In effect, this approach would provide better control than the k-means-based one, where introducing a third cluster may yield uncertain outcomes in detecting an intermediate ideological class (Fig. 1).

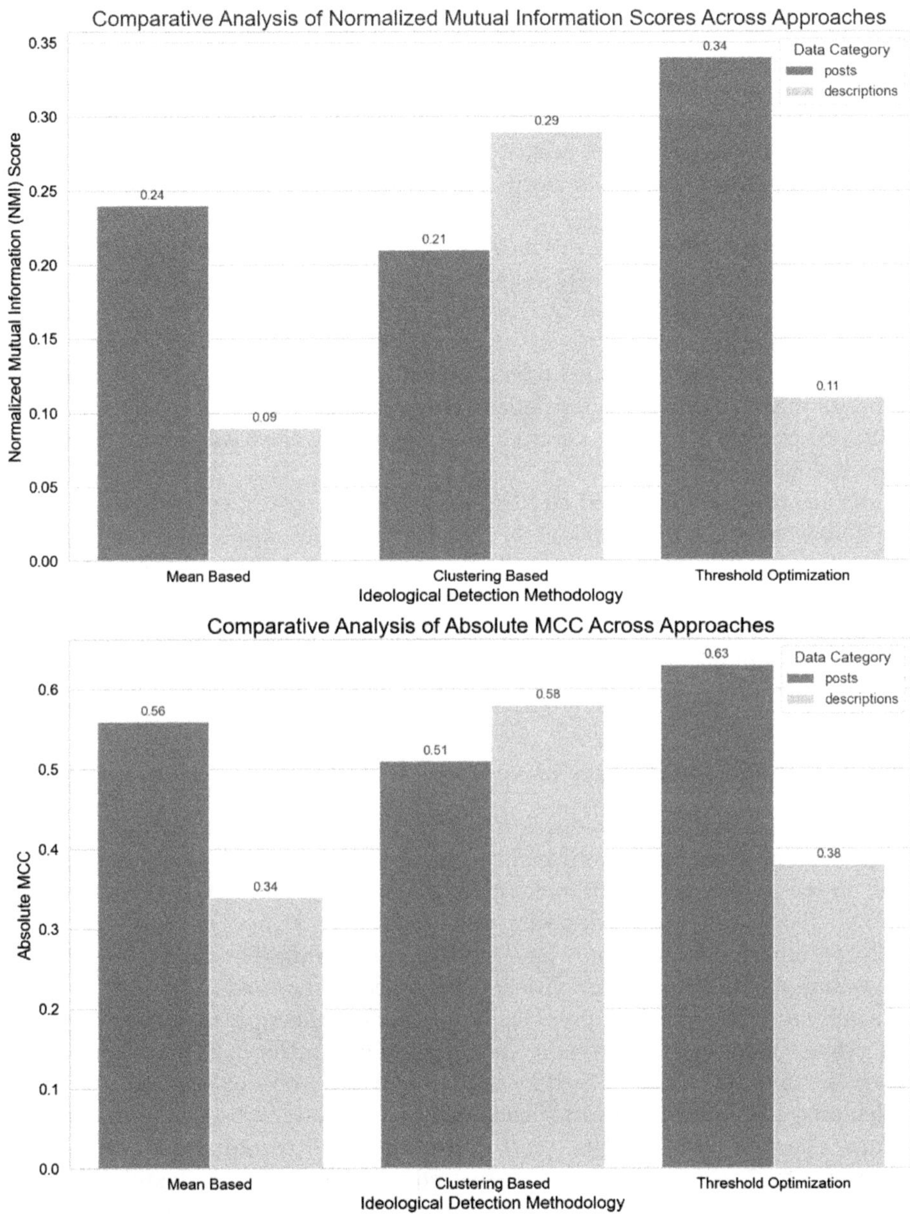

Fig. 1. Comparing approaches for categorizing ideological polarity via word embeddings. Clustering-based methodology provides the most balanced results across diverse data types, but threshold optimization of projected embedding scores on posts performs best.

4.2 Ideological Continuous Scoring Performance

As for the continuous scores generated through dimensionality reduction of the embeddings, Content score showed a significantly stronger PB correlation (0.66) with the true label than the Self-dscription score (0.22). Following score dichotomization, the performance gap between content produced and self-presentation further increases, with the NMI consistently higher for the Content score in both mean-based and optimization-based methods when compared to actor status.

However, when dividing users into two distinct ideological categories through k-means performed on the entire set of features (i.e., the 1531 embedding columns), the self-description of the users proved effective in estimating uers' ideological polarity, albeit with lower performance compared to the projection-based approach with threshold optimization on posts. In this sense, both the self-presentation of the climate debate stakeholders on social media and their public rethoric on Facebook showed some degree of effectiveness in predicting their ideological polarity.

Overall, the content-based embedding generated the best predictive performance and resilience to variations in the classification methodology, albeit with varying results. On the other hand, although clustering-based approach with Facebook pages descriptions yielded slightly lower classification performance compared to the optimal post-based classification, it still achieved satisfactory performance while effectively managing computational resources as a result of the reduction of the demanded data volume.

4.3 Influential Factors on Ideological Classification Performance

After comparing ideological detection with different data types and dichotomization processes, given the meaningful results obtained in ideological classification based on the public rhetoric of social media users, we examined how its predictive and classificational capabilities vary across different temporal scopes. This in-depth examination stems from the assumption that different time windows, due to their unique durations and political occurrences, may yield varying amounts of posts and linguistic nuances, potentially influencing the approach's performance.

First, we recomputed the score for individual months (e.g., the first month, the second month, etc.). Subsequently, we cumulatively aggregated posts and recalculated ideological scores for each aggregated data subset (the first month of data, the first two months, etc.). After that, we recomputed performance parameters for each time frame and assessed the differences in classification and predictive performance of the Content Score. We employed the PB Correlation between our score and the dummy ground truth as a metric for score performance (that we called *PB Performance*) and then, we calculated the Pearson Correlation between the number of posts within each time window and the PB Performance of each time-specific Content Score (Fig. 2).

Overall, different time periods have shown different levels of performance: the most recent months shown stronger association between the Content score and

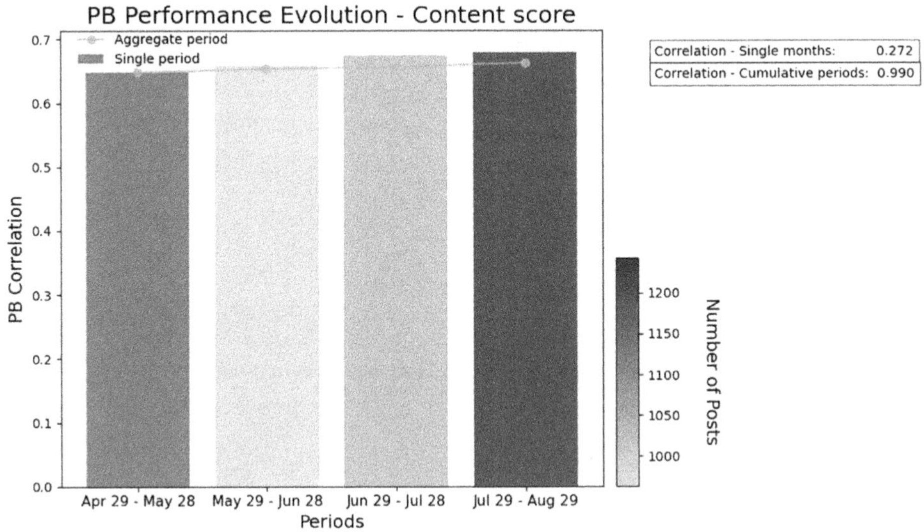

Fig. 2. PB Performance of Content Score over months. Bars represent PB Performance for each month, while dots indicate cumulative PB Performance (first month, first two months, etc.). Bar color tones reflect the number of posts per period.

stakeholders' true labels. Furthermore, the number of released posts in a given month partially affected the performance of the Content score: when the dataset is segmented into four individual monthly periods, the correlation between the number of posts in a month and score performance is 0.27; when the correlation between the number of posts in each cumulative aggregation of months and the model performance, the correlation coefficient reaches 0.99. These findings suggest that the model performance is affected by features of the examination period. However, further investigations over prolonged time intervals are required to validate this observation.

Subsequently, our analysis focused on our dichotomized Content score, to measure the elements affecting its efficacy in aligning with the ground truth in estimating the ideological polarity of each Facebook page. We considered three potential factors affecting the probability of correct classification: the posting frequency of each page, the average length of the posts, and the diversity of the published content (Figs. 3 and 4).

Our results indicate that, after Content Score dichotomization, rhetoric-based ideological polarity prediction among social media users was resilient to variations in terms of factors such as posting frequency of the Facebook pages: although misclassified pages exhibited a skewed distribution toward lower values, no statistically significant differences were observed in average tendencies. The approach maintained substantial resilience to variety of the produced content and average post length, though both higher posts homogeneity and average

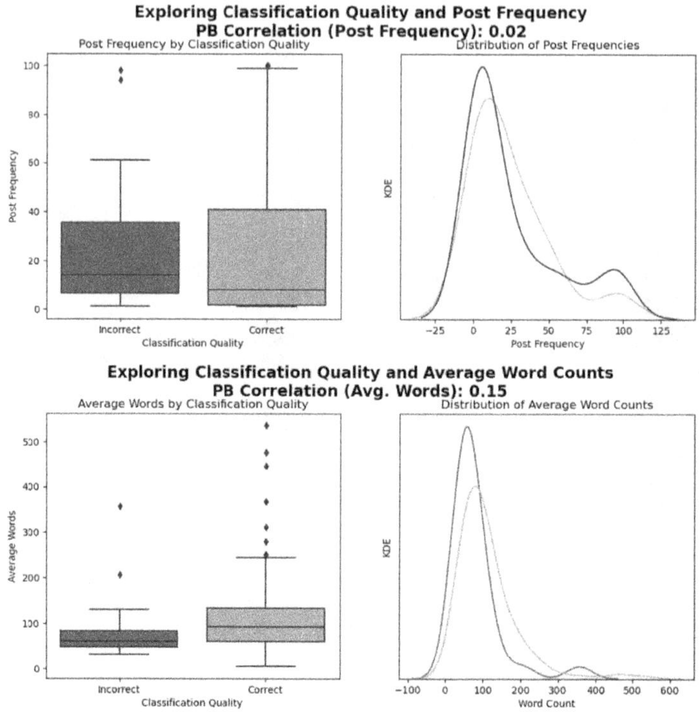

Fig. 3. Classification quality and user activity: The Content Score (post-based) shows that posting frequency does not significantly affect ideological polarity estimation on Facebook, while there is a weak positive correlation with average post length.

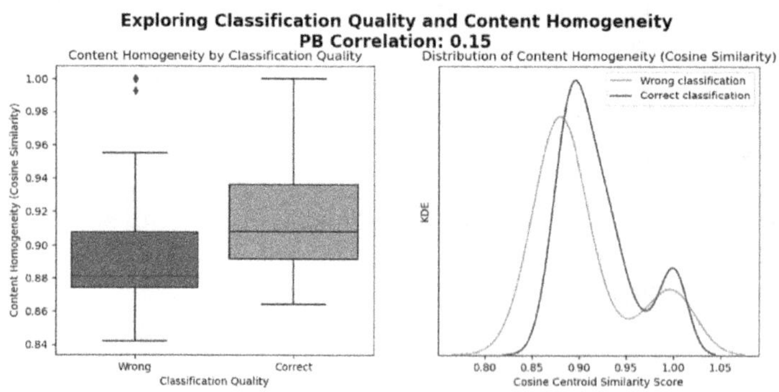

Fig. 4. Classification quality vs Content Homogeneity - The match between dichotomized Content Score and ideological ground truth is slightly affected by the diversity level of published content by each climate debate stakeholder

length exhibited a slight positive correlation with the effective categorization of the pages.

5 Discussion and Conclusion

In conclusion, our study provided compelling insights about the use of word embedding to determine the ideological polarity of social media actors in the context of the climate change debate. Our goal was to broaden the current knowledge of word embedding for ideological classification by examining a specific polarizing issue, the climate debate, which has been under-explored in the analysis of embedding-based methods for ideology detection. Our approach involved applying the OpenAI "ada" model, a promising yet underutilized method in this context, to assess the performance of embedding-based ideological classification with different textual data sources. In particular, we collected both users' page descriptions and their released posts over a four-month period. This allowed us to capture and compare two distinct dimensions of social media users' textual production: their public rethoric (given by their published posts), and their self-presenation (given by their page descriptions) to determine which better predicts users' ideological polarity. Then, starting from the word embedding outcomes, we implemented three distinct approaches to bifurcate users into two polarized categories: two employing dimensionality reduction for continuous ideological scoring and one integrating embedding with k-means clustering for a two-cluster model representing the ideological polarization among the users. Our analysis revealed notable findings for both self-presentation and public-rhetoric in the context of ideological classification. In particular, the top-performing approach was the integration of word embeddings with dimensionality reduction applied to users' posts. The clustering-based approach applied to user descriptions was the second best method in terms of performance.

Overall, approaches relying on users' public rethoric (posts) showed higher resilience to methodological variations but were more computationally demanding. This textual embedding-based approach gains significance when the lack of interactional data prevents researchers from mapping the network dynamics among political actors or social media users as well as when the lack of homophilic patterns in social media interactions might cause network-based ideological estimation to fail. In light of the significance of both the produced content and the self-description of the pages in estimating ideology, the selection of the best approach should be guided by the available computational resources and the interest in the use of a continuous score to place actors on an ideological scale.

Concerning the limitations, data constraints prevented including a neutral third group between pro-climate actors and counter-actors, which could have enriched our ideological classification. Moreover, the rhetoric-based ideological detection varied with the dataset's temporal span, as unique periods with specific discussion themes and posting frequencies impacted Content Score performance. Moreover, longer time spans boosted the quality of ideological predictions. At the same time, content-based ideology detection showed resilience to content variation.

In terms of future developments, this analysis can be replicated using other embedding models traditionally employed in ideology detection and social media studies, such as BERT and Word2Vec. This approach may allow us to compare the effectiveness of different embedding models in ideological classification of climate debate stakeholders based on either their self-presentation or their public rhetoric. Furthermore, given the variety of textual content that users produce on social media, including different tones, styles and register, it may be relevant to measure and explore how the subjectivity level of the produced content (Yaqub et al. 2018) can affect correct text-based ideological classification: understanding this relationship could increase the precision and reliability of ideology detection and may provide a better understanding of the interplay between discursive tone and the expression of specific ideological stances.

6 Supplementary Material

Due to privacy and data protection reasons, the collected data through Crowdtangle during the current study are not publicly accessible. However, the code employed in this study is available in our repository at *github.com/arminioluigi/ideology_ estimation*.

References

Alter, K.J., Madsen, M.R.: The International adjudication of mega-politics. Law Contemp. Probs. **84**, 1 (2021). https://doi.org/10.2139/ssrn.3961417

Arbuckle, M.B.: The interaction of religion, political ideology, and concern about climate change in the United States. Soc. Nat. Resourc. **30**(2), 177–194 (2017). https://doi.org/10.1007/s11109-015-9323-7

Bafumi, J., et al.: Practical issues in implementing and understanding Bayesian ideal point estimation. Politic. Anal. **13**(2), 171–187 (2005). https://doi.org/10.1093/pan/mpi010

Baly, R., et al.: We can detect your bias: predicting the political ideology of news articles. In: Proceedings of the 2020 Conference on Empirical Methods in Natural Language Processing (EMNLP), pp. 4982–4991 (2020). https://doi.org/10.18653/v1/2020.emnlp-main.404

Barberá, P.: Birds of the same feather tweet together: Bayesian ideal point estimation using Twitter data. Politic. Anal. **23**(1), 76–91 (2015). https://doi.org/10.1093/pan/mpu011

Bond, R., Messing, S.: Quantifying social media's political space: estimating ideology from publicly revealed preferences on Facebook. Am. Politic. Sci. Rev. **109**(1), 62–78 (2015). https://doi.org/10.1017/S0003055414000525

Boughorbel, S., Jarray, F., El-Anbari, M.: Optimal classifier for imbalanced data using Matthews Correlation Coefficient metric. PloS One **12**(6), e0177678 (2017). https://doi.org/10.1371/journal.pone.0177678

Buechner, B.M., et al.: Political ideology and cultural consumption: the role of flexibility in shaping liberal and conservative preferences for global-local experiences. J. Assoc. Consum. Res. **7**(3), 266–275 (2022). https://doi.org/10.1080/08941920.2016.1209267

Casañ, R.R., et al.: Online polarization and cross-fertilization in multi-cleavage societies: the case of Spain. Soc. Netw. Anal. Mining **12**(1), 79 (2022). https://doi.org/10.1007/s13278-022-00909-5

Ceron, A.: Intra-party politics in 140 characters. Party Politics **23**(1), 7–17 (2017). https://doi.org/10.1177/1354068816654325

Chen, X., et al.: A comparative study of demographic attribute inference in twitter. In: Proceedings of the International AAAI Conference on Web and Social Media, vol. 9(1), 590–593 (2015). https://doi.org/10.1609/icwsm.v9i1.14656

Chinn, S., Hart, P.S., Soroka, S.: Politicization and polarization in climate change news content, 1985–2017. Sci. Commun. **42**(1), 112–129 (2020). https://doi.org/10.1177/1075547019900290

Conover, M.D., et al.: Predicting the political alignment of twitter users. In: 2011 IEEE Third International Conference on Privacy, Security, Risk and Trust and 2011 IEEE Third International Conference on Social Computing, pp. 192–199. IEEE (2011). https://doi.org/10.1109/PASSAT/SocialCom.2011.34

Dassonneville, R., Fréchet, N., Liang, B.: Linguistic cleavages in public opinion. In: Comparative Public Opinion, pp. 284–309. Routledge (2022). https://doi.org/10.4324/9781003121992-18

Del Valle, M.E.: Homophily and polarization in twitter political networks: a cross-country analysis. Media Commun. **10**(2), 81–92 (2022). https://doi.org/10.17645/mac.v10i2.4948

Forbes, A.D.: Classification-algorithm evaluation: five performance measures based on confusion matrices. J. Clin. Monit. **11**, 189–206 (1995). https://doi.org/10.1007/BF01617722

Geana, M.V., Rabb, N., Sloman, S.: Walking the party line: the growing role of political ideology in shaping health behavior in the United States. SSM-Populat. Health **16**, 100950 (2021). https://doi.org/10.2139/ssrn.3961417

Gurciullo, S., Mikhaylov, S.J.: Detecting policy preferences and dynamics in the un general debate with neural word embeddings. In: 2017 International Conference on the Frontiers and Advances in Data Science (FADS), pp. 74–79 (2017). https://doi.org/10.18653/v1/S19-2146

Hanteer, O., et al.: From interaction to participation: the role of the imagined audience in social media community detection and an application to political communication on twitter. In: 2018 IEEE/ACM International Conference on Advances in Social Networks Analysis and Mining (ASONAM), pp. 531–534. IEEE (2018). https://doi.org/10.1109/ASONAM.2018.8508575

Hjorth, F., et al.: Computers, coders, and voters: comparing automated methods for estimating party positions. Res. Politics **2**(2), 2053168015580476 (2015). https://doi.org/10.1177/2053168015580476

Hornsey, M.J., et al.: Resolving the small-pockets problem helps clarify the role of education and political ideology in shaping vaccine scepticism. Br. J. Psychol. **112**(4), 992–1011 (2021). https://doi.org/10.1111/bjop.12500

Jahn, D.: The rise and decline of new politics and the Greens in Sweden and Germany: resource dependence and new social cleavages. Eur. J. Politic. Res. **24**(2), 177–194 (1993). https://doi.org/10.1111/j.1475-6765.1993.tb00375.x

Jiang, Y., et al.: Team bertha von suttner at semeval-2019 task 4: hyperpartisan news detection using elmo sentence representation convolutional network. In: Proceedings of the 13th International Workshop on Semantic Evaluation, pp. 840–844 (2019). https://doi.org/10.18653/v1/S19-2146

Jurafsky, D., Martin, J.: Speech and Language Processing: An Introduction to Natural Language Processing, Computational Linguistics, and Speech Recognition, vol. 2. Prentice Hall (2008)

Kamboj, M., et al.: Multimodal political deception detection. IEEE MultiMedia **28**(1), 94–102 (2020). https://doi.org/10.1109/MMUL.2020.3048044

Kareem, N.: Deep-translator (Version 1.11.4). In: PyPI (2023). pypi.org/project/deep-translator/. Accessed 10 Jan 2024

Kornbrot, D.: Point biserial correlation. In: Wiley StatsRef: Statistics Reference Online (2014). https://doi.org/10.1002/0470013192.bsa485

Laver, M., Benoit, K., Garry, J.: Extracting policy positions from political texts using words as data. Am. Politic. Sci. Rev. **97**(2), 311–331 (2003). https://doi.org/10.1017/S0003055403000698

Lipset, S.M., Rokkan, S.: Cleavage structures, party systems, and voter alignments: an introduction. In: Party Systems and Voter Alignments: Cross-national Perspectives, pp. 1–64. The Free Press (1967)

Lonsdale, A.: Sociodemographic Inequalities and political cleavages: an investigation of recent elections, nativist movements, and individual-level voting shifts. CID: 20.500.12592/p2ngm6b. PhD thesis. Ecole des Hautes Etudes en Sciences Sociales (2023)

Matthews, B.W.: Comparison of the predicted and observed secondary structure of T4 phage lysozyme. Biochim. Biophys. Acta (BBA)-Protein Struct. **405**(2), 442–451 (1975). https://doi.org/10.1016/0005-2795(75)90109-9

Neelakantan, A., et al.: Text and code embeddings by contrastive pre-training. arXiv preprint arXiv:2201.10005 (2022)

Pelinka, A.: Language as a political category: the viewpoint of political science. J. Lang. Politics **6**(1), 129–143 (2007). https://doi.org/10.1075/156921507781509626

Proksch, S.O., Slapin, J.B.: How to avoid pitfalls in statistical analysis of political texts: the case of Germany. German Politics **18**(3), 323–344 (2009). https://doi.org/10.1080/09644000903055799

Ramaciotti Morales, P.: Multidimensional online American politics: mining emergent social cleavages in social graphs. In: International Conference on Complex Networks and Their Applications, pp. 176–189. Springer, Cham (2022). https://doi.org/10.1007/978-3-031-21127-0_15

Rao, A., Spasojevic, N.: Actionable and political text classification using word embeddings and LSTM. arXiv preprint arXiv:1607.02501 (2016)

Rheault, L., Cochrane, C.: Word embeddings for the analysis of ideological placement in parliamentary corpora. Political Anal. **28**(1), 112–133 (2020). https://doi.org/10.1017/pan.2019.26

Rogowski, J.C., Sutherland, J.L.: How ideology fuels affective polarization. Polit. Behav. **38**, 485–508 (2016). https://doi.org/10.1007/s11109-015-9323-7

Ruedin, D., Morales, L.: Estimating party positions on immigration: assessing the reliability and validity of different methods. Party Politics **25**(3), 303–314 (2019). https://doi.org/10.1177/1354068817713122

Van der Maaten, L., Hinton, G.: Visualizing data using t-SNE. J. Mach. Learn. Res. **9**(11), 2579–2605 (2008)

Wallis, H., Loy, L.S.: What drives pro-environmental activism of young people? A survey study on the Fridays For Future movement. J. Environ. Psychol. **74**, 101581 (2021). https://doi.org/10.1016/j.jenvp.2021.101581

Yaqub, U., et al.: Analysis and visualization of subjectivity and polarity of Twitter location data. In: Proceedings of the 19th Annual International Conference on Digital Government Research: Governance in the Data Age, pp. 1–10 (2018). https://doi.org/10.1145/3209281.3209313

Multidisciplinary

Five Patterns of Vaccine Misinformation on Telegram

Fabio Malini[1], Francis Sodré[2], Athus Cavalini[3](✉), Gabriel Herkenhoff[2], and Fabio Goveia[1]

[1] Department of Communication, Federal University of Espírito Santo, Vitória, Brazil
fabiomalini@gmail.com
[2] Postgraduate Program in Public Health, Federal University of Espírito Santo, Vitória, Brazil
francisodre@uol.com.br
[3] Postgraduate Program in Computer Science, Federal University of Espírito Santo, Vitória, Brazil
athus.cavalini@ifes.edu.br

Abstract. The high production of information during the COVID-19 pandemic, characterized by the WHO as an "infodemic", has fueled uncertainties and anti-science discourse, particularly on social media. This phenomenon has significantly contributed to vaccine hesitancy, evident in Brazil's low childhood vaccination rates against COVID-19, where only 10% were fully vaccinated. To address this challenge, this study analyzed over 1.8 million messages shared in 248 Telegram anti-vaccine channels between July 1, 2022 and January 31, 2023. Using a methodological approach based on the use of data science and data analysis techniques, we identified five common types of vaccine misinformation: 1) pseudo-protective (propagating pseudo-protective measures against 'experimental vaccines'); 2) immunological (building clinical stories that would demonstrate the reduction of immunity due to immunization); 3) false utility (disseminating legal and argumentative instructions for parents not to vaccinate their children); 4) medicalizing (highlighting medical reports to legitimize false or inaccurate facts); and 5) conspiracist (spreading conspiracy theories based on the idea that global corporations want to subjugate people with eugenic policies). These findings offer valuable insights for health professionals and managers to strategically develop targeted communication campaigns to counter these narratives and address vaccine hesitancy.

Keywords: Misinformation · Telegram · Vaccine · COVID-19

1 Introduction

The high production of information in the context of the COVID-19 pandemic, with the accelerated dissemination of news about the virus, recommendations on how to avoid contagion, reports of contamination, and comments on the

L. M. Aiello et al. (Eds.): ASONAM 2024, LNCS 15213, pp. 181–196, 2025.
https://doi.org/10.1007/978-3-031-78548-1_15

disease, led the World Health Organization (WHO) to characterize the situation as an infodemic. Infodemic is a pandemic of information, including those which are false or misleading, in physical or digital environments, caused during an epidemiological outbreak [18]. An infodemic can intensify or extend pandemic events, as they generate uncertainties about forms of protection, conduct, and behavior. The use of digital media and social networks further the circulation of information that can be harmful for disease control, accentuating the problem of misinformation [17].

The objective of this work is to identify patterns of vaccine misinformation on Telegram mapping the denialist narratives present in 248 channels and groups, which mobilize a superbase made up of 3.903.049 subscribers. By misinformation we mean false content that is deliberately intended to harm or cause damage. It can be fabricated or decontextualized content - designed to deceive, spread fear or distrust among the population -, or even the use of expert speeches, which are taken out of context, to generate confusion and undermine the credibility of official authorities [7]. Misinformation can circulate according to a political agenda or it can simply be inaccurate information amplified with manipulative intent [17]. In the specific case of the COVID-19 pandemic, the amount of inaccurate or false information concerns the origin of the virus, the treatments, and the mechanisms of propagation [13].

The vaccine against COVID-19 was one of the main resources used by public health institutions to mitigate the effects of the pandemic. In this context, the emergence of groups resistant to vaccination is noteworthy. [11] demonstrated that, in Brazil, despite the social recognition of the historical importance of vaccines (which, between the years 2001 and 2015, achieved coverage of more than 70% of the population [1]), the rapid dissemination of false and inaccurate information indicated a favorable scenario for the growth of anti-science discourse, especially in social networks, provoking a historical increase of vaccine hesitancy, particularly against COVID-19. Data from [6] reveals that only 16% of children were fully vaccinated against coronavirus in Brazil.

It is possible to state that, if "in the case of the COVID-19 pandemic, infodemics and misinformation are closely related phenomena" [5], the same process dangerously spreads in the field of immunization, generating noise and changes in habits that lead to vaccine hesitancy in Brazil.

In this context, this study aims to map anti-vaccination narratives that reinforce a process of health misinformation. These narratives often present claims as established facts, despite lacking scientific evidence [4]. In our view, the narrative is conceived as "the power we have to imagine complex sets of ideas into being, and to then share these among the community, which creates both the societies we live in and the cultures that provide the meanings for our lives." [16].

2 Methods

The mapping and analysis of messages that produce misinformation in Telegram groups and channels is based on the combination of data science techniques with social network analysis. Therefore, our methodology is divided into four phases: (1) delimitation of Telegram channels and groups, (2) extraction of techno-linguistic data, (3) data mining, and (4) data visualization and interpretation.

2.1 Delimitation and Collection Data

The approach used to obtain the network of Telegram channels and groups involved in the vaccine misinformation process was based on the "snowball" method, often used in research on this kind of platform [15]. In this method, from the messages collected from a certain channel, those that were forwarded from other channels are identified in order to add such channels to the collection as they become relevant sources. This allows to generate a network based on "connections" between channels and groups and procedurally identify more "hidden" actors, for whom access is less intuitive and requires the researcher to be immersed in the network [8].

To apply this method, it is necessary to obtain an initial base of representative channels and groups to start composing the network that is to be obtained at the end of the process. In the case of this work, initially, it was selected channels and groups characterized as ultraconservative and listed in the Supreme Federal Court (the highest court of law in Brazil) survey for disseminating disinformation against the Brazilian democratic order, in addition to those found by the application's search interface when searching by the terms "vaccine" and "covid". Other channels and groups associated with the anti-vaccine movement were also searched on Twitter and Facebook pages.

The initial base was formed by 108 chats. After applying the snowball method, the base was expanded to 248 Telegram chats. Among these, 129 are channels and 119 are groups. Together they are composed of 3,903,049 users registered as followers.

In the channels, communication is massive and works on a top-down model, since only the administrator can send messages to all subscribers. The groups work in a different way, as an interpersonal communication, since the participants are at the same time message producers and commentators. Telegram treats both as chats because they can house textualities formed by posts, comments, forwarding, and emojis, serving as a digital public arena.

2.2 Extracting Techno-Linguistic Data

Once the sample of channels and groups was defined, the next methodological procedure was to export the messages collected from them to a text file, to form the initial dataset. In internet social media analysis, datasets have communicational and linguistic functions, as they allow the researcher to detect and calculate the degree of relationships between actors and between textual entities.

The extraction of posts in these chats was performed by the Telegram Observatory software, or Obtel [3], which communicates directly with the Application Programming Interface (API) of Telegram. Hence, it was possible to download massive data from hundreds of channels and groups. Obtel has captured 2,564,886 publications and detected 408,939 forwarded messages posted in the 248 Telegram chats between July 1, 2022, and January 31, 2023. This time frame was chosen due to the high number of political acts that took place in the period, such as general elections and their subsequent consequences (e.g. the invasion of the seats of the Legislative, Judiciary, and Executive branches by ultraconservative demonstrators). The correlation between these political events and the issue of vaccine misinformation is because, during the COVID-19 pandemic, the stance of the Brazilian far-right representatives (including former president Jair Bolsonaro) was to encourage disbelief in the safety and the efficiency of the vaccines against coronavirus. Therefore, it is fair to say that there was in Brazil a politicization of pandemics that led ultra-conservative groups to stand against the vaccination.

Posts consisting of only links or emojis were excluded from the database. After this refinement, the data was reduced to 1,842,211 textual messages, which then formed the macro dataset of our work. Following the macro dataset generation, it was necessary to filter the messages that contained words associated with the context of vaccination. For this purpose, we have adopted as inclusion criteria only messages with the term "vaccin" (in Portuguese, "vacin"). We used this radical to extract messages that contained derivations of the word, such as "vaccine", "vaccination", "vaccinated", "vaccining", etc. After this processing, 48,313 messages were selected, forming the micro dataset.

Macro and micro datasets form the techno-linguistic corpus of this article. The macro dataset reveals the ideological issues and interests that stand out in the exchanges of messages performed by the members of the 248 chats of ultra-conservative groups. The micro dataset, for its part, makes it possible to link the ideological dimensions of the groups to the arguments, facts, sources of information, and types of misinformation conveyed in the messages that constitute the anti-vaccine discourse.

2.3 Data Mining: Generating Graph Archives

In possession of the datasets, in text extension format, each of these datasets was processed in the softwares Ford [14] and Obtel to export files in .gdf format (supported by the complex network visualization software, Gephi). This phase is called data mining, which is used to answer questions and unravel complex network patterns.

Our first focus was to find a semantic network analysis process based on graphs [10], in which words are the nodes and the association between them are the edges. We initially started with the 150 most frequent words among the 48,313 messages. Then, using the same corpus, Ford calculated the 30 words most associated with each of those 150 in the Telegram messages set. Thus, this mining resulted in a matrix with 2,200 unique words, which formed 40,000 pairs.

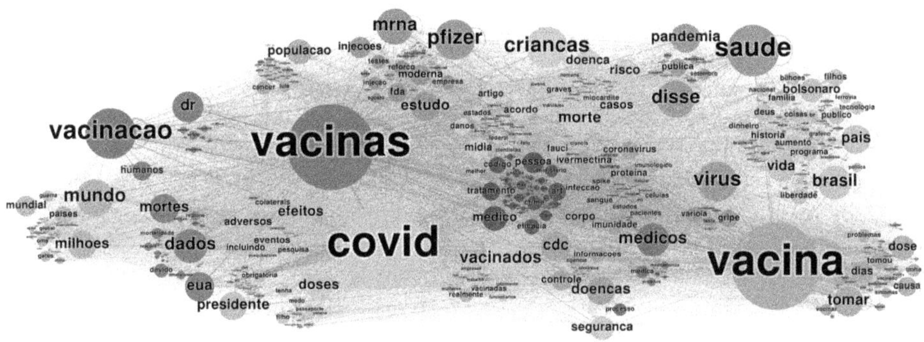

Fig. 1. General Narrative Network based on messages of the Brazilian anti-vaccine movement on Telegram. Each hue in the graph represents a cohesive group of words that represent a part of the overall narrative. The lines signify the ties between the words. The thicker the line, the more times the terms (nodes) appear together.

2.4 Visualizing and Interpreting Data

The next step was to make the relationships between the words visible. To achieve this goal, we created a network in Gephi software [2]. In it, we applied the "modularity" metric to detect the most cohesive groupings of words. In total, 23 groups, called "clusters", were identified and marked with colors (Fig. 1). We then applied the "degree" statistic to the nodes. This metric calculates the number of relationships maintained by a node (in this case, a word) in the network, or the number of times a word was paired with another present in the graph. The more pairs a word generates, the greater its importance in the semantic network. With this information, it was possible to assign the size of the point in the network according to the value of its degree.

The final phase in Gephi was to apply the "Circle Pack Layout" algorithm, randomly distributing the clusters of words, allowing a better perception of which words are grouped more intensely with which ones. We call this visualization the General Narrative Network, as it allows us to detect the topics that each group of words denotes. Therefore, we were able to map all the subjects covered by the 48,313 messages on Telegram, as we can see in Fig. 1.

Each word group in this graph is conceptualized as a perspective [10], because it contains a specific topic on the object of study (vaccine). The idea of a lexical cluster as a perspective has allowed us to visualize a diversity of meanings that disinformation is shaped. In this method, digital tools taken from data science are activated to detect groups of perspectives (clusters) that derive from the association of words that are linked to denote different points of view on a narrated or argued subject [9]. Thus, "every network is understood as an overlapping of layers-worlds-network that deepen relationships with each other" [12].

3 Results and Discussion

When viewing Fig. 1, it is clear that the movement of vaccine misinformation on Telegram has given greater weight to the word "covid", mentioned in 10,219 of the 48,313 filtered messages. This is a much higher number than the citation of other diseases for which there are immunizations - e.g. there are only 605 mentions of "flu" ("gripe") and 546 of "Monkeypox" among the messages.

It is also worth highlighting the other 9 terms most cited in messages about vaccination: "children" ("crianças"), present in 1,983 messages; "take" ("tomar"), in 1,950; "Pfizer", in 1,711; "virus", in 1,667; "world" ("mundo"), in 1,559; "doctors" ("doutores"), in 1,353; "mRNA", in 1,256; "Dr", in 1,253; and "death" ("morte"), in 1,157. Each of these words is located in a different perspective in Fig. 1, and they are associated with particular discursivities in the web of posts and comments that mark vaccine misinformation on Telegram.

3.1 Pseudo-protective

By analyzing the universe of each of the 23 perspectives present in the General Narrative Network (Fig. 1), we detected five predominant patterns in the messages with vaccine misinformation on Telegram. The first narrative pattern has a protective bias, inducing people to protect themselves against "adverse reactions", "side effects", and "adverse events" caused by the act of "taking" an "experimental vaccine", an expression used by the Telegram groups studied to pejoratively qualify the vaccine.

Figure 2 shows the groupings with the rhetoric that offers a kind of guiding line for the anti-vaccine movement on Telegram (AMT). It is the main driving force of the movement. The terms "vaccine" ("vacina"), "take" ("tomar"), "they took" ("eles tomaram"), "experimental", "experiment" ("experimento"), "problem(s)" ("problema"), "cause" ("causa"), "symptoms" ("sintomas"), "events" ("eventos"), "adverse" ("adverso"), "effects" ("efeitos"), "side effects" ("efeitos colaterais") and "reaction" ("reação") form two cohesive and associated groups of topics. This linguistic community is derived from messages that correlate severe symptoms of illness with immunization. In this sense, the main point of view of the anti-vaccination movement is to associate the vaccine with a permanent state of illness.

One of the most viewed messages in the ultraconservative chats, with 13,894 views and 700 forwards, explores the impossibility of reversing any symptom manifested by a patient with COVID-19, implying that they will persist and be constant. In the anti-vaccination ruse, every immunization causes irreversible reactions ("[...] Vaccinated people will no longer be able to continue treating the symptoms of the vaccine. Vaccinated people will have to bear the consequences because they will no longer be able to be cured [...] because the genetic defect is eternal").

The basis of AMT manipulation is therefore to plant seeds of doubts grounded on pseudo-scientific certainties that generate panic, correlating the immunization with severe side effects. These side effects are reported in the

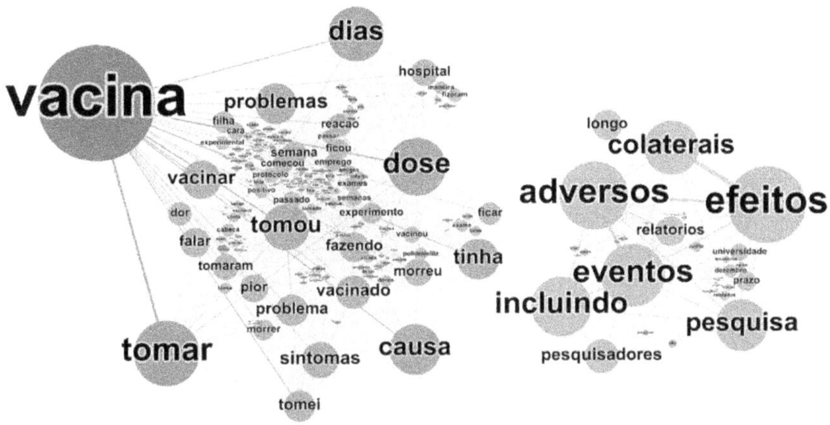

Fig. 2. Two clusters originated from the General Network Narrative. The cohesion of the terms means refusal of vaccination, preaching that the vaccine is experimental and that it generates side effects and adverse events

first person, however, the patient is often a third person ("my daughter", "my mother", "my father") who is experiencing serious reactions caused by the vaccine, such as: "acute allergic reaction", "acute myocardial infarction", "cerebral ischemic attacks", "hypertension", "cardiac hypertrophy", "genetic defect", "acute headache".

Testimony is taken, in this context, as clinical evidence. It is the particular that intends to generalize, like a virus, to spread fear and mistrust about the vaccine, thus operating as the main veridiction of misinformation language. In conspiracy logic, the qualitative data would be the generating principle of truth, while the statistician would be what manipulates society. Thus, in the anti-vaccine rhetoric, the humanized narrative of the clinical case overrules the cold and impersonal discourse of statistics. Between the particular and the general, the misinformation language adopts the first one as a convincing strategy ("I will never stop talking. Think of a healthy man, who at the age of 51, an athlete, had three cardiac arrests after 20 days of the third dose. A tragedy (sic) in my family [...] On the day he told me he had taken the third dose of the experimental vaccine, my body froze because I knew something was going to happen [...]").

The anti-vaccine rhetoric captures the virality of the digital world to serve as symptomatological evidence, turning an episodic reaction (mild vaccine effects) into unsuspected proof that they will be adverse and permanent, generating uncertainty about the safety and effectiveness of vaccines. The irreversibility of pain, the irreversibility of a health problem, and the irreversibility of death are commonly celebrated by "anti-vax" as a strategy to simultaneously generate doubt (which brings out its experts) and monetization on social networks (which will guarantee their financing).

The cause of health problems and deaths, reported in the anti-vaccine Telegram groups, is associated with the "experimental vaccine" ("vacina experimental", in Portuguese), as said, an expression used as a metaphor to refer pejoratively to any vaccine against covid, as well as "experimental shot" or "experimental drug". The denialist narrative propagates that the vaccine, instead of immunizing against diseases, makes people "sick again", with "serious sequelae". "Or worse, they kill!" The vaccine is portrayed as a "poison", sold as "safe and mandatory by the press", pushed and "injected into guinea pigs" by pharmaceutical companies, generating more risks for people than coronavirus itself. Thus, AMT proposes as the first therapy that people should increase their immunity in a natural manner, "not trusting an experimental vaccine".

In an abstract way, the term "experimental vaccine" has become an ultraconservative identity token to make believe that vaccines against COVID-19 are immunizations whose formulas would still be in the clinical trial phase. But, in a concrete way, it originally served as a discursive bulkhead to justify the delay in vaccination against covid during the government of President Jair Bolsonaro[1]. The rhetoric of scientificity, when used, serves as a shield for political negligence, reinforcing the AMT's political belief more than the credulity in the shared misinformation ("An experimental vaccine that does not immunize, does not prevent contagion or dissemination. They used our children as guinea pigs and, just like the Nazis, they made propaganda for the people to accept it!"; "[...] People are falling to the ground like flies because of heart-related problems that are, in virtually all cases, caused by an adverse reaction to mRNA injection").

AMT's rhetoric, however, is not limited to denial. Far from being a discursive element that would only work to deny vaccination, the expression "experimental vaccine" operates in a positive rotation, by propagating messages whose content is prevention against adverse reactions and/or risks of death, and problems allegedly derived from the vaccine. In the building of vaccine misinformation, the protection of the other is the subliminal text that the messages carry syllogistically. Instead of being written with a denialist tone, misinformation about vaccines is shaped to generate the impression of care intention, which makes its dismantling more complex. In addition to spreading fear and panic, the anti-vaccine narrative seeks to suggest care for the other, inverting the scientific sense of vaccines, as it convinces its members that adverse reactions to the vaccine are the norm, when they are an exception.

The formula "experimental vaccine" unfolds, in short, in three narrative structures. First, it is used to refute the seriousness of a disease and its vaccine. Second, it contributes to inventing or overestimating risks and damages. Third, it is strengthened by exemplifying testimonial stories that would have occurred within the family.

[1] The theme was extensively reported by Brazilian press, e.g.: Bolsonaro contraria dados do governo e diz que vacina está em "fase experimental". Estadão. https://www.estadao.com.br/politica/bolsonaro-contraria-dados-do-governo-e-diz-que-vacina-esta-em-fase-experimental. Accessed: 09-05-2023.

3.2 Immunological

The second pattern of vaccine misinformation creates a narrative centered on the idea that there is a loss of immunological defenses when someone is vaccinated. The irreversibility of "side effects" and "adverse events" is connected to the second narrative pattern of vaccine misinformation, expressed in the idea that messenger RNA (mRNA) vaccine technology, developed by Pfizer/BioNTech and Moderna, triggers in people a loss of immunity, leaving them susceptible to autoimmune diseases. These vaccines were the first to use RNA technology in history to be permitted for use in humans. This unprecedented development was labeled "experimental" by denialist groups, corroborated by official speeches by far-right politicians. An example is Brazilian former president Jair Bolsonaro, who amplified distrust about the Pfizer vaccine, in particular, saying, in a tone of irony, that people could turn into anything when taking a dose of this type of vaccine: "Turning into an alligator"[2].

The declared political position against mRNA vaccines became a password for ultra-conservative supporters to question the safety and effectiveness of the vaccines. In Fig. 3 it is possible to detect two semantic groupings that demonstrate how the topics "mRNA" and "spike protein" ("proteína spike") drive the fabrication of false controversies about immunization. It once again reveals how the anti-vaccine movement on Telegram, in Brazil, focuses its efforts predominantly on undermining the vaccine against COVID-19, since both "mRNA" and "spike protein" are terms related to the most advanced technology in vaccine production. Contrary to what propagates AMT, the spike protein is responsible for the entry of the virus into the human cell, coupling to the ACE217 receptor. The mRNA vaccine produces antibodies against the spike, preventing the coronavirus from entering the body, neutralizing it[3]. The villainization of the scientific method adopted in the production of mRNA vaccines became an AMT strategy to spread the feeling that the pharmaceutical industry was conducting genetic manipulation of human bodies. The movement does this by embodying the rhetoric of the weak against the oppressors. They collectively embark on a crusade against what they consider to be an intellectual elite (scientists), a media elite (mainstream press), and an economic elite (big pharma). Thus, the vaccine denialist message emulates the social processes of political resistance, inverting them, by encouraging its members to refute the benefits of vaccination, as a way of not bending to power structures. On one side, there would be citizens who resist all coercion; on the other, the manipulated unwary people who get vaccinated. Denialism presents itself as a movement of "real people" skeptical of what comes from the establishment. And, because of this, they seek their own therapies.

---

[2] On this subject: Bolsonaro é provavelmente o primeiro líder político da história a desencorajar vacinação, diz especialista francês. 2021. BBC News Brasil. https://www.bbc.com/portuguese/brasil-55939354. Accessed: 09-05-2023.

[3] About that, it is worth reading: Como imunizantes contra covid-19 podem alavancar vacinas que combatem câncer. 2022. BBC News Brasil. https://www.bbc.com/portuguese/geral-59814298. Accessed: 09-05-2023.

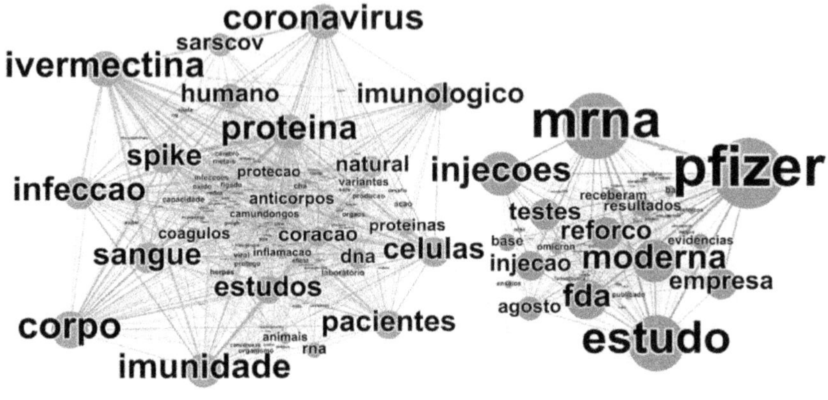

Fig. 3. Two clusters of the General Narrative Network mined from 1.9 million messages from the anti-vaccine movement on Brazilian Telegram. It is possible to note that the cohesion of the terms shows the incidence of the misinformative narrative that links the mRNA/spike protein vaccine, from Pfizer and Moderna, as causing the loss of human immunological defense

Figure 3 reveals some very frequent expressions that appear in the messages in the second pattern, such as: (1) "natural-protection" ("proteção-natural") and "natural-immunity" ("imunidade-natural"), which works as the opposite of vaccine immunization; (2) "human-body-inflammation" ("corpo-humano-inflamação"), "inflammation-human-body-dna" ("humano-inflamação-corpo-dna") and "infection-immunological" ("infecção-imunológica"), semantic marks linked to messages that support stories that would prove the weakening of the immune system triggered by spike protein; and (3) "ivermectin-patients" ("ivermectina-pacientes"), reported continuous as drug for preventing COVID-19 massively. Stories and expressions shared in groups serve to create community bonds, revealing a way of thinking and behaving in terms of health. In this case, within the AMT logic, the vaccination would imply being dominated by the system and hostage to a governability that would place bodies in a state of continuous risk by making them susceptible to "inflammations", "infections", "reactions", "effects". In sum, a permanent state of lack of defenses. Some messages highlight this: "COVID vaccines are affecting T cell memory"; "the spike protein reactivates the herpes zoster virus, weakens the immune system and causes several side effects such as clots".

3.3 False Utility

The narrative of public utility is what characterizes the third pattern of misinformation about vaccines. After COVID-19, the word most associated with "vaccine(s)" ("vacina") is "children" ("crianças"). Together they appeared in 1,983 messages. In them, there are several emotional appeals for members of AMT groups to refrain from vaccinating their "children", "babies" ("bebês"), and "young people" ("jovens"), correlating immunization with the development of serious diseases.

The messages associate the vaccine with: child intoxication ("Covid vaccines do not protect children at all, they are toxic poisons", says one of the comments); autism ("Vaccines have greatly increased the number of children with autism nowadays is everywhere"); heart attack and myocarditis ("vaccine, I mean experimental genetic therapy (sic) without registration, even 7-year-old children having heart attacks in schools"; "woman rebels against journalist and says that children are dying of myocarditis due to vaccines"); sudden illness ("8-year-old child dies after sudden illness. The killer 'covid vaccine' claims another victim"); and even polio ("Health [institution] suspect of "inadequate vaccination' in case of paralysis in Pará. 3-year-old child presented symptoms compatible with polio in a city in the interior of the state. The Ministry of Health is investigating").

This small extract from the messages reproduces what Fig. 4 shows: children are the preferred target of narratives that connect deaths to vaccination. That is why in this graph the words "death" ("morte"), "risk(s)" ("risco"), "syndrome" ("síndrome"), "serious cases" ("casos sérios"), "myocarditis" ("miocardite"), "thrombosis" ("trombose"), "paralysis" ("paralisia"), "liver disease" ("doença de fígado"), "heart disease" ("doença de coração"), "brain disease" ("doença no cérebro"), "embolism" ("embolismo"), "sudden death" ("morte súbita"), "hospitalization" ("hospitalização") belong to the same class as "babies" ("bebês") and "children".

The misinformation strategy therefore affects the population that most intensely needs to seek out vaccination rooms and, at the same time, those that are most susceptible to doubt, since fathers, mothers, or guardians are more attentive to any reported danger against the children.

Fig. 4. Cluster of words belonging to the General Narrative Network. The cohesion of the terms shows the incidence of the misinformative narrative that links vaccines with severe diseases in babies and children

Messages focused on children do not just use the rhetoric of a health alert, but above all a political alert. The architecture of misinforming messages chooses the target to protect (children), to avoid (serious diseases) and the position to defend. AMT posts suggest how parents should proceed to oppose the mandatory vaccination card required by schools or the National Immunization Program (PNI) or even how to argue, using false studies or instructions from false doctors, when confronting pressure from other parents for collective vaccination. The logic of misinformation on Telegram is not only supported by messages that share false collective concerns and false care for others, but it is a narrative on how to behave, and how to act. In other words, it produces protocols, which serve as an orientation guide for child care. Messages like this illustrate the tactic: "A POISON vaccine will never be mandatory...fight react, take your child out of school but don't vaccinate your child [...] read the ADVERSE effects OF THIS EXPERIMENT".

Thus, these messages are instructive materials on how to acquire "healthy behaviors", which, in the end, serve as a gateway to resistance to any and all vaccines. Vaccine misinformation presents itself as public utility information, self-instructive knowledge, and easy to understand and replicate.

3.4 Medicalizing

One of the strengths of misinformation is providing instructions on how to act, a kind of tutorial. However, the use of an argument from authority helps to change the habit of getting vaccinated. This is why there is the fourth pattern of misinformation in vaccine contexts: the medical expert narrative. It is important to point out that several messages are carried out by professionals who participate in a network called "Doctors for Life" ("Médicos pela Vida", in Portuguese). The strategy is reporting medical guidelines in videos posted on Telegram channels. This is why the word "video" appears intensely connected to the term "Dr" in Fig. 5.

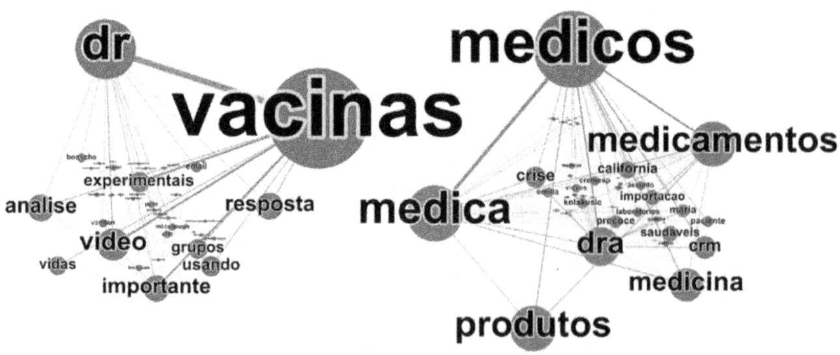

Fig. 5. Cluster of words belonging to the General Narrative Network. There is a cohesion of terms associated with medical reports

In about 20% of messages that we analyzed, there is the presence of the lexicons "dr" (divided into "dr" and "dra", in Portuguese), "doctor" ("doutor"), 'study" ("estudo"), "research" ("pesquisa"), "products" ("produtos"), "CFM" (an acronym for Conselho Federal de Medicina or Federal Council of Medicine, in English), "medicines" ("medicamentos"), "products" ("produtos") as markers of authorities that legitimize misinformation. This indicates that medical discourse shapes the truth that supports arguments against the mandatory PNI vaccines, against vaccination records in schools, and even in conversation streams.

In a context where there are doubts about the manufacture, safety and effects of the vaccine, the anti-vaccine movement uses the reported speech of a "medical expert" ("especialista médico") as a persuasion strategy. The doctors' speech serves as an authentication of the facts, in which the case report is considered a scientific fact. The reality comes from medical misinformation, in which the doctor explains, tells, speaks, studies, affirms, prescribes, indicates to an audience ready to share the recommendations of these professionals or even looking for reports to avoid getting vaccinated ("Dr. Robert Malone explains that the FDA knew for a long time that vaccines caused harm such as Myocarditis, Blood Clots, reactivation of viruses in the form of Herpes Zoster, Epstein-Barr, among others and did their best to obfuscate."; "Dr Maria Emilia explains the serious risks of applying vaccines to young people. spectacular live").

3.5 Conspiracism

Finally, there is the fifth pattern of vaccine misinformation, which is the conspiracist bias. This perspective can be seen in Fig. 1, in the light green cluster, where the words "world" ("mundo"), "global" ("global"), "countries" ("países"), "war" ("guerra") appear. This grouping also has key terms for understanding the conspiracy theories generated by vaccine misinformation, they are: "Bill Gates", "WHO" ("OMS", according to the acronym in Portuguese), "China", and "reduction" ("redução"). These are clue words, whereas they function as clues for detecting conspiracy theories.

Firstly, such theories serve to provide easy answers to complex problems, generating socio-cognitive relief for those who disseminate them, since they create certainty for the group, freeing its components of concerns arising from events without immediate explanations. Secondly, they serve to give AMT a global identity, since several of the theories already circulate in other languages (especially English) and are tropicalized and adapted for Brazil, revealing that the dissemination of misinformation follows dynamics orchestrated by internationalized groups, making it more complex for national health policies. ("Bill Gates is now pumping millions of dollars into a new brainwashing program that manipulates people into getting future experimental COVID-19 vaccines").

In the conspiracy narrative, the word "Gates" refers to Bill Gates. The term works as a force of attraction for the other terms previously mentioned. Gates would be investing millions of dollars in a plan to genetically modify humanity to create a new way of reducing global population, together with countries like China and pharmaceutical corporations, through mutations that could do both: to manipulate people into smoothly accepting new experimental vaccines, as well as to allow the "depopulation of the world" (the "reduction" in the number of inhabitants on the planet), due to deaths caused by adverse reactions to vaccines.

The conspiracy about the American millionaire is based on the image of Gates as a generator of patents and financier of research and scientific initiatives in different parts of the world. And this would become proof of his global influence on ways of thinking and living. In the conspiracist narrative, Gates would function as a pervasive power, capable of creating a liquid chip to be inserted into the vaccine or buying Coca-Cola to ship "RNA in the recipe, just like they put in the vaccine". This same theory is also attributed to other billionaire foundations (Rockefeller, George Soros, etc.). These conspiracy theories reproduce a logic based on the warlike and dichotomous discourse of the "war" of natural versus artificial, the war of rich against poor, and the war of death against life. In practice, AMT thus adopts storytelling as a resource to popularize a narrative of the war against vaccines (which would be one of the technological elements to combat in contemporary life).

In practice, as [16] analyzed, these stories, transformed into conspiracism, convert abstract theories about changes in the world into concrete scenarios, translating deep philosophical reflections into clear experiences and images for cognition. This is because shared stories ("the global elite", "the vaccine chip", "the experimental vaccine") link the private sphere of our individual consciousness to the public domain of politics, as if vaccine refusal were also resistance to policies, commercials of large global corporations, the subjugation to the guidelines of more technologically advanced countries, the programmatic definitions of partisan groups that dominate the political establishment. The causal link of vaccine conspiracy theories has a historical basis of veridiction, making it seductive and persuasive. After all, international inequalities and asymmetries of power are felt by the vast majority of the population in a context of globalization.

4 Final Remarks

This work sheds light on the key strategies employed by ultra-conservative Telegram channels to persuade their audiences. We observed that, in addition to the narratives about the COVID-19 vaccine, these groups are intensely influenced by Brazilian political polarization. In this sense, the first conclusion is that the efforts of these groups are focused on debunking the population on the COVID-19 vaccine. On the other hand, the number of mentions regarding other vaccines was significantly lower. This presents a strategic opportunity for health professionals and managers to focus communication campaigns on vaccines that are not yet entangled in societal polarization.

Furthermore, the investigation identified five patterns of vaccine misinformation on Telegram: protection of others; genetic risk of vaccines; concern for children; doctors against vaccines; and conspiracy against the global order. For each of these categories, there are specific narrative strategies and clue words that were evident in the messages analyzed. Our attention was drawn to the need to formulate health communication actions that make greater use of reporting clinical cases in arguments defending vaccination. This is because the anti-vaccine discourse uses the example of personal stories to verify its thesis. Using the same tactic with users of the healthcare system seems to be more effective in persuading people to get vaccinated than disseminating numerical and purely scientific arguments.

Lastly, the analysis underscores the power of narratives that leverage medical discourse and doctors as trusted authorities to persuade audiences on Telegram. This highlights the critical need for ongoing communication from medical experts. By engaging in a daily dialogue with the population, these experts can effectively dispute the war of narratives around adherence to vaccination schedules and promote evidence-based information.

Acknowledgments. This work was partially supported by the following funding agencies: *Coordenação de Aperfeiçoamento de Pessoal de Nível Superior – Brasil* (CAPES) – Finance Code 001; Fundação de Amparo à Pesquisa do Espírito Santo (FAPES) – 368/2022 - P: 2022-NGKM5, 129/2021 - P: 2021-GL60J; and Instituto Capixaba de Ensino Pesquisa e Inovação em Saúde (ICEPI).

References

1. Albuquerque, F.: Brasil atingiu em 2021 menor cobertura vacinal em 20 anos (2023). https://agenciabrasil.ebc.com.br/saude/noticia/2023-08/brasil-atingiu-em-2021-menor-cobertura-vacinal-em-20-anos
2. Bastian, M., Heymann, S., Jacomy, M.: Gephi: an open source software for exploring and manipulating networks. In: Proceedings of the International AAAI Conference on Web and Social Media, vol. 3, no. 1, pp. 361–362 (2009). https://doi.org/10.1609/icwsm.v3i1.13937
3. Cavalini, A., Malini, F., Gouveia, F., Comarela, G.: Politics and disinformation: analyzing the use of telegram's information disorder network in brazil for political mobilization. First Monday, May 2023. https://doi.org/10.5210/fm.v28i5.12901
4. Chou, W.Y.S., Oh, A., Klein, W.M.P.: Addressing health-related misinformation on social media. JAMA **320**(23), 2417 (2018). https://doi.org/10.1001/jama.2018.16865
5. Floss, M., Tolotti, G., Rossetto, A.d.S., Camargo, T.S.d., Saldiva, P.H.N.: Linha do tempo do "tratamento precoce" para covid-19 no brasil: desinformação e comunicação do ministério da saúde. Interface - Comunicação e, Saúde, Educação e **27**, e210693 (2023). https://doi.org/10.1590/interface.210693
6. Fundação e Osvaldo Cruz: Covid-19: apenas 16% das crianças de 3 e 4 anos tomaram duas doses da vacina (2023). https://agencia.fiocruz.br/covid-19-apenas-16-das-criancas-de-3-e-4-anos-tomaram-duas-doses-da-vacina
7. Giordani, R.C.F., Donasolo, J.P.G., Ames, V.D.B., Giordani, R.L.: A ciência entre a infodemia e outras narrativas da pós-verdade: desafios em tempos de pandemia. Ciência & Saúde Coletiva **26**(7), 2863–2872 (2021). https://doi.org/10.1590/1413-81232021267.05892021
8. Johnson, T.P.: Snowball Sampling: Introduction. Wiley (2014). https://doi.org/10.1002/9781118445112.stat05720
9. Latour, B., Jensen, P., Venturini, T., Grauwin, S., Boullier, D.: The whole is always smaller than its parts. Br. J. Sociol. **63**(4), 590–615 (2012). https://doi.org/10.1111/j.1468-4446.2012.01428.x

10. Malini, F.: Um método perspectivista de análise de redes sociais: cartografando topologias e temporalidades em rede. Encontro Anual da Compós (2016). http://www.labic.net/wp-content/uploads/2016/06/compos_malini_2016.pdf
11. Massarani, L., Leal, T., Waltz, I.: O debate sobre vacinas em redes sociais: uma análise exploratória dos links com maior engajamento. Cad. Saude Publica **36**, e00148319 (2020). https://doi.org/10.1590/0102-311X00148319
12. Milonopoulos, A., Teixeira, R.R.: Potências da fome, potências da vida: uma cartografia das práticas alimentares expressas no Instagram. Master's thesis, Universidade de São Paulo (2023). https://doi.org/10.11606/T.5.2023.tde-25052023-162329
13. OPAS: Entenda a infodemia e a desinformação na luta contra a covid19 (2020). https://iris.paho.org/bitstream/handle/10665.2/52054/factsheetinfodemic_por.pdf
14. Passos, N.A.R.deA.: Uma ferramenta para o jornalismo de dados, a análise de redes sociais e a cartografia do ciberespaço. Master's thesis, Department of Communication Studies. Federal University of Espírito Santo (2018)
15. Peeters, S., Willaert, T.: Telegram and digital methods: mapping networked conspiracy theories through platform affordances. M/C J. **25**(1) (2022). https://doi.org/10.5204/mcj.2878
16. Seargeant, P.: The art of political storytelling: Why stories win votes in post-truth politics **27** (2020). https://doi.org/10.5040/9781350107427
17. Tomes, N., Parry, M.S.: What are the historical roots of the covid-19 infodemic? Lessons from the past. WHO Regional Office for Europe, Copenhagen (2022)
18. WHO: Health topics: Infodemic (2022). https://www.who.int/health-topics/infodemic

MLCDG: Multi-Level Contrastive Graph Clustering in Dynamic Graphs

Mohamed Mahmoud Amar, Mohamed Bouguessa(✉), and Abdoulaye Banire Diallo

University of Quebec at Montreal, Montreal, QC, Canada
amar.mohamed_mahmoud@courrier.uqam.ca,
{bouguessa.mohamed,diallo.abdoulaye}@uqam.ca

Abstract. Dynamic graphs are widespread in social networks, biological networks, and recommendation systems. Community detection in dynamic graphs presents several challenges because these graphs continually change over time. Traditional methods often tackle the problem by clustering graphs separately at each timestep and matching these communities. Unfortunately, this strategy does not account for the temporal continuity and can lead to instability in community detection. Moreover, most current methods do not consider the scenario where nodes have attributes. To address these challenges, we introduce a multi-level contrastive graph clustering approach for Dynamic Graphs (MLCDG), a novel methodology employing deep clustering in the context of dynamic graph neural networks. MLCDG innovatively incorporates contrastive learning to generate clustering-oriented latent representations that effectively capture both node-level and temporal-level community structures. Our approach stabilizes community detection by maintaining temporal coherence and minimizing clustering disruptions caused by dynamic changes. The methodology outperforms existing state-of-the-art techniques, demonstrated by our experiments on real-world and synthetic datasets, validating our approach's effectiveness in enhancing community detection in attributed dynamic graphs.

Keywords: Dynamic Graphs · Deep Clustering · Contrastive Learning · Community detection

1 Introduction

Many real-world graphs, such as those in social networks [31], biological networks [10], and recommendation systems [1], commonly exhibit community structure. Community detection is essential for understanding real-world graphs, identifying important nodes, grouping similar nodes, and predicting future behavior. Despite the attention given to community detection in static graphs, relatively less research has focused on the dynamic scenario in which communities evolve. This problem is particularly challenging as slight changes can significantly impact

L. M. Aiello et al. (Eds.): ASONAM 2024, LNCS 15213, pp. 197–210, 2025.
https://doi.org/10.1007/978-3-031-78548-1_16

the results of graph clustering methods [2]. Community detection in dynamic graphs is commonly approached as a problem with two distinct steps. First, the graphs are clustered independently at each time step using a static clustering algorithm. Then, the communities discovered at different time steps are matched [4,6,39]. This set of approaches is based on the assumption that communities at a time step are influenced solely by the state of the graph at that particular time. However, the inherent instability of clustering algorithms can hurt the quality of clustering results [2]. Discerning between changes in the community structure stemming from community evolution and those arising from algorithm instability can be challenging.

Several solutions have been suggested to address, or at least alleviate, the instability problem. They all share the goal of creating a more gradual evolution of communities. One commonly employed strategy for addressing instability involves iteratively leveraging historical information to identify the community structure at the time-step of interest. These methods allow us to identify communities at a specific point in time without diverging significantly from the community structure seen in previous time steps. In [2], the authors modified the Louvain [3] algorithm to stabilize the community structure. Another research direction [8,24] explored two distinct facets. The first aspect is the current partition's quality, while the second is temporal coherence. This research initiative focuses on optimizing a quality function that integrates both aspects. Lin *et al.* [24] formulated the problem through non-negative matrix factorization, which unifies communities' factorization and temporal evolution. In [8], Folino *et al.* proposed a multi-objective evolutionary approach based on genetic algorithms. Unfortunately, all these methods remain shallow at their core, heavily depending on similarity measures built around the topological similarities of the clusters, making them unreliable in the case of attributed graphs.

We argue that these methods fail to provide the expressive capabilities of neural networks. Neural networks, specifically graph neural networks, have demonstrated unprecedented achievements in learning graph representations by capturing information from the graph structure using message passing [37]. Over the past few years, deep clustering [13,40] has exhibited remarkable effectiveness in grouping a dataset into clusters. This paradigm consists of two training phases: pretraining for solving a pretext task and fine-tuning based on pseudo labels. The objective of the fine-tuning phase is to transform the latent space so that it becomes clustering-oriented. This clustering approach has been employed with different types of datasets, including graph data [18], [28]. Despite deep learning achieving notable success in dynamic graphs across several downstream tasks, including node classification and edge prediction [19,23,26,35], the application of deep clustering techniques for community detection in dynamic graphs remains unexplored.

Contrastive learning [25] has emerged as an effective approach for learning representations in machine learning, especially within unsupervised or self-supervised contexts. The goal of contrastive learning is to learn a meaningful representation of data by encouraging the model to pull similar data points closer together in the latent space while pushing dissimilar data points farther apart. This helps in creating a more discriminative and structured latent representation of the data. In unsupervised learning, negative and positive pairs are created by applying data augmentation techniques to generate multiple views of the same data point. One advantage of using contrastive learning with graphs is the capability to customize the learning approach to focus on a desired level of abstraction (node, cluster, time).

The problem of instability has been explored in [2]. The researchers selected three traditional static clustering algorithms and assessed their outcomes' stability when the network undergoes minor modifications. To achieve this, they compared the modularity and Normalized Mutual Information (NMI) scores across consecutive outcomes. In a network comprising 9,377 nodes and 24,107 edges derived from a co-authorship network, eliminating a single node resulted in NMIs between successive outcomes of 0.99, 0.9, and 0.75 for each algorithm, respectively. The impact becomes even more evident when considering the number of individual changes: a solitary node modification led to approximately 500, 2,000, and 3,000 individual community changes for the respective algorithms. Several solutions have been proposed to address, or at least alleviate, the instability issue. They all share the goal of creating a more gradual evolution of communities.

We introduce a novel method for clustering dynamic graphs, which we call Multi-Level Contrastive Graph Clustering for Dynamic Graphs (MLCDG). Our deep clustering approach uses contrastive learning principles to train a Dynamic Graph Neural Network (DGNN) to learn clustering-oriented latent representations. Specifically, we present three contrastive loss components that constitute the overall loss function to be minimized. During the pretraining phase, the loss function comprises two terms. The first term consists of node-level contrastive learning, which is used to learn a general-purpose graph representation. The second component is a temporal-level contrastive loss, designed to enforce a temporal smoothness constraint and address the community instability problem. During the clustering phase (second phase), a novel loss term is incorporated to refine the clusters identified at the end of the pretraining stage.

Contributions. (1) We establish the first deep clustering model for dynamic graphs that tackles the instability problem in static community detection algorithms. **(2)** We introduce a novel multi-level contrastive loss function to improve node representation learning by harnessing data from multiple abstraction levels. **(3)** In the fine-tuning stage, we incorporate a contrastive clustering loss aimed at gradually transforming the latent space to become clustering-oriented. This transformation capitalizes on the nodes assigned with high confidence. We conduct experiments on real-world and synthetic datasets. The results illustrate that the proposed model exceeds state-of-the-art methods' performance.

2 Related Work

Graph Neural Networks (GNNs) have gained significant attention and demonstrated their effectiveness in learning powerful representations for complex graph datasets. However, when dealing with dynamic graphs, solely using static GNNs may not be sufficient for capturing relevant temporal patterns. Multiple research efforts have been conducted to address this limitation, primarily focusing on extending static GNNs to dynamic GNNs by incorporating a sequential model.

In [34], the authors proposed a first dynamic graph neural network model by combining a GNN with a recurrent neural network such as a vanilla RNN [27], a GRU [7] or an LSTM [16]. The authors proposed two variants. The first variant employs a combination of Convolution layers and Recurrent Neural Networks. The second variant adopts a convolutional LSTM that replaces fully connected operations in the recurrent neural network definition with convolutional operations. In [22], Kumar *et al.* proposed *JODIE*, a coupled recurrent neural network model designed to acquire trajectories of node embeddings. They employed a recurrent neural network to update a node embedding given historical information. Additionally, to capture the trajectory of future embeddings, they introduced a projection operator that can learn a node's embedding at any given point. The work in [33] calculates node representations by employing simultaneous self-attention mechanisms that consider neighborhood structural information and temporal dynamics. EvolveGCN [30] employs a graph convolution network (GCN) and a recurrent neural network. Unlike prior methods, this approach utilizes the RNN to update the GCN parameters instead of the node embeddings. The authors of [14] proposed a variational graph recurrent neural network model aimed at capturing changes in dynamic graph topology and node attributes. Additional notable examples of dynamic graph neural networks have been introduced [19,23,26,35]. These techniques are primarily developed for supervised tasks such as node classification and edge prediction, but they overlook the problem of clustering in dynamic graphs.

Deep clustering has demonstrated remarkable performance in partitioning graph data. Contrastive learning has been applied within deep clustering frameworks. A node-level loss function is employed during the pretraining stage to develop a versatile node representation. Inspired by image augmentation techniques, graph augmentations like edge modification and node feature masking are employed by models to uncover deeper semantic information. This is achieved by creating more complex pretext tasks challenging the model to learn effectively. In the clustering phase, the contrastive method moves a particular node closer to its corresponding centroid and further from other centroids, thereby reshaping the representation space to make the clusters separable.

3 Methodology

We introduce MLCDG, a novel graph neural network model designed to address the problem of detecting communities in evolving graphs. We have adopted the contrastive learning paradigm as the foundation of our approach. The model architecture consists of T (the number of timesteps) Graph Neural Networks interconnected with GRU units. The tth GNN takes the A^t (the tth adjacency matrix) and X^t (The tth attribute matrix) as input and projects the graph into a low dimensional latent space-this projection in denoted H^t. H^t and H^{t-1} represent the inputs of a GRU unit, which produces the graph's embedding at the tth timestep, denoted Z^t. The architecture of MLCDG is illustrated in Fig. 1.

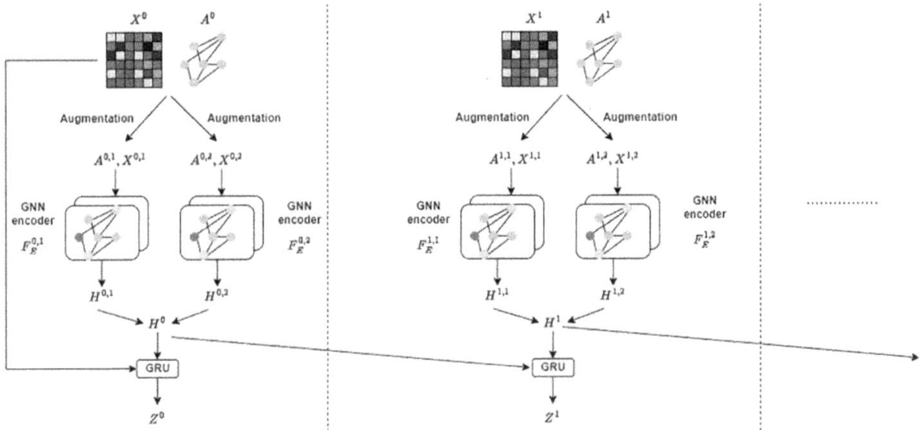

Fig. 1. The MLCDG Architecture.

MLCDG is based on contrastive learning. Given the unsupervised nature of the clustering problem, creating two different graph augmentations is necessary. These augmentations or graph views are denoted $(A^{t,1})$ and $(A^{t,2})$. We denote by F_E^t (the view index is omitted to simplify notation) the GNN encoder at the tth timestep and L the number of layers. We employ a Convolutional Graph Network (GCN), as described in [20].

$$H_L^t = F_E^t(A^t, X^t) = ReLU(\tilde{\Delta^t}^{-\frac{1}{2}} \tilde{A}^t \tilde{\Delta^t}^{-\frac{1}{2}} H_{L-1}^t W_L^t) \tag{1}$$

$$Z^t = GRU(H^t, H^{t-1}) \tag{2}$$

where W_L^t represent the learning parameters of the GCN F_E^t, ReLU is the rectified linear unit activation function corresponding to the output layer of the GCN, $\tilde{\Delta^t}^{-\frac{1}{2}} \tilde{A}^t \tilde{\Delta^t}^{-\frac{1}{2}}$ represents the normalized adjacency matrix where $\tilde{A}^t = A + I$, I is the identity matrix and $\tilde{\Delta} = A\mathbf{1}_N$.

Mutual Information (MI): Mutual information measures the statistical dependence or information shared between two random variables. It quantifies how much knowing the value of one variable tells you about the other. In other words, it measures the reduction in uncertainty about one variable when you know the value of the other. Mathematically, the mutual information between two random variables X and Y is defined as:

$$I(X,Y) = \mathrm{E}_{p(x,y)}(\log(p(x,y))/p(x),p(y)) \tag{3}$$

where $p(x,y)$ represents the joint density of X and Y, $p(x)$ and $p(y)$ denote, respectively, the marginal density of X and Y.

Over the past few years, numerous research studies have demonstrated significant achievements in representation learning through the optimization of mutual information between two different data views [5,15,36]. Given the difficulty of estimating mutual information [32], it is expected to maximize MI by first deriving a lower bound and then optimizing this lower bound. For instance, InfoNCE [29] (Noise Contrastive Estimator) is frequently used as the lower bound of MI.

$$I(X,Y) \geq I_{NCE} = \mathrm{E}\Big(\frac{1}{M}\sum_{i=1}^{M}\frac{\exp(\theta(x_i,y_i))}{\frac{1}{M}\sum_{j=1}^{M}\exp(\theta(x_j,y_j))}\Big) \tag{4}$$

$\{x_i, y_i\}_{i=1}^{M}$ represent M independent samples from the joint density $p(x,y)$, θ is a critic function. This function aims to assign large scores to pairs (x_i, y_i) drown together (positive pairs) and small scores to negative pairs $(\{x_i, y_j\}_{i \neq j})$. The MLCGD is trained in two distinct phases: first, a pretraining phase is performed to solve a pretext task, followed by a clustering phase with the goal of refining clusters using pseudo-labels.

First Phase (Pretraining): Before delving into the primary task, which is pseudo-supervision or clustering, an initial pretraining phase is conducted. The pretraining allows the model to acquire high-level representations from which pseudo-labels can be derived. At the heart of this phase, two contrastive loss functions are introduced. The first loss function is a contrastive loss applied at the node-level. We pick two graph augmentations, yielding two different views of the graph at every time step. At the time step t, the two views of the graph (A^t, X^t) are denoted $(A^{t,1}, X^{t,1})$ and $(A^{t,2}, X^{t,2})$. These augmentations are fed to two GNNs producing latent representations to the two graph views denoted $H^{t,1}$ and $H^{t,2}$. The ith row in $H^{t,l}$, denoted $h_i^{t,l}$, represents the latent representation of of the node i in the graph view l at the tth timestep. Given a node i, the positive samples associated to $h_i^{t,l}$ ($l \in \{1,2\}$)are $\{h_i^{t,k}\}_{k \neq l}$ and the negative samples are sampled from the set $\{h_j^{t,k}\}_{j \neq i}$. We opt for the Noise Contrastive Estimator (I_{NCE}) as our choice of a contrastive loss function. At the timestep t, the node-level contrastive loss is defined as follows:

$$\mathcal{L}_{NL} = \sum_{l=1}^{2}\sum_{i=1}^{N} -\log\Big(\frac{exp((h_i^{t,l})^T h_i^{t,k})}{\sum_{k=1}^{2}\sum_{j=1}^{N} exp((h_i^{t,l})^T h_j^{t,k})}\Big) \tag{5}$$

The critic function in (4) is a dot product: $\theta(x,y) = x^T y$. The loss L_{NL} is minimized when the representations of a specific node in both of its distinct views are embedded in such a way that they are in close proximity to each other while simultaneously being positioned far apart from the representations of other nodes. Conversely, this loss is maximized when the embeddings of the two views of the same node are situated far from each other and in close proximity to other nodes.

We leverage two graph augmentations: (i) Node Feature Masking (NFM), Which is a commonly employed attribute-based augmentation technique [17, 42]. Specifically, NFM involves the random masking of features for a subset of nodes in the provided graph. The masking can be applied either row-wise or column-wise. (ii) Graph Diffusion (GD) [9] which is a topological augmentation. The concept behind graph diffusion involves enhancing the augmented graph by incorporating global structural information by adding connections between nodes and their indirect neighbors.

In the context of evolving graphs, characterized by continuous modifications involving both the addition and removal of nodes as well as alterations in their interactions over time, our first main assumption is that these dynamic changes occur in a gradual and smooth manner, thereby avoiding sudden and abrupt shifts in the underlying community structure within the graph. To encourage the preservation of temporal smoothness and then tackle the problem of community instability, we propose a *time-level* contrastive loss function denoted $\mathcal{L}_{TL}$. Given a node i, the goal is to ensure that the latent representation h_i^t doesn't significantly diverge from h_i^{t-1}. This loss function is defined as follows:

$$\mathcal{L}_{TL} = \sum_{i=1}^{N} -\log\left(\frac{\exp((z_i^t)^T z_i^{t-1})}{\sum_{j=1}^{N}\exp((z_i^t)^T z_j^{t-1})}\right) \tag{6}$$

In the pretraining phase, the the self-supervision loss is calculated as a combination of node-level and time-level losses, with appropriate balancing hyperparameters.

$$\mathcal{L}_{SS} = \lambda_N \mathcal{L}_{NL} + \lambda_T \mathcal{L}_{TL} \tag{7}$$

where λ_N and λ_T denote the balancing hyperparameters in the self-supervised loss $\mathcal{L}_{SS}$.

Second Phase (Clustering): The loss terms mentioned above contrast individual entities such as node representations, enabling the learning of the community structure. Rather than focusing solely on low-level entities like nodes in contrastive learning, we will focus on higher-level entities, specifically cluster centroids. The core idea is based on the gradual transformation of the latent representational space to be well-suited for clustering. Following the pretraining phase, we employ a clustering algorithm, like K-means, on the latent codes. At the timestep t, the cluster membership matrix Φ^t is defined by:

$$\Phi_{i,j}^t = \frac{(1 + \|z_i^t - \mu_j^k\|^2)^{-1}}{\sum_{k=1}^{K}(1 + \|z_i^t - \mu_k^t\|^2)^{-1}} \tag{8}$$

where z_i^t denotes the embedding of the node i at the timestep t, $\{\mu_k\}_{0\leq k\leq K}$ represent the K centroids of the clusters.

Let $\mathcal{H}_\epsilon^t$ be the set of nodes assigned to their clusters with high confidence, meaning that:

$$i \in \mathcal{H}_\epsilon^t \Leftrightarrow \Phi_{i,c_{max}} \geq \epsilon$$

Where $\Phi_{i,c_{max}}$ is the highest value in the ith row of Φ^t and ϵ is a confidence threshold. Let $\{\Omega_k^t\}_{0\leq k\leq K}$ be the centroids of the nods belonging to $\mathcal{H}_\epsilon^t$. We introduce a contrastive clustering loss that aims to pull the nodes assigned with high confidence to their corresponding high confident centroids, thus creating a letent space with well separated clusters. We periodically increase the hyperparameter ϵ every M iterations, aiming to refine the community structure within the latent space by enlarging the circle of high-confidence.

$$\mathcal{L}_{CL} = \sum_{i\in\mathcal{H}_\epsilon^t} -\log\left(\frac{\exp((z_i^t)^T\Omega_{k_i}^t)}{\sum_{k=1}^{K}\exp((z_i^t)^T\Omega_k^t)}\right) \tag{9}$$

$\Omega_{k_i}^t$ is the centroid corresponding to the node i. Finally the overall loss function is a weighted sum of the previously defined loss terms.

$$\mathcal{L} = \lambda_{NL}\mathcal{L}_{NL} + \lambda_{TL}\mathcal{L}_{TL} + \lambda_{CL}\mathcal{L}_{CL} \tag{10}$$

where λ_{NL}, λ_{TL} and λ_{CL} denote the balancing hyperparameters in the overall loss $\mathcal{L}$.

Training Algorithm: Our model is pretrained for I_1 iterations. During this phase, the parameters of the graph neural networks, as well as the GRU units, are updated to minimize the loss function $\mathcal{L}_{SS}$ using the Adam optimizer. Subsequently, the clustering is performed on the latent codes using K-means. Afterward, the model is fine-tuned by minimizing the loss function $\mathcal{L}$. Algorithm 1 provides the details of the clustering phase of MLCDG.

4 Experiments

Baselines: We assess the effectiveness of our approach by conducting a comprehensive comparison with a range of methodologies. Specifically, our evaluation includes traditional clustering algorithms, deep graph clustering techniques, and dynamic graph representation learning methods, which serve as the baseline for our analysis.

- **Traditional Clustering methods**: We opted for considering two prominent traditional clustering algorithms: K-means and spectral clustering. The K-means algorithm typically operates under the assumption that clusters exhibit a spherical or concave shape, meaning data points are expected to fall within a certain radius around the cluster centroid. On the contrary, spectral clustering proves beneficial when dealing with datasets characterized by non-linear and complex cluster structures.

Algorithm 1. MLCDG Clustering

Input: Dynamic graph $\{(A^t, X^t)\}_{0 \leq t \leq T-1}$; number of clusters K; threshold ϵ; number of epochs N_{epochs}; update interval M; initial cluster membership matrix $\{\Phi^t\}_{0 \leq t \leq T-1}$
Output: Final cluster membership matrix $\{\Phi^t\}_{0 \leq t \leq T-1}$

```
epoch ← 0
for 0 ≤ t ≤ T − 1 do
  Compute two graph views (A^{t,1}, X^{t,1}) and (A^{t,2}, X^{t,2})
  for epoch ≤ N_epochs do
    if epoch%M = 0 then
      Update ε
    end if
    H^{t,1} ← F_E^{t,1}(A^{t,1}, X^{t,1})
    H^{t,2} ← F_E^{t,2}(A^{t,2}, X^{t,2})
    H^t = (H^{t,1} + H^{t,2})/2
    Z^t = GRU(H^t, H^{t−1})
    Compute H_ε^t
    Compute and backpropagate L
  end for
  Compute and store Φ^t using Z^t
end for
return {Φ^t}_{0≤t≤T−1}
```

- **Deep Graph Clustering**: We evaluate our approach by comparing it to three deep graph clustering methods. The first model in this comparison is a graph autoencoder (**GAE**) [21], which first learns node embeddings and then employs a clustering algorithm. The second model is a variational variant of GAE(**VGAE**) [21]. **DAEGC** [38] is an other deep graph clustering approach. DAEGC utilizes an attention network to capture the significance of adjacent nodes with respect to a central target node.
- **Dynamic Graph Neural Networks Models**: As far as we know, our method represents the first deep graph clustering approach designed for dynamic graphs. Previous dynamic graph neural network models were primarily oriented toward different downstream tasks such as node classification, edge detection, and edge prediction. We have selected several notable examples of these prior methods as baselines for our approach. **DynGEM** [12] employs autoencoders to progressively produce node embeddings at time t, utilizing the graph snapshot from time $t-1$. **DynAERNN** [11] employs a deep architecture comprising dense and recurrent layers to capture relevant knowledge from the temporal transitions within the network. **EvolveGCN** [30] leverages both a graph convolution network and a recurrent neural network to update the learnable parameters of the GCNs over time. **VGRNN** [14] is a variational graph recurrent neural network that enhances the GRNN model by introducing high-level latent random variables into its architecture.

Datasets: Due to the scarcity of labeled dynamic graph datasets, our experiments were carried out using two distinct datasets. We collected one real-world dataset and generated one synthetic dynamic graph dataset. Despite the initial challenge posed by the limited availability of datasets, the positive outcomes of our preliminary experiments serve as a strong incentive for us to incorporate additional datasets into our future research.

Table 1. Clustering results on two dynamic graph datasets. Best methods in bold and second best underlined.

	Baseline	DBLP-T		SynSBM	
		ACC	F1-Sc	ACC	F1-Sc
Classical Methods	K-Means	58.83	<u>75.22</u>	41.30	49.02
	Spectral Clustering	58.93	73.22	26.00	23.66
Static Deep Graph Clustering	GAE	57.87	74.55	45.69	53.21
	VGAE	<u>62.76</u>	75.10	75.12	87.32
	DAEGC	59.46	73.82	81.43	84.47
Dynamic Graph Models	DynGEM	61.63	74.88	93.12	94.21
	DynAERNN	60.73	73.20	90.32	92.75
	EvolveGCN	61.02	74.52	<u>95.75</u>	<u>96.53</u>
	VGRNN	59.17	73.58	91.13	93.79
	MLCDG	**72.31**	**79.01**	**98.10**	**98.10**

- **DBLP-T** [41]: This dataset is obtained from the DBLP website, which offers accessible bibliographic datasets for prominent computer science journals and conferences. The authors are represented by nodes and the co-authorship by edges. Each year corresponds to a single time step, with co-author connections being established in the year a collaborative paper is published. The two classes represent the research domain (*computer networks* or *machine learning*). The node class has the potential to undergo changes over time.
- **SynSBM**: We slightly modify the static stochastic block (SBM) model to generate dynamic graphs exhibiting community structure. After generating an initial graph using a static SBM, a number of nodes are randomly selected and assigned to other communities. The connections of these nodes are added according to the two parameters of the SBM: α and β represent, respectively, the probabilities of a node making connections within its own community and with nodes in other communities.

Hypermapameters: In the MLCDG algorithm, we set the number of epochs to 500 during the second phase. Additionally, we incorporate a gradual increase in ϵ value, starting from 0.4 and incrementing by 0.1 every 100 epochs until it reaches 0.8.

We assess our model's performance using two commonly employed metrics: Accuracy (ACC) and the F1-score (F1-Sc). Let y_{true} be the true labels vector and y_{pred} denote the predicted labels. Table 1 displays the preliminary results of our model alongside the results from the baseline models. The results represent the mean performance metrics across all snapshots. Table 1 shows that MLCDG surpasses all comparing models, notably achieving a remarkable improvement of 13% in ACC and 5% in F1-Sc when compared to the top-performing baseline on the DBLP-T dataset. Regarding the SynSBM synthetic dataset, our model maintains its superiority by achieving an increase of 2% and 1.6% in ACC and F1-Sc, respectively, when compared to the top-performing baseline. The lower performance of the baselines can be attributed to the inability of static and dynamic graph neural networks to effectively capture temporal dependencies using basic graph and feature reconstruction losses. This leads to a failure in learning latent representations suitable for clustering. Furthermore, we observed a consistent improvement in clustering performance as time advances, indicating that our approach benefits from the communities identified earlier to discover new communities effectively.

We argue that the effectiveness of our model stems from two key factors. First, the combination of a recurrent neural network and the temporal loss allowed our model to capture relevant historical information and mitigate the negative impact of the well-known problem of community instability. Second, by incorporating the proposed clustering term into the loss function, which relies on highly confident clustering assignments, the latent representational space can be gradually reshaped to enhance the separability of the clusters.

5 Conclusion

In this paper, we presented MLCDG, a novel deep clustering framework for dynamic attributed graphs, leveraging multi-level contrastive learning to address the challenges of community detection in evolving networks. Our method integrates node-level, temporal-level, and cluster-level contrastive losses to foster stable and meaningful clustering over time. By employing a Dynamic Graph Neural Network architecture, MLCDG effectively captures the evolving nature of graph data, maintaining temporal coherence and enhancing the quality of community detection. By combining the strengths of contrastive learning and deep graph neural networks, MLCDG represents an advancement in dynamic graph clustering. It opens up new possibilities for understanding complex systems in which the underlying graph data evolve over time. In line with this perspective, we are collecting additional real datasets to further investigate our methodology under various real-world scenarios. In addition, we plan to develop an analytical tool to track and predict structural changes and explain the evolution of communities over time. Our goal is to track and predict the evolution of community structures and the variation of the role of nodes within communities over short and long-term scales.

References

1. Ahmadian, S., Joorabloo, N., Jalili, M., Meghdadi, M., Afsharchi, M., Ren, Y.: A temporal clustering approach for social recommender systems. In: 2018 IEEE/ACM International Conference on Advances in Social Networks Analysis and Mining (ASONAM), pp. 1139–1144. IEEE (2018)
2. Aynaud, T., Guillaume, J.L.: Static community detection algorithms for evolving networks. In: 8th International Symposium on Modeling and Optimization in Mobile, Ad Hoc, and Wireless Networks, pp. 513–519. IEEE (2010)
3. Blondel, V.D., Guillaume, J.L., Lambiotte, R., Lefebvre, E.: Fast unfolding of communities in large networks. J. Stat. Mech. Theory Exp. **2008**(10), P10008 (2008)
4. Bourqui, R., Gilbert, F., Simonetto, P., Zaidi, F., Sharan, U., Jourdan, F.: Detecting structural changes and command hierarchies in dynamic social networks. In: 2009 International Conference on Advances in Social Network Analysis and Mining, pp. 83–88. IEEE (2009)
5. Chen, T., Kornblith, S., Norouzi, M., Hinton, G.: A simple framework for contrastive learning of visual representations. In: International Conference on Machine Learning, pp. 1597–1607. PMLR (2020)
6. Chen, Z., Wilson, K.A., Jin, Y., Hendrix, W., Samatova, N.F.: Detecting and tracking community dynamics in evolutionary networks. In: 2010 IEEE International Conference on Data Mining Workshops, pp. 318–327. IEEE (2010)
7. Chung, J., Gulcehre, C., Cho, K., Bengio, Y.: Empirical evaluation of gated recurrent neural networks on sequence modeling. arXiv preprint arXiv:1412.3555 (2014)
8. Folino, F., Pizzuti, C.: An evolutionary multiobjective approach for community discovery in dynamic networks. IEEE Trans. Knowl. Data Eng. **26**(8), 1838–1852 (2013)
9. Gasteiger, J., Weißenberger, S., Günnemann, S.: Diffusion improves graph learning. Adv. Neural Inf. Process. Syst. **32** (2019)
10. Girvan, M., Newman, M.E.: Community structure in social and biological networks. Proc. Natl. Acad. Sci. **99**(12), 7821–7826 (2002)
11. Goyal, P., Chhetri, S.R., Canedo, A.: dyngraph2vec: capturing network dynamics using dynamic graph representation learning. Knowl.-Based Syst. **187**, 104816 (2020)
12. Goyal, P., Kamra, N., He, X., Liu, Y.: DynGEM: deep embedding method for dynamic graphs. arXiv preprint arXiv:1805.11273 (2018)
13. Guo, X., Gao, L., Liu, X., Yin, J.: Improved deep embedded clustering with local structure preservation. In: IJCAI, vol. 17, pp. 1753–1759 (2017)
14. Hajiramezanali, E., Hasanzadeh, A., Narayanan, K., Duffield, N., Zhou, M., Qian, X.: Variational graph recurrent neural networks. Adv. Neural Inf. Process. Syst. **32** (2019)
15. Hjelm, R.D., et al.: Learning deep representations by mutual information estimation and maximization. arXiv preprint arXiv:1808.06670 (2018)
16. Hochreiter, S., Schmidhuber, J.: Long short-term memory. Neural Comput. **9**(8), 1735–1780 (1997)
17. Hu, W., et al.: Strategies for pre-training graph neural networks. arXiv preprint arXiv:1905.12265 (2019)
18. Hui, B., Zhu, P., Hu, Q.: Collaborative graph convolutional networks: unsupervised learning meets semi-supervised learning. In: Proceedings of the AAAI Conference on Artificial Intelligence, vol. 34, pp. 4215–4222 (2020)

19. Jin, W., Qu, M., Jin, X., Ren, X.: Recurrent event network: autoregressive structure inference over temporal knowledge graphs. arXiv preprint arXiv:1904.05530 (2019)
20. Kipf, T.N., Welling, M.: Semi-supervised classification with graph convolutional networks. arXiv preprint arXiv:1609.02907 (2016)
21. Kipf, T.N., Welling, M.: Variational graph auto-encoders. arXiv preprint arXiv:1611.07308 (2016)
22. Kumar, S., Zhang, X., Leskovec, J.: Predicting dynamic embedding trajectory in temporal interaction networks. In: Proceedings of the 25th ACM SIGKDD International Conference on Knowledge Discovery & Data Mining, pp. 1269–1278 (2019)
23. Li, J., et al.: Predicting path failure in time-evolving graphs. In: Proceedings of the 25th ACM SIGKDD International Conference on Knowledge Discovery & Data Mining, pp. 1279–1289 (2019)
24. Lin, Y.R., Chi, Y., Zhu, S., Sundaram, H., Tseng, B.L.: FacetNet: a framework for analyzing communities and their evolutions in dynamic networks. In: Proceedings of the 17th International Conference on World Wide Web, pp. 685–694 (2008)
25. Liu, Y., et al.: Graph self-supervised learning: a survey. IEEE Trans. Knowl. Data Eng. **35**(6), 5879–5900 (2022)
26. Manessi, F., Rozza, A., Manzo, M.: Dynamic graph convolutional networks. Pattern Recogn. **97**, 107000 (2020)
27. Medsker, L.R., Jain, L.: Recurrent neural networks. Des. Appl. **5**(64–67), 2 (2001)
28. Mrabah, N., Bouguessa, M., Touati, M.F., Ksantini, R.: Rethinking graph autoencoder models for attributed graph clustering. IEEE Trans. Knowl. Data Eng. **35**, 9037–9053 (2022)
29. Oord, A.v.d., Li, Y., Vinyals, O.: Representation learning with contrastive predictive coding. arXiv preprint arXiv:1807.03748 (2018)
30. Pareja, A., et al.: EvolveGCN: evolving graph convolutional networks for dynamic graphs. In: Proceedings of the AAAI Conference on Artificial Intelligence, vol. 34, pp. 5363–5370 (2020)
31. Pizzuti, C.: GA-Net: a genetic algorithm for community detection in social networks. In: Rudolph, G., Jansen, T., Beume, N., Lucas, S., Poloni, C. (eds.) PPSN 2008. LNCS, vol. 5199, pp. 1081–1090. Springer, Heidelberg (2008). https://doi.org/10.1007/978-3-540-87700-4_107
32. Poole, B., Ozair, S., Van Den Oord, A., Alemi, A., Tucker, G.: On variational bounds of mutual information. In: International Conference on Machine Learning, pp. 5171–5180. PMLR (2019)
33. Sankar, A., Wu, Y., Gou, L., Zhang, W., Yang, H.: DySAT: deep neural representation learning on dynamic graphs via self-attention networks. In: Proceedings of the 13th International Conference on Web Search and Data Mining, pp. 519–527 (2020)
34. Seo, Y., Defferrard, M., Vandergheynst, P., Bresson, X.: Structured sequence modeling with graph convolutional recurrent networks. In: Cheng, L., Leung, A.C.S., Ozawa, S. (eds.) ICONIP 2018. LNCS, vol. 11301, pp. 362–373. Springer, Cham (2018). https://doi.org/10.1007/978-3-030-04167-0_33
35. Taheri, A., Gimpel, K., Berger-Wolf, T.: Learning to represent the evolution of dynamic graphs with recurrent models. In: Companion Proceedings of the 2019 World Wide Web Conference, pp. 301–307 (2019)
36. Veličković, P., Fedus, W., Hamilton, W.L., Liò, P., Bengio, Y., Hjelm, R.D.: Deep graph infomax. arXiv preprint arXiv:1809.10341 (2018)
37. Vignac, C., Loukas, A., Frossard, P.: Building powerful and equivariant graph neural networks with structural message-passing. Adv. Neural. Inf. Process. Syst. **33**, 14143–14155 (2020)

38. Wang, C., Pan, S., Hu, R., Long, G., Jiang, J., Zhang, C.: Attributed graph clustering: a deep attentional embedding approach. arXiv preprint arXiv:1906.06532 (2019)
39. Wang, Y., Wu, B., Pei, X.: CommTracker: a core-based algorithm of tracking community evolution. In: Tang, C., Ling, C.X., Zhou, X., Cercone, N.J., Li, X. (eds.) ADMA 2008. LNCS (LNAI), vol. 5139, pp. 229–240. Springer, Heidelberg (2008). https://doi.org/10.1007/978-3-540-88192-6_22
40. Xie, J., Girshick, R., Farhadi, A.: Unsupervised deep embedding for clustering analysis. In: International Conference on Machine Learning, pp. 478–487. PMLR (2016)
41. Yao, Y., Joe-Wong, C.: Interpretable clustering on dynamic graphs with recurrent graph neural networks. In: Proceedings of the AAAI Conference on Artificial Intelligence, vol. 35, pp. 4608–4616 (2021)
42. Zhu, Y., Xu, Y., Yu, F., Liu, Q., Wu, S., Wang, L.: Deep graph contrastive representation learning. arXiv preprint arXiv:2006.04131 (2020)

Dissecting the Advocacy Discourse Behind the #StopAsianHate Movement on X/Twitter

Yuze Sha[1], Nicholas Micallef[2(✉)], and Yan Wu[3]

[1] Department of Linguistics and English Language, Lancaster University, Lancaster, UK
y.sha2@lancaster.ac.uk

[2] Department of Computer Science, Swansea University, Swansea, Wales, UK
nicholas.micallef@swansea.ac.uk

[3] Department of Literature, Media and Language, Swansea University, Swansea, Wales, UK
y.wu@swansea.ac.uk

Abstract. This paper takes a multidisciplinary approach and conducts a three-dimensional analysis of #StopAsianHate tweets from January 1, 2021, to December 31, 2022 by combining computer science, applied linguistics, and cultural studies. It employs a 'funnel approach', from a broad examination to specific sentimental and linguistic dimensions within the top 10% most engaged tweets. The analysis reveals that the #StopAsianHate hashtag is primarily used for counter-discourse against Anti-Asian hate crime, expressing collective in-group identity and inclusionary out-group solidarity against racism. A key finding is the representation of Asian people as the 'model minority', derived from combined analyses of sentiments, politeness, toxicity, and Corpus-Assisted Critical Discourse Analysis of the tweets. The #StopAsianHate movement is characterised as moderate, evidenced by the large number of tweets with positive sentiment scores and frequent relational identification, which refers to anti-racism supporters as 'friends', 'folks', and 'family'. Though negative sentiment scores are also prevalent, they are found non-toxic and can be explained by tweet genre's rare use of polite expressions, as well as the prominence of #StopAsianHate thematic words such as 'hate', 'racism', and 'crime', serving as tools to challenge racism. Most notably, the study provides fresh insights into the growing self-reflective collective awareness of the negative impacts of 'model minority' stereotypes within the Asian communities and discusses ongoing opportunities and challenges in #StopAsianHate movement.

Keywords: #StopAsianHate · model minority · social media · digital activism · sentiment analysis · corpus-assisted critical discourse analysis

L. M. Aiello et al. (Eds.): ASONAM 2024, LNCS 15213, pp. 211–229, 2025.
https://doi.org/10.1007/978-3-031-78548-1_17

1 Introduction

The utilisation of X/Twitter as a political tool has become increasingly prevalent among various societal actors, including social elites, advocacy groups, and grassroots activists, for advancing political agendas. Notably, in recent years, there has been a significant surge in the use of X/Twitter to advocate social justice concerning gender, sexuality, and race related inequalities. While existing research is predominantly grounded in computational techniques, it often overlooks the importance of qualitative evaluation, particularly concerning the discursive nuances embedded within Tweets through cultural studies lens.

Taking a triangulation approach, we use both quantitative and qualitative methods in analysing X/Twitter activism via a case study of #StopAsianHate tweets during the COVID-19 global pandemic. This study features three variations of knowledge production that extend across computer science, applied linguistics, and cultural studies while combining methods from Natural Language Processing (NLP) and Corpus-Assisted Critical Discourse Analysis (CACDA) to explore the discursive features of the #StopAsianHate movement on X/Twitter.

While NLP facilitates the processing of large-scale data, traditionally utilising sentiment analysis (e.g., [36]) and topic modelling (e.g., [30]) to reveal general lexicon and semantic patterns, CACDA offers a framework for integrating quantitative and qualitative analyses at textual level and interprets discursive features within social contexts. By integrating NLP with CACDA, our study not only explores anti-Asian racism during COVID-19 global pandemic as an individual phenomenon, but also investigate how such discourse is supported by a dominant social structure that might be expressed in overt or covert terms.

Employing a 'funnel approach', this paper follows a data-inductive strategy, moving from broad patterns to specific features, thereby providing a comprehensive and in-depth analysis of the general characteristic, sentimental landscapes, and discourse practice associated with #StopAsianHate discourse on X/Twitter. Such an approach allows us to provide a comprehensive analysis of the discursive essence of the movement and its implications for and limits to causing social changes. Through the lens of cultural racism, we also investigate how anti-Asian hate online reflects the problematic 'model minority' ideologies and how such ideologies shape the counter discourse to anti-Asian racism.

2 Related Work

In this section, we firstly assess the concept of 'model minority', a form of cultural racism against Asian communities in the US. We then review related work on the technological affordances of X/Twitter in mobilising people for racial justice. Both the Black Lives Matter (BLM) movement and the #StopAsianHate movement utilise hashtags and real-time engagement to raise awareness, challenge stereotypes, and advocate for social justice. Identifying current research gaps, we propose a multidisciplinary approach integrating natural language processing (NLP) and computational-aided analysis of communication dynamics (CACDA) to analyse the nascent form of digital resistance to anti-Asian racism.

2.1 'Model Minority' and 'Medical Scapegoating' Racism

'Model Minority' as collective identity label has been attached to ethnic minority communities with Chinese, and other East Asian (such as Japanese, Korean) and sometimes other Southeast Asian heritages in the US. This term was firstly used by American press in the 1960s to laud the academic and economical success of such minority communities and their propensity of not causing "problems" [16,23,32]. Despite its seemingly positive connotation, the model minority stereotypes usually imply such community's lack of visibility and subordination in the society where 'the model minority has become integrated, modernised, and civilised' [5]. Model minority stereotype promotes the myth of the US as a colourblind post-racial society and become a hegemonic discourse to silence Asian Americans from voicing their grievance and isolate them from standing in solidarity with other oppressed minority communities in striving for racial equality and social justice. As Frank Wu argues, 'Asian Americans have been excluded by the very terms used to conceptualise race. People speak of "Americans" as if it means "white", and "minority" as if it means "black.". In that semantic formula, Asian Americans, neither black nor white, consequently are neither American nor minority' [51].

A particular form of anti-Asian racism is what Trauner (1978) terms 'medical scapegoating', which describes how the authorities blamed the ethnic Chinese during the public health crisis in the late 19th Century to early 20th Century to cover up their health governance failure [48]. The 'medical scapegoating' strategies have been similarly adopted during the COVID-19 global pandemic, which spurred a wave of Sinophobic violence in the US and worldwide. Asian descent worldwide had been scapegoated for the public health crisis and suffered from heightened hostility, exacerbated stereotyping, verbal harassment, physical attacks, and online hatred ranging from blatant racist postings to racism memes [52], affecting both their safety and mental well-being [15]. Meanwhile, politicians and media played significant roles in activating and intensifying anti-Asian hatred [34]. In the US, 'rhetorical labelling of the virus as "Chinese virus," "Wuhan virus," and "Kung flu" by influential and prominent individuals sharing their racism openly via both traditional media and social media platforms to an audience of millions strongly suggest[ed] that the virus' harms and dangers emanate from a specific community and location' [27].

2.2 Social Media for Racial Justice: from #BLM to #StopAsianHate

Information Communication Technologies has forever revolutionised the way in which social movement is initiated, coordinated, organized, and escalated [13,25]. In particular, social networking services (SNSs) such as X/Twitter, play a key role in digital social movement for social justice in recent years. The "affordances of technology" of SNSs which includes searchability; replicability; scalability, the ability to capture and archive content; and the broad range of dissemination (boyd p. 7 in [7]) make them ideal tools for social movement. X/Twitter can mobilise and widely engage users 'by incorporating the use of hashtags, or conversation streams to label the meanings they express' [54].

SNSs played a crucial role during the Black Lives Matter (BLM) movement [6]. 'Black Twitter' convenes public discourse centering around the black identity construction and racial inequities felt by the American black community [19,45]. Through the use of Twitter for real-time engagement, users experience 'heightened temporality' that creates a sense of belonging and solidarity among Twitter users and movement participants, contributing to movement mobilisation and escalation [6]. Meanwhile, Yang contends that the temporal unfolding of the postings under a common hashtagged word or phrase develops a narrative form and agency, creating a powerful sociopolitical agenda for the BLM movement [53]. Worth noticing is also the use of emotions and sentiment in postings in building a following on social media. Keib et al. (2018) observe that tweets that express emotion, or tweets include content about policy or action or social actor are more often retweeted than neutral tweets [28]. Tweets contain all these components, i.e. emotion content about a policy or action or established group, were more often retweeted.

The anti-Asian hatred during the COVID-19 global pandemic reached its peak on 16th March 2021 when 21-year-old Robert Aaron Long carried out a shooting spree at two spas and a massage parlour in Atlanta, Georgia. Eight people were killed and a ninth was wounded and out of the eight death, six were women of Asian descent. The shooting triggered a wave of protest against anti-Asian crime and racism across the country. #StopAsianHate, #StopAAPI-Hate, and other hashtags trended online [43], calling law enforcement to tackle hate crime and advocating for social justice for Asian communities. Such hashtags help to raise awareness, break stereotypes, and challenge racial injustice and discrimination [35]. A study of 46,058 Tweets featuring #StopAsianHate and #StopAAPIHate reveals that the movement attracts more participation from women, younger adults, Asian and Black communities. About half of the Twitter users show direct support, but 5.43% show a negative attitude towards the movement. The Black and White communities blame each other for anti-Asian hate crimes [36]. In recent work, Wheeler et al. (2022) studied both negative (e.g.,#kungflu) and positive (e.g.,#StopAAPIHate) hashtags and keywords related to anti-Asian prejudice [50]. Through descriptive analyses, they observed differences in the frequency of negative and positive keywords based on geographic location [50]. Using burst detection, they identified distinct increases in negative and positive content in response to key political tweets and events [50].

2.3 Research Gaps

It is evident that by delivering real-time streaming communication, X/Twitter is instrumental in activating, mobilising, escalating social movement for social justice. However, there are still debates as to how this plays out in particular contexts, political practice, and activist organisations (see [10–12,41,42,46]). Comparing to #BLM, #StopAsianHate is a less 'impactful' digital social movement. We therefore aim to examine the assembly of technical system in the specific contexts which led to its differentiating forces and outcomes. We are interested in investigating the linguistic strategies and rhetorical devices used by #StopAsianHate activists on X/Twitter and consider in what ways, the sentiment, politeness, and toxicity features within #StopAsianHate Tweets facilitate or constrain the computational expression of the advocacy for social justice.

Methodologically, previous NLP studies examining #StopAsianHate sentiments often focused narrowly on a singular sentimental perspective [36,50], which can result in contextual blindness and oversimplification. This limitation is particularly significant in social movement datasets, where the presence of extreme lexicons such as hate and racism may lead to false positives. Additionally, these studies typically concentrate solely on statistical patterns, neglecting the contextualisation of these features or overlooking the deeper, specific patterns of interaction that are crucial for understanding social movements. To bridge such research gaps, this study integrates NLP and CACDA methods, takes a multidisciplinary approach and offers a comprehensive analysis that connects statistical data with broader discourse dynamics in the study of digital resistance to a particular form of cultural racism.

3 Methodology

Corpus-assisted Critical Discourse Analysis (CACDA) is an interdisciplinary methodology that combines qualitative Critical Discourse Analysis (CDA) methods and quantitative Corpus Linguistics (CL) methods [4]. CDA is viewed as a social movement, analysing discourse from a critical perspective to 'understand, expose, and ultimately challenge social inequality' [49]. CDA methods are predominantly qualitative, interpreting results within the broader social, cultural, and political contexts [4]. On the other hand, CL often involves quantitative analyses performed on large collections of machine-readable texts. It aims to provide methodologically rigorous analysis to elucidate underlying structures and patterns, and to highlight differences between texts, genres, and languages [38]. In CACDA studies, CL typically identifies areas of interest for closer analysis with CDA; meanwhile, findings from both approaches mutually triangulate and reinforce each other [20].

3.1 Data Collection and Corpus Construction

We collected data on #StopAsianHate, which was the most popular hashtag used to advocate racial equality and social justice for Asian communities on social media platforms [9,30,31,47]. Similar to previous research that collected Tweets [14,39,40], we utilised a keyword-based method to collect tweets that mention the #StopAsianHate term. Since the streaming API has data collection limitations [2], we used the X Search API [1] to retrieve and collect data from January 1st, 2021 to December 31st, 2022. This method returned 721,531 tweets. We employed a 'funnel approach' to explore the research topic across three dimensions (to be explained in the following subsection), each of which required distinct datasets. After outlining the general characteristics in the first dimension, the second dimension (i.e., sentimental landscapes) analysed the English-language tweets, yielding a subset of 424,346 tweets. We further isolated the 10% most popular tweets (42,435 in total) according to retweets and favourites to compare features between these and the general dataset. Finally, the third dimension (discourse practices) focused on these popular English-language tweets for a more detailed analysis. Each tweet was manually cleaned, and its multi-modal textual content was retained for analysis. They were then uploaded to Sketch Engine, an online platform for corpus construction and analysis [29]. The finalised corpus comprises 591,583 tokens.

3.2 Analytical Framework and Procedures

Taking Fairclough's (2013) [18] perspective that views language as both discourse and social practice, we employed a three-dimensional approach to analyse the #StopAsianHate tweets, starting from broader perspectives and narrowing down to specific linguistic details.

The first dimension focuses on dataset characteristics. Specific focus is given to mapping the distribution of languages and geographical locations represented in the #StopAsianHate tweets. In addition, we calculate how many tweets share images, videos, and links to YouTube videos and the level of engagement (in terms of retweets and favourites) gathered by the tweets.

In the second layer, we used Natural Language Processing (NLP) methods to examine the sentimental landscapes in the dataset, analysing sentiment, politeness, and toxicity scores. We use state-of-the art classifiers which have been widely adopted by previous research [3,39]. For sentiment study, we use the widely deployed SentimentIntensityAnalyzer python module, which generates a positive score to represent positive valence and a negative score to represent negative valence [26]. For politeness study, we use ConvoKit python module to calculate the probability that tweets are polite based on 'lexical and syntactic features operationalizing key components of politeness theory, such as indirection, deference, impersonalisation, and modality' [17]. Higher scores indicate polite tweets, while scores close to 0 indicate impolite discourse. To measure toxicity, we use the detoxify python module [22] that is also a widely used classifier to predict harmful content within discourse [21]. Scores close to 1 indicate toxic discourse, while scores close to 0 indicate non-toxic discourse. Moreover, our analysis uses He et al. (2021) classifier to determine the percentage of tweets that can be considered hateful, counter-speech, or neutral [24].

In the last section, we employed CACDA and narrowed our focus to the discourse practices of the 10% most popular English-language tweets. Utilising Sketch Engine [29], we began with Keyword Analysis to identify words that occur more frequently than expected by chance [37]. Keywords are words that occur statistically more frequently in the target corpus compared to another reference corpus [44]. Brookes and Collins (2023) points out that the reference corpus is typically equal to or larger in size than the target corpus and serves as a standard for the type of language under study [8]. For the present study, The reference corpus used for generating the keyword list was the English Web 2021 (enTenTen21), an English-language corpus of 52 billion words collected from the Internet. We then semantically categorised the top 20 keywords, using extended concordances (i.e., the context before and after each example) to delineate the discourses [4] present in the focused dataset.

Following the CDA tradition, which extends analysis beyond the text to encompass various contextual levels, we interpret the results in relation to the social and cultural contexts of the #StopAsianHate movement through the lens of cultural racism.

4 Findings

Our findings reveal dataset characteristics, sentimental landscapes, and an investigation of the linguistic features and discourses in the 10% most popular English-language tweets in the dataset. Overall, the #StopAsianHate movement is characterised with positivity, politeness, a moderate tone, few toxic expressions, and an avoidance of condemning perpetrators of anti-Asian racism.

4.1 Dataset Characteristics

To determine locations, we extract the location configured by users in their X/Twitter profiles. Since this information is self-reported, we cannot verify the authenticity of this information and only analyse instances where such information is provided. It is evident that the technological affordance of X/Twitter enables the global reach of #StopAsianHate movement, and the use of hashtags enabled the focused public discussion about racially instigated hatred towards Chinese and Asian communities worldwide. However, the density of tweets containing this hashtag is biased towards the US. The highest percentage of tweets (16%) are originated in the US while global cosmopolitan cities such as Los Angeles (1.2%), New York (1.1%), San Francisco (1%), Toronto (<1%), London (<1%) etc. that feature sizeable Asian communities are also the major places where such virtual forms of public debate take place (see Fig. 1).

As for the languages of the tweets, we use the language attribute provided by the API and find that most tweets are in English (59%). Other popular languages of the rest of the tweets include Asian languages (traditional Chinese–12%, Korean–2%, Thai–1%, Japanese–1%) and Latin languages (Spanish–6%, Portuguese–2%). The rest of the tweets have either an undetected language or a less popular language (<1%).

We calculated the number of tweets that contain an image, video, or link to a YouTube video. Interestingly, the most popular tweets are twice as much more likely to link to multi-modal content (*All tweets* with images: 6%; *Most popular tweets* with images: 15%; *All tweets* with videos: 0.7%; *Most popular tweets* with videos: 2.3%), which could be one of the reasons why these tweets are more popular. Surprisingly, a low proportion of tweets (less than 1%) share links to YouTube videos, which indicates that sharing YouTube video is not a

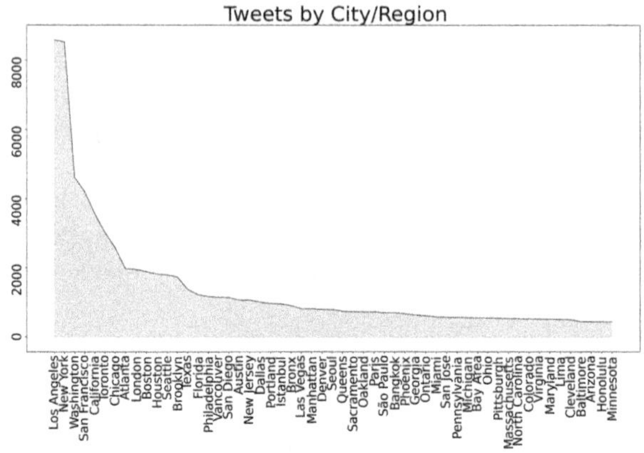

Fig. 1. Most popular cities/regions in our dataset (claimed by tweeters).

common practice within the Asian community. In this case there is no difference between *All tweets* (1%) and the *Most popular tweets* (0.9%).

Most tweets do not receive much engagement, with 77% of tweets not being retweeted and 58% of tweets not receiving any favourites. A considerable number of tweets receive moderate engagement, i.e., between 1 and 100 retweets or favourites (retweets: 22%, favourites: 41%), with only a very small percentage being widely diffused across the platform, i.e., receiving more than 100 retweets or favourites (retweets:0.63%, favourites: 1.42%). This finding shows that these tweets are receiving some traction within the platform, although they might not be receiving the widespread diffusion gathered by other social movements [45].

Moreover, 27% of the users that used the #StopAsianHate hashtag have more than 1,000 followers with 44% having between 100 and 1,000 followers. This finding shows that although these tweets do not seem to receive much traction, they might still have been viewed by a considerable audience. Most users (98%) only posted between 1 to 10 times with accounts dedicated to addressing the racism topic posting several times (RespondToRacism having 2,789 posts and ActToChange having 294). This finding suggests that, with a few exceptions, posting about #StopAsianHate was a sporadic collective action.

4.2 Sentimental Landscapes

We conduct this analysis using only English-language tweets, since the classifiers used are most effective with English corpora. All tweets refer to all English tweets (424,346), while *Most popular tweets* refer to English tweets that received most retweets and favourites (42,435). The classifier developed by He et al. (2021) [24] found that 99% of our dataset is counter-speech discourse.

Figure 2a shows the density distribution of the tweet sentiment from negative (−1), to positive (1), with neutral (0) in the middle (*All tweets*: mean = −0.27, standard deviation = 0.60; *Most popular tweets*: mean = −0.16, standard devia-

Table 1. Sentiment, Politeness, and Toxicity within tweets.

Tweet	Type
@XYZ Thank you Senator XYZ for standing up for the AAPI community and for fighting against hate. Please continue to spread the word in <location> (ps, thanks also for giving work opportunities to students in your office-awesome staff	Positive Sentiment
Hate needs to stop now Be kind to one another let's make the world a better place #StopAsianHate <link>	Neutral Sentiment
It's not just that Asian's are being attacked- it's elders! Fuck is wrong with people? #StopAsianHate <link>	Negative Sentiment
@XYZ @XYZ Great coverage of today's event and cute green dress thanks for sharing people's stories #StopAsianHate	Polite
I spoke with CNN to keep this important conversat going. #StopAsianHate #StopAAPIHate <link>	Moderately Polite
WOW WPs behaving badly. Really wild stuff downtown. A Stop Asian Hate rally is clashing with a Pro-Uighur drive by. The pro-Uighur group is shouting "F—China!" The Asian rally is responding by calling them "racist." <link>	Impolite
so it's #StopAsianHate. how about we make a change to that, just stop being a hate-filled fucker to others? ...no matter who they damn are. Sure I am an asshole and disagree with others but I at least do not hate nor disrespect ppl for petty shit such as skin color	Toxic
Learn more from @XY in an hour than a lifetime in the US school system #StopAsianHate <link>	Moderately Toxic
and it's not just her story that makes me upset, it's the THOUSANDS of people sharing their stories. #StopAsianHate	Non-toxic

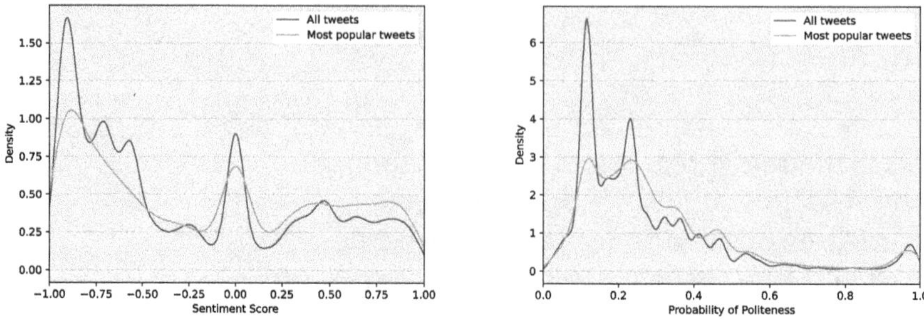

(a) (b)

Fig. 2. (a) Density distribution of Sentiment across our dataset. (b) Density distribution of Politeness across our dataset.

tion = 0.62). For both datasets the negative sentiment is the most prevalent (*All tweets*: 66.4%, *Most popular tweets*: 53.4%), which could be explained by the negative terms, such as "hate", "racist", and "protest" that are generally used in such discourse (see row 3 in Table 1). There are still a considerable number of tweets that use positive terms (*All tweets*: 24.3%, *Most popular tweets*: 35.8%), meaning that despite the negative terms inherent to this topic people still find ways to spread positive messages to their followers (see row 1 in Table 1). The rest of the tweets are neutral (*All tweets*: 9.3%, *Most popular tweets*: 10.8%). Refer to Table 1 for examples of positive, neutral, and negative tweets. When comparing the two datasets, we find that the *Most popular tweets* have a significantly higher percentage of positive tweets. This finding could indicate that a positive sentiment might lead to more engagement, but also people/organisations with more followers might be more careful in crafting positive messages. Building on Wheeler et al.'s (2022) analysis of #StopAsianHate dataset [50], our study addresses an important research gap, which has been previously unexplored, by delving deeper into the levels of politeness and toxicity within these tweets.

Figure 2b shows the density distribution of the probabilities of the tweets being polite (*All tweets*: mean = 0.27, standard deviation = 0.21; *Most popular tweets*: mean = 0.31, standard deviation = 0.21). Most tweets are scored as impolite because they were classified with a score less than 0.4 (see row 6 in Table 1). This finding confirms the previously described finding that most tweets contain a negative sentiment. However, a good number of tweets contain some polite discourse (between 0.4 and 0.6) and only very few tweets are polite (scores more than 0.8). Refer to Table 1 for examples of impolite, neutral, and polite tweets. Our in-depth analysis of which terms are used by polite and impolite tweets revealed that the most common terms used in impolite tweets are: "discrimination", "racial", "racist", "unprofessional", "hate", "condemn", "stand", "violence", "stop", "apologize". Some of these terms are also common in neutral tweets, which in addition contained terms such as "support", "together", "respect", "community", "live", "represent", "comment". Polite tweets consisted of the usual terms such as "racism", "hate", "stop", "condemn", but also have a prevalence of polite and positive terms, mainly: "thank", "together", "respected", "support", "voice", "proud". Moreover, although the distributions show similar patterns for *All tweets* and *Most popular tweets*, Fig. 2b shows that *Most popular tweets* contain slightly more polite discourse. This finding could be explained by these tweets being written by professionals who manage accounts on social media platforms for organisations and public figures.

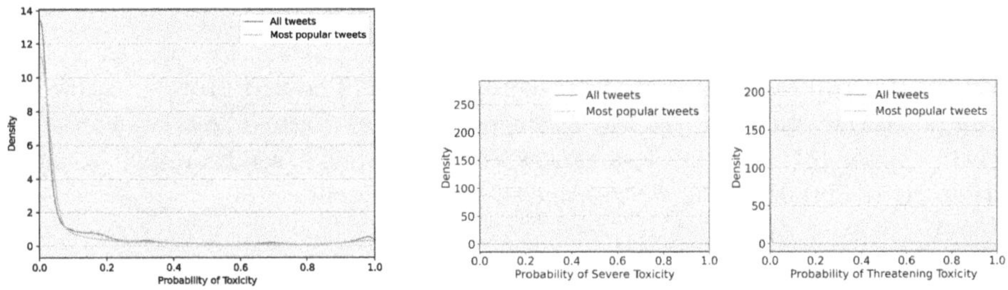

(a) (b)

Fig. 3. (a) Density distribution of Toxicity across our dataset. (b) Density distribution of Insulting, Obscene, Severe, and Threatening Toxicity.

Figure 3a shows the density distribution of the probabilities of the tweets being toxic, meaning having harmful terms (*All tweets*: mean = 0.12, standard deviation = 0.25; *Most popular tweets*: mean = 0.09, standard deviation = 0.22). Interestingly, we find that most tweets are not toxic. Despite most tweets being negative and impolite, further analysis (see Fig. 3b) shows that in general the discourse around #StopAsianHate does not contain insulting, obscene, severe, or threatening terminology. Our in-depth analysis of which terms are used in toxic tweets revealed that profane terms are common in these tweets, e.g., "f**k", "f**king", "*ss", "sh*t", "wtf". This finding shows that the discourse around this topic is mostly moderate and does not get out of control and radicalised. Refer to Table 1 for examples of toxic, moderate, and non-toxic tweets.

4.3 Textual Level

The textual level analysis focuses on the discourses in the 10% most popular English-language tweets in our dataset. Table 2 displays the top 20 keywords. Based on the keyword and concordance analyses, the top 10% most popular English-language tweets can be interpreted from four perspectives: (1) counter-discourse against anti-Asian hate, (2) self-positioning and attack dynamics, (3) tweet dissemination, and (4) expressions of attitudes and sentiments.

Table 2. Top 20 keywords organised by semantic group

No.	Top 20 Keywords (Lemma)
(1)	(1)anti-asian, (2)aapi, (3)asians, (4)asian, (7/13)asian-american(s), (18)islander; (5)racism, (9)xenophobia, (11)hate; (12)xiaojie, (16)daoyou, (17)delaina, (20)yaun;
(2)	(10)supremacy;
(3)	(6)carrd; (8)bts, (15)rt, (19)tacha;
(4)	(14)solidariy;

Counter-Discourse Against Anti-Asian Hate. The first group of keywords is related to the counter-discourse against anti-Asian hate. These keywords frequently appear in the form of slogans within the tweets (see Example 1), articulating the demands of the #StopAsianHate movement.

Example 1:
Racism is Not an opinion.
Stand against racism.
We condemn racism and xenophobia in all forms.

Asians, particularly Asian Americans, are positioned as central social actors within the target corpus, as evidenced by the top keywords like *anti-Asian*, *AAPI*, *Asian(s)*, and *Asian-American(s)*. Notably, keywords such as *AAPI* and *Asian-American(s)* indicate a strong connection to the US context in these most popular English-language tweets.

The associated concordances highlight a strong sense of community and solidarity. *AAPI community* appeared 831 times in the dataset, with the singular form being much more prevalent than the plural, suggesting a unified group solidarity. The keyword *solidarity* was prominent, appearing 930 times, with *support*, *love*, *stand in solidarity*, and *rally* frequently collocating with it. Additionally, the sense of community is reinforced with relational identification strategies [33] in portraying these Asian social actors. They are referred to as *friends*, *folks*, *family*, and *groups*, further emphasising collective in-group identity and inclusionary out-group solidarity. In Example 2, the tweeter employs the pronoun 'we' to refer to the social group advocating #StopAsianHate and promoting unity and love, and encourages the readers (i.e., 'you') to join this community by appealing to the intimacy of being their 'friends, family, and fellow human beings'.

Example 2:
#StopAsianHate it's really as simple as that. Ive said it before its much easier to love one another and focus on what brings ***us*** *together instead of focusing on what divides us!!* ***We*** *are your* ***friends, family, and fellow human beings.*** *Spread love not hate*

The second focus within Group 1 is racism and hate crimes, highlighted by the keywords *racism*, *hate*, and *xenophobia*. A notable incident emerged from the keywords is the Atlanta Shooting on March 16th, 2021, where six of the eight victims were women of Asian descent, classifying the event as a hate crime. Many among the top 10% most popular tweets commemorate the victims (e.g., *Xiaojie, Daoyu, Delaina, & Yaun*).

Self-positioning. A bottom-up self-positioning of Asian people is prominent in the discourse, condemning long-standing neglect of anti-Asian racism and portraying an unequal power relationship between Asians and racists. In the concordances, Asian individuals are frequently depicted as the patient of actions (i.e., social actors being the recipient of/affected by the actions), with 848 instances of the phrase *against Asians* (see Example 3). In terms of how racists are represented, notable keywords include *supremacy* (ranked 10th), appearing 447 times predominantly in *white supremacy*, stating the subordinating position of Asian in American society.

> ***Example 3:***
> *He was one of the victims of violence against Asian people, especially the elderly ones.*

The tweet in Example 4 further highlights this long-standing power inequality in American society. When eight people, including six Asian women, were killed, the sheriff's comment that the shooter had 'a really bad day' demonstrates the minimisation and deflection of racial crimes. This is summarised by the tweeter as 'what white supremacy looks like', demonstrating how hate crimes against Asians are often downplayed due to bigotry.

> ***Example 4:***
> *The Cherokee County sheriff said the Atlanta spa shooter killed 8 people, 6 of them Asian women, because "it was a really bad day for him," and that's it's not about race. This is what* ***white supremacy*** *looks like. #StopAsianHate #StopAAPIHate*

Dissemination. The third keyword group sheds light on the dissemination and promotion of these tweets. This can be found through providing *carrds* (ranked 6th) providing further information, sources, and links, along with requests for retweets (*RT & retweet*) and adding *hashtags*.

Celebrity endorsements emerge among the keywords, notably with K-pop boy band *BTS* (ranked 8th). The concordances of BTS reveal expressions of support and acknowledgement for BTS's anti-racism speech at the White House (see Example 5). Additionally, *Tacha* (19th), *Anita Natacha Akide*, a prominent social media Key Opinion Leader (KOL) who expressed her support for the #StopAsianHate movement, also emerged as a keyword.

> ***Example 5:*** *....it's a reality now BTS PAVED THE WAY YOU HAVE OUR BACK SO PROUD OF YOU BTS BTS IN WHITE HOUSE BTS SOUTH KOREA'S PRIDE ...*

For the dissemination discourse, we further went down the keyword list to investigate potential associations with previous movements. We discovered that *BLM* (Black Lives Matter) (101st) appeared in 50 instances, which is not very frequent considering the size of our dataset. This suggests a relatively weak connection between BLM and the most popular #StopAsianHate tweets.

Expressions of Attitudes/Sentiments. The three aforementioned keyword groups cover most of the top 20 keywords, with *solidarity* notably expressing support towards #AntiAsianHate movement and condemnation of anti-Asian crimes. Extending our analysis to the top 50 keywords, we categorised them into two groups. The first group includes responses to hate crimes, such as *heart-broken* (ranked 31st), *heartbreaking* (45th), while the second group describes the crimes or the perpetrators using terms like *senseless* (37), *horrific* (41st), *disgusting* (42nd), and *sickening* (44th). However, an interesting pattern emerges when examining the concordances: the majority of these descriptors are applied with the abstraction strategy (i.e., a detachment from the agent of social actions), such as *shootings*, *acts*, *killings*, and *violence* (see Example 6), rather than to specific individuals such as killers, racists, perpetrators, or men. This echoes the linguistic trends observed in the first keyword group, where abstracted slogans rarely criticise or condemn individuals directly but the racist actions and behaviours. This prevalent use of abstraction suggests a deliberate linguistic practice to avoid direct attacks on individual racists, mirroring the patterns identified in the second layer of analysis.

Example 6:
.. the racism and violence against asians (especially recently) is disgusting and i hate that i live in constant fear. please let us be heard

5 Discussion

Through an examination of the #StopAsianHate tweets from January 1, 2021, to December 31, 2022 on X/Twitter, this study consists of three analytical dimensions and employs a 'funnel approach', starting with a broad examination of the #StopAsianHate movement and progressively delving into the specific sentimental and linguistic dimensions within the top 10% most popular tweets. The analysis shows #StopAsianHate to be predominantly used in the US, with English being the mostly employed language. This suggests this movement is originated from English-speaking countries and targeting at a broader, English-speaking global audience to raise awareness and rally support. Overall, #StopAsianHate is predominantly used as counter-discourse.

A key insight from the study of #StopAsianHate discourse is how 'model minority' is represented as both a trope and a myth within the Asian community on X/Twitter. This conclusion is derived from combining NLP's sentiment, politeness, and toxicity analyses with Corpus-Assisted Critical Discourse Analysis results. Firstly, the movement appears moderate from a sentimental perspective, as evidenced by many tweets receiving positive sentiment scores, highlighted by expressions of solidarity at the textual level. The popular tweets frequently use relational identification strategy, referring to those countering racism as 'friends', 'folks', and 'family'. Although the politeness analysis indicates that many tweets may not seem polite (likely due to the brevity of tweet texts, which typically feature fewer polite lexicons than longer communication genres) and the qualitative analysis shows that tweets with negative sentiment scores often contain theme words like hate, racism, and crime, these are not harmful tweets. Instead, these tweets directly challenge racism and hate crime, which is evidenced by the extreme low probability of toxicity.

A significant linguistic feature further supporting the 'model minority' conclusion is the avoidance of direct condemnation or confrontation against individual racists. Even when negative attitudinal lexicons like 'disgusting' and 'senseless' are used, as shown in concordance examples, they target events or actions such as shootings and killings rather than the perpetrators. This prevalent use of abstraction suggests a deliberate linguistic strategy to focus on challenging the institutional racism (in this case, law enforcement agencies' lack of understanding and prosecution of anti-Asian hate crime) while avoiding direct conflicts with individuals, further revealing the movement's moderate tone.

Lastly, the discourse reveals #StopAsianHate activists' self-positionings as lacking influential power and as being vulnerable in the movement. This is evident in the frequent use of keywords such as (white) supremacy and descriptions of Asians as passive recipients of actions. Such self-positionings expose the unequal power dynamics within the discourse, while indicating a growing self-reflecting collective awareness of the negative impacts of model minority stereotypes on Asian communities.

6 Conclusion

In conclusion, this paper takes an innovative approach by integrating NLP and CACDA methodologies in the study of X/Twitter activism via a case study of #StopAsianHate tweets during the COVID-19 global pandemic. NLP techniques identify statistical patterns related to the emotional and social appropriateness of tweets, a capability where CACDA methods are comparatively less effective. Conversely, CACDA provides a three-dimensional framework that incorporates textual features, discursive patterns, and social practices, thereby enriching NLP findings by situating them within the social context and interpreting them through linguistic features and strategies. Findings suggest that the moderate tone, communal sentiment and the vulnerability self-positionings reveal Asian communities' general subordinate position in the political power structure, and their lack of visibility and experience in activism. However, data at the same time provides fresh insights into the growing self-reflective collective awareness within the Asian communities of the negative impacts of 'model minority' stereotypes. The use of #StopAsianHate as a response to Atlanta shooting hate crime demonstrates the growing knowledge and skill sets of digital activism within the Asian community. Celebrity endorsement contributes to the global reach, dissemination and solidarity of the movement, giving momentum to Asian communities' ongoing efforts in fighting racism in all forms.

Acknowledgements. This work was supported by the Morgan Advanced Studies Institute, Swansea University under Grant number MID1001-104.

References

1. Standard search API. https://developer.twitter.com/en/docs/twitter-api/v1/tweets/search/api-reference/get-search-tweets
2. Understanding X limits—X Help. https://help.twitter.com/en/rules-and-policies/x-limits
3. Ajao, O., Bhowmik, D., Zargari, S.: Sentiment aware fake news detection on online social networks. In: 2019 IEEE International Conference on Acoustics, Speech and Signal Processing (ICASSP), pp. 2507–2511, May 2019. https://doi.org/10.1109/ICASSP.2019.8683170
4. Baker, P., Gabrielatos, C., KhosraviNik, M., Krzyżanowski, M., McEnery, T., Wodak, R.: A useful methodological synergy? Combining critical discourse analysis and corpus linguistics to examine discourses of refugees and asylum seekers in the UK press. Discourse Soc. **19**(3), 273–306 (2008). https://doi.org/10.1177/0957926508088962
5. Bascara, V.: Model-Minority Imperialism. U of Minnesota Press
6. Bonilla, Y., Rosa, J.: #Ferguson: digital protest, hashtag ethnography, and the racial politics of social media in the United States: #Ferguson. Am. Ethnol. **42**(1), 4–17 (2015). https://doi.org/10.1111/amet.12112
7. Boyd, D.: Social Network Sites as Networked Publics: Affordances, Dynamics, and Implications (2010)
8. Brookes, G., Collins, L.C.: Corpus Linguistics for Health Communication: A Guide for Research. Taylor & Francis, New York (2023)
9. Cao, J., Lee, C., Sun, W., De Gagne, J.C.: The #StopAsianHate movement on Twitter: a qualitative descriptive study. Int. J. Environ. Res. Public Health **19**(7), 3757 (2022). https://doi.org/10.3390/ijerph19073757
10. Castells, M.: The Rise of the Network Society: The Information Age: Economy, Society, and Culture. Wiley, New York (1996)
11. Castells, M.: The Network Society: A Cross-Cultural Perspective. Edward Elgar Publishing, Incorporated, October 2004
12. Castells, M.: Communication Power, 2nd edn. Oxford University Press, Oxford (2013)
13. Castells, M.: Networks of Outrage and Hope: Social Movements in the Internet Age. Wiley, New York (2015)
14. Chen, E., Lerman, K., Ferrara, E.: Tracking social media discourse about the Covid-19 pandemic: development of a public coronavirus twitter data set. JMIR Public Health Surveill. **6**(2), e19273 (2020). https://publichealth.jmir.org/2020/2/e19273
15. Cheng, H.L.: Xenophobia and racism against Asian Americans during the COVID-19 pandemic: mental health implications. J. Interdisc. Perspect. Scholarsh. **3**(1), 3 (2020)

16. Chou, R.S., Feagin, J.R.: Myth of the Model Minority: Asian Americans Facing Racism, 2nd edn. Routledge, New York (2015). https://doi.org/10.4324/9781315636313
17. Danescu-Niculescu-Mizil, C., Sudhof, M., Jurafsky, D., Leskovec, J., Potts, C.: A Computational Approach to Politeness with Application to Social Factors, June 2013. https://doi.org/10.48550/arXiv.1306.6078
18. Fairclough, N.: Critical discourse analysis. In: The Routledge Handbook of Discourse Analysis, pp. 9–20. Routledge (2013)
19. Florini, S.: Tweets, Tweeps, and Signifyin': communication and cultural performance on "Black Twitter." Telev. New Media **15**(3), 223–237 (2014). https://doi.org/10.1177/1527476413480247
20. Gabrielatos, C., Baker, P.: Fleeing, sneaking, flooding: a corpus analysis of discursive constructions of refugees and asylum seekers in the UK Press, 1996–2005. J. English Linguist. **36**(1), 5–38 (2008). https://doi.org/10.1177/0075424207311247
21. Hanu, L., Thewlis, J., Haco, S.: How AI Is Learning to Identify Toxic Online Content. https://www.scientificamerican.com/article/can-ai-identify-toxic-online-content/
22. Hanu, L., Unitary, T.: Detoxify, November 2020. https://doi.org/10.5281/zenodo.7925667, https://github.com/unitaryai/detoxify
23. Hartlep, N.D.: The Model Minority Stereotype: Demystifying Asian American Success, 2nd edn. IAP, April 2021
24. He, B., Ziems, C., Soni, S., Ramakrishnan, N., Yang, D., Kumar, S.: Racism is a virus: anti-asian hate and counterspeech in social media during the COVID-19 crisis. In: Proceedings of the 2021 IEEE/ACM International Conference on Advances in Social Networks Analysis and Mining, ASONAM 2021, pp. 90–94. Association for Computing Machinery, New York, NY, USA, January 2022. https://doi.org/10.1145/3487351.3488324
25. Hill, S.: Digital revolutions: activism in the internet age. New Internationalist (2013)
26. Hutto, C., Gilbert, E.: VADER: a parsimonious rule-based model for sentiment analysis of social media text. Proc. Int. AAAI Conf. Web Soc. Media **8**(1), 216–225 (2014). https://doi.org/10.1609/icwsm.v8i1.14550
27. Ittefaq, M., Abwao, M., Baines, A., Belmas, G., Kamboh, S.A., Figueroa, E.J.: A pandemic of hate: social representations of COVID-19 in the media. Anal. Soc. Issues Public Policy **22**(1), 225–252 (2022). https://doi.org/10.1111/asap.12300
28. Keib, K., Himelboim, I., Han, J.Y.: Important tweets matter: predicting retweets in the #BlackLivesMatter talk on twitter. Comput. Hum. Behav. **85**, 106–115 (2018). https://doi.org/10.1016/j.chb.2018.03.025
29. Kilgarriff, A., et al.: The Sketch Engine: ten years on. Lexicography **1**(1), 7–36 (2014). https://doi.org/10.1007/s40607-014-0009-9
30. Lee, C.S., Jang, A.: Questing for justice on Twitter: topic modeling of #StopAsianHate discourses in the wake of Atlanta shooting. Crime Delinquency **69**(13–14), 2874–2900 (2023). https://doi.org/10.1177/00111287211057855
31. Lee, J.J., Lee, J.: #StopAsianHate on TikTok: Asian/American women's space-making for spearheading counter-narratives and forming an Ad Hoc Asian Community. Soc. Media + Soc. **9**(1), 205630512311575 (2023). https://doi.org/10.1177/20563051231157598
32. Lee, S.J.: Unraveling the "Model Minority" Stereotype: Listening to Asian American Youth, 2nd edn. Teachers College Press (2015)

33. van Leeuwen, T.: Discourse and Practice: New Tools for Critical Analysis. Oxford University Press, Oxford (2008). https://doi.org/10.1093/acprof:oso/9780195323306.001.0001
34. Li, Y., Nicholson Jr., H.L.: When "model minorities" become "yellow peril"–othering and the racialization of Asian Americans in the COVID-19 pandemic. Sociol. Compass **15**(2), e12849 (2021). https://doi.org/10.1111/soc4.12849
35. Lloret-Pineda, A., He, Y., Haro, J.M., Cristóbal-Narváez, P.: Types of racism and Twitter users' responses amid the COVID-19 outbreak: content analysis. JMIR Formative Res. **6**(5), e29183 (2022). https://formative.jmir.org/2022/5/e29183/
36. Lyu, H., Fan, Y., Xiong, Z., Komisarchik, M., Luo, J.: State-level racially motivated hate crimes contrast public opinion on the #StopAsianHate and #StopAAPIHate movement, April 2021. https://doi.org/10.48550/arXiv.2104.14536
37. McEnery, T.: Swearing in English: Bad Language, Purity and Power from 1586 to the Present. Routledge, London (2005). https://doi.org/10.4324/9780203501443
38. McEnery, T.: Corpus Linguistics. Edinburgh University Press (2019)
39. Micallef, N., He, B., Kumar, S., Ahamad, M., Memon, N.: The role of the crowd in countering misinformation: a case study of the COVID-19 infodemic. In: 2020 IEEE International Conference on Big Data (Big Data), pp. 748–757. IEEE (2020). https://ieeexplore.ieee.org/abstract/document/9377956/
40. Micallef, N., Sandoval-Castañeda, M., Cohen, A., Ahamad, M., Kumar, S., Memon, N.: Cross-platform multimodal misinformation: taxonomy, characteristics and detection for textual posts and videos. Proc. Int. AAAI Conf. Web Soc. Media **16**, 651–662 (2022). https://doi.org/10.1609/icwsm.v16i1.19323
41. Morozov, E.: To save everything, click here: the folly of technological solutionism. PublicAffairs (2013)
42. Murthy, D.: Twitter: social communication in the Twitter age. Digital media and society, Polity, Cambridge (2013)
43. Press, A.: #StopAsianHate trends as Asian-Americans grieve after Atlanta attack, March 2021. https://www.scmp.com/news/world/united-states-canada/article/3126040/stopasianhate-trends-online-asian-americans-grieve
44. Scott, M.: PC analysis of key words–and key key words. System **25**(2), 233–245 (1997). https://doi.org/10.1016/S0346-251X(97)00011-0
45. Sharma, S.: Black Twitter? Racial hashtags, networks and contagion. New Formations **78**(78), 46–64 (2013)
46. Tilly, C.: Social Movements, 1768–2004. Routledge, New York (2019). https://doi.org/10.4324/9781315632063
47. Tong, X., Li, Y., Li, J., Bei, R., Zhang, L.: What are people talking about in #BackLivesMatter and #StopAsianHate? Exploring and categorizing Twitter topics emerged in online social movements through the Latent Dirichlet Allocation Model. In: Proceedings of the 2022 AAAI/ACM Conference on AI, Ethics, and Society, AIES 2022, pp. 723–738. Association for Computing Machinery, New York, NY, USA (2022). https://doi.org/10.1145/3514094.3534202
48. Trauner, J.B.: The Chinese as medical scapegoats in San Francisco, 1870–1905. Calif. Hist. **57**(1), 70–87 (1978). https://doi.org/10.2307/25157817
49. Van Dijk, T.A.: Critical discourse analysis. In: The Handbook of Discourse Analysis, pp. 466–485. Wiley (2015). https://doi.org/10.1002/9781118584194.ch22
50. Wheeler, B., Jung, S., Barioni, M.C.N., Purohit, M., Hall, D.L., Silva, Y.N.: #WashTheHate: understanding the prevalence of anti-Asian prejudice on Twitter during the COVID-19 pandemic. In: 2022 IEEE/ACM International Conference on Advances in Social Networks Analysis and Mining (ASONAM), pp. 484–491, November 2022. https://doi.org/10.1109/ASONAM55673.2022.10068578

51. Wu, F.H.: Yellow: Race in America Beyond Black and White. Basic Books (2002)
52. Wu, Y., Wall, M.: COVID-19 and viral anti-Asian racism: a multimodal critical discourse analysis of memes and the racialization of the COVID-19 pandemic. J. Contemp. Chin. Art **8**(2), 107–127 (2021). https://doi.org/10.1386/jcca_00040_1
53. Yang, G.: Narrative agency in hashtag activism: the case of #BlackLivesMatter. Media Commun. **4**(4), 13–17 (2016). https://doi.org/10.17645/mac.v4i4.692
54. Zappavigna, M.: Discourse of Twitter and Social Media: How We Use Language to Create Affiliation on the Web. Continuum International Publishing Group, London (2012)

Evaluating LLMs Capabilities Towards Understanding Social Dynamics

Anique Tahir[1], Lu Cheng[2], Manuel Sandoval[3](✉), Yasin N. Silva[3], Deborah L. Hall[4], and Huan Liu[1]

[1] School of Computing and AI, Arizona State University, Tempe, AZ, USA
{artahir,huanliu}@asu.edu

[2] University of Illinois Chicago, Chicago, IL, USA
lucheng@uic.edu

[3] Loyola University Chicago, Chicago, IL, USA
{msandovalmadriga,ysilva1}@luc.edu

[4] School of Social and Behavioral Sciences, Arizona State University, Glendale, AZ, USA
d.hall@asu.edu

Abstract. Social media discourse involves people from different backgrounds, beliefs, and motives. Thus, often such discourse can devolve into toxic interactions. Generative Models, such as Llama and ChatGPT, have recently exploded in popularity due to their capabilities in zero-shot question-answering. Because these models are increasingly being used to ask questions of social significance, a crucial research question is whether they can understand social media dynamics. This work provides a critical analysis regarding generative LLM's ability to understand language and dynamics in social contexts, particularly considering cyberbullying and anti-cyberbullying (posts aimed at reducing cyberbullying) interactions. Specifically, we compare and contrast the capabilities of different large language models (LLMs) to understand three key aspects of social dynamics: language, directionality, and the occurrence of bullying/anti-bullying messages. We found that while fine-tuned LLMs exhibit promising results in some social media understanding tasks (understanding directionality), they presented mixed results in others (proper paraphrasing and bullying/anti-bullying detection). We also found that fine-tuning and prompt engineering mechanisms can have positive effects in some tasks. We believe that a understanding of LLM's capabilities is crucial to design future models that can be effectively used in social applications.

Keywords: large language models · social media · anti-bullying

1 Introduction

In today's digitally connected world, social media platforms have become integral arenas for communication, allowing individuals from diverse backgrounds to engage in discussions, share ideas, and express their opinions. However, this increased connectivity has brought with it a range of challenges, including the emergence of toxic online behavior, cyberbullying, and the propagation of harmful content [2]. As the online landscape becomes more complex and dynamic, the need to understand and address these issues at scale becomes ever more pressing.

L. M. Aiello et al. (Eds.): ASONAM 2024, LNCS 15213, pp. 230–244, 2025.
https://doi.org/10.1007/978-3-031-78548-1_18

While previous research has made significant strides in specific social interaction settings such as identifying and combating online toxicity and bullying [21], a notable gap remains in understanding the positive forces at play within digital conversations. Generative models such as Llama and ChatGPT are increasingly being viewed as a panacea for arbitrary problems. However, employing them in social settings can have detrimental consequences. Thus, this paper aims to investigate the extent to which language models can classify and provide explanations for their decisions when social factors are involved. Since the area of social dynamics is diverse, we use bullying and anti-bullying as exemplars for our analysis. We define anti-bullying as interactions where online mediators intervene in toxic discussions with the goal of counteracting cyberbullying behaviors. In essence, these mediators are engaging in bystander intervention [39]. Our analysis seeks to pave the way toward providing interpretable insights into online discourse patterns from the perspective of generative language models, further contributing to the comprehension of human interaction in the digital realm.

Recently, large language models (LLMs) have gained traction for their general problem-solving capabilities. LLMs, such as ChatGPT, have been claimed to show signs of intelligence [4] and outperform human crowdsourcing [17]. In contrast, several efforts have studied the limitations of LLMs, such as their tendency to hallucinate [43], regurgitate obsolete information which is attributed to the staleness of their training data, and to have racial, gender, and religious biases [35].

In the context of social dynamics, LLMs can potentially provide a pathway to identify and explain behaviors on a large scale. In addition, they could also explain the rationale behind social discourse and, in turn, enable the promotion of positive interactions. This work aims to identify current strengths and limitations of LLMs to identify and explain social behaviors. Our evaluation framework considers multiple enhancement techniques and aims to answer the following research questions:

- RQ1: Do LLMs understand language in the social context?
- RQ2: Which evaluation dimensions expose the weaknesses of LLMs in a social analysis setting?
- RQ3: Can LLMs understand directionality in the social context?
- RQ4: Can LLMs identify behaviors involved in social dynamics, such as cyberbullying and anti-bullying?

2 Related Work

There is considerable prior research that studies social dynamics [33]. While this paper aims to evaluate general LLMs capabilities in social contexts, we use cyberbullying and anti-bullying as case studies in our analysis.

2.1 LLMs in Social Media Analysis

LLMs have been extensively used in prior work related to social media analysis. Transformer based models [38], such as BERT [10], encode words into tokens and learn attention weights signifying relationship between sequences. Although these models were originally developed for machine translation, they have been repurposed for classification tasks.

Sentiment Analysis and Emotion Detection. One prominent application of LLMs in social media analysis is sentiment analysis, which involves determining the emotional tone behind a piece of text. Prior researchers have utilized models like BERT and GPT-2 to gauge sentiment polarity in tweets [1,6]. Emotion detection has also been explored, with models being fine-tuned to recognize emotions such as joy, anger, and sadness [12, 13].

Irony and Sarcasm Detection. Identifying irony and sarcasm presents a unique challenge in online discourse due to the absence of vocal cues and facial expressions. LLMs, including RoBERTa, have been leveraged to detect instances of irony and sarcasm in tweets [27,44]. These models employ contextual understanding to differentiate between literal and figurative language, contributing to a deeper comprehension of social media interactions.

Hate Speech and Offensive Language Detection. The issue of hate speech and offensive language on social media platforms has spurred efforts to develop automated tools for detection and mitigation. Researchers have employed LLMs to create models capable of identifying hate speech [3,24] and offensive content [9,32,42]. The contextual awareness of LLMs aids in distinguishing between genuine expressions of opinion and harmful speech.

2.2 Improving Generative Language Models: Fine-Tuning vs. Prompt Engineering

For a few years now, standard practice has been to pretrain a model using abundantly available textual data, and then fine-tune the model to perform a more specific task using a new task-specific objective function and/or task-specific dataset. This approach gives the model an opportunity to learn the various general-purpose features of the pretraining data, and avoid biases that could result from being trained on a small dataset. In practice, the task-specific dataset can be difficult to collect, in short supply, or expensive to produce. Prompt engineering addresses the need for these datasets by having the model instead learn from prompts, a text template engineered specifically to get a better response from the language model. The drawback to prompt engineering is the complexity of identifying how to correctly structure the templates for each task the model will perform [22].

The expanding repertoire of LLM applications in social media analysis underscores their potential in unraveling the nuanced behaviors on these platforms. We aim to shed light on LLMs' strengths and areas in need of additional work to effectively understand intricate social dynamics.

3 Prompt Generation for Detecting and Explaining User Behavior

3.1 Background

Many conversations in social media revolve around certain topics of interest. For instance, information on Reddit is structured in a hierarchical format where sub-reddits are centered around a large scale topic and each thread in a subreddit is a subset of the larger topic. The comments in a thread are centered around the thread topic. Similarly, X (formerly Twitter) uses hashtags to convey several topics of interest. Posters on X can reply to a comment, creating a hierarchical discourse structure. Instagram and 4chan use similar discourse structures. We refer to the combination of a topic and its responses as a session. Prior work in social media analysis has extensively studied the extraction of useful information from social media sessions, ranging from fake news detection [8, 11, 23] to echo chamber formation [5, 14, 26].

Given our interest in finding the capabilities of LLMs to detect and explain cyberbullying and anti-bullying behavior, we formalize next our problem setting. Given a dataset of social media discourse containing text-based conversations, posts, comments, or interactions from various social media platforms, a goal of our work is to analyze whether it is feasible to develop an LLM-based framework for identifying behavior within these online discussions. We define cyberbullies as individuals who demonstrate intentionally hostile or toxic behavior directed toward other users. In contrast, we define anti-bullies as individuals who, in response to cyberbullying, actively seek to counter or combat the harmful content, e.g., by steering conversations in a positive direction or fostering constructive interactions. The problem can be formally stated as follows:

Input. The input dataset $\mathcal{D} = \{S_1, S_2, \cdots, S_N\}$ consists of N instances of social media sessions, where each instance $d_{ij} \in \mathcal{S}_i$ is as a sequence of text tokens. Each instance d_i includes the j'th comment of session S_i and the previous 1 to $j-1$ comments. This representation allows the model to rely on the context of a session.

Output. The objective is to produce the following output components:

1. **Behavior Identification (Quantitative):** For each instance d_{ij}, let $b_{ij} \in \{0, 1\}$ represent a binary classification of the behavior as a cyberbully. Similarly, we let $m_{ij} \in \{0, 1\}$ represent a binary classification of behavior as anti-bullying.
2. **Behavior and Motivation Explanations (Qualitative):** For each identified cyberbullying and anti-bullying comment, $b_{ij}, m_{ij} \in \mathcal{S}_i$ respectively, generate human-interpretable explanations that shed light on the behavior, actions, and motivations behind the comment.

Generative LLMs have demonstrated general purpose capabilities in question answering, surpassing humans in some cases [15]. Several methods focused on enhancing the output of these models have been proposed. For our task of social behavior analysis, we created a taxonomy of methods for enhancing responses. The most common approach for querying foundational LLMs is to use zero-shot prompting. However, because these models are trained on general data, they may fail to perform niche tasks. In addition, because natural language is ambiguous, the models may respond in a manner that is not expected for a specific use-case. For instance, a model trained as a 'comedy bot' might respond in a humorous style, whereas such a style would not be relevant across a broader range of applications.

There are several approaches to guide prompting for enhancing generation. The most common ones are chain-of-thought (CoT) prompting [41] and exemplar-based prompting [25]. In chain-of-thought prompting, the prompt is constructed such that the response consists of the rationale for the decision before responding with the final decision in the generation. Exemplar-based generation provides the model with example generations, in essence guiding the model to continue generations in similar fashion.

3.2 Prompt Generation

Because social media sessions have a specific structure, prompt design needs to take four major factors into consideration: (i) The task objective, i.e., identifying and explaining behavior; (ii) context, i.e., what is the topic of the subset of the conversation we are analysing; (iii) the classification instance d_{ij}, i.e., how can we differentiate the query objective from the context; and (iv) the response format, i.e., how do we want the LLM to structure its generated response.

Given that a social media conversation is sequential in nature, we use the tabular format such that various characteristics of the conversation are encoded alongside the comments, e.g., the post id, comment author, and a label specifying which comment is the target of classification. In addition, the instruct-tuning prompt format [36] allows the instruction and input to be structured together in a prompt query. Because we are interested in classification, we expect the LLM to return the output in a structured manner such that class labels, b_{ij} and m_{ij}, can be extracted from the generated open-ended response. We employ templating of the procedural generation through constraints in the generation process. Figure 1 shows one of the sample prompts used in our analysis. The first question, which asks for a summary of the conversation around the post of interest, acts as a CoT mechanism.

3.3 Fine-Tuning

In addition to prompt and response engineering, we also studied the effect of fine-tuning on the veracity of the responses. For this work, each phase of the fine-tuning process is performed using the **Low Rank Adaptation** (LoRA) strategy [19]. This strategy allows us to target particular layers in the model for fine-tuning. Consider a weight matrix, $W_0 \in \mathbb{R}^{m \times n}$, representing the original weights. We use a new set of corresponding weights, $A \in \mathbb{R}^{m \times r}$ and $B \in \mathbb{R}^{r \times n}$, where $r << min(m, n)$. Our models' corresponding weight is now $W_0 + AB$. Thus, backpropogating on A and B, we can fine-tune on a significantly smaller number of parameters.

Machine learning models are vulnerable to mode collapse and catastrophic forgetting when training on a certain type of data. That is, training on structural understanding might make the model forget about previously learned knowledge in lieu of learning structure. To prevent such a hazard, we use a combination of task-specific data and general instruction data. For the instruction data, we make use of Alpaca [36] instructions, which are combined with the task data with a certain probability.

4 Do LLMs Understand Language in Social Context?

Large language models are pretrained mostly on formal and semi-formal text corpora. Given our motivation to analyze social media discourse, we compare the capabilities of LLMs to understand informal social media comments. Our hypothesis is

Prompt

Below is an instruction that describes a task, paired with an input that provides further context. Write a response that appropriately completes the request.

Instruction:
You have been provided with a CSV file containing a social media conversation. For this task, you should only make assumptions about posters based on the provided CSV input. The posts in the input are in the order of date posted i.e. replies do not occur before posts being replied to.
Answer the following questions:
a) Summarize the discourse revolving around the active post ID.
b) Is the active post considered bullying?
c) Is the active post considered anti-bullying?
d) If it is bullying, explain why?
e) If it is anti-bullying, explain why?

Input:
post_id, author, comment, active
1, User Information, Post description, 0
2, sabrinatee10, I fuckin love you Marcus, 0
3, lameesam, Hopsin u da shit!!, 1

Response:
a) The post with the active ID ([POSTID]) [SUMMARY]
b) [BULLYING], the active post is [BULLYINGEXTRA]
c) [ANTIBULLYING], the active post is [ANTIBULLYINGEXTRA]
d) The reason the post is [BULLYREASON1] bullying is because [BULLYREASON2]
e) The reason the post is [REASONANTIBULLY1] anti-bullying is because [REASONANTIBULLY2]

Constraints

```
len(POSTID) < 10
AND regex(POSTID, r"([0-9]|\))+")
AND stops_before(POSTID, ")")
AND len(SUMMARY) < 100
AND stops_before(SUMMARY, "b)")
AND BULLYING IN ["Yes", "No"]
AND len(BULLYINGEXTRA) < 100
AND len(BULLYINGEXTRA) > 1
AND stops_before(BULLYINGEXTRA, "c)")
AND ANTIBULLYING IN ["Yes", "No"]
AND len(ANTIBULLYINGEXTRA) < 100
AND stops_before(ANTIBULLYINGEXTRA, "d)")
AND REASONBULLY1 IN ["not considered", "considered"]
AND len(REASONBULLY2) < 200
AND stops_before(REASONBULLY2, "e)")
AND REASONANTIBULLY1 IN ["not considered", "considered"]
AND len(REASONANTIBULLY2) < 200
AND stops_before(REASONANTIBULLY2, "f)")
```

Fig. 1. We use constrained generation for our analysis. Constrained prompt generation allows extraction of categorical labels and explanations while preventing the responses from deviating from expectation. The prompt placeholders are highlighted in yellow. (Color figure online)

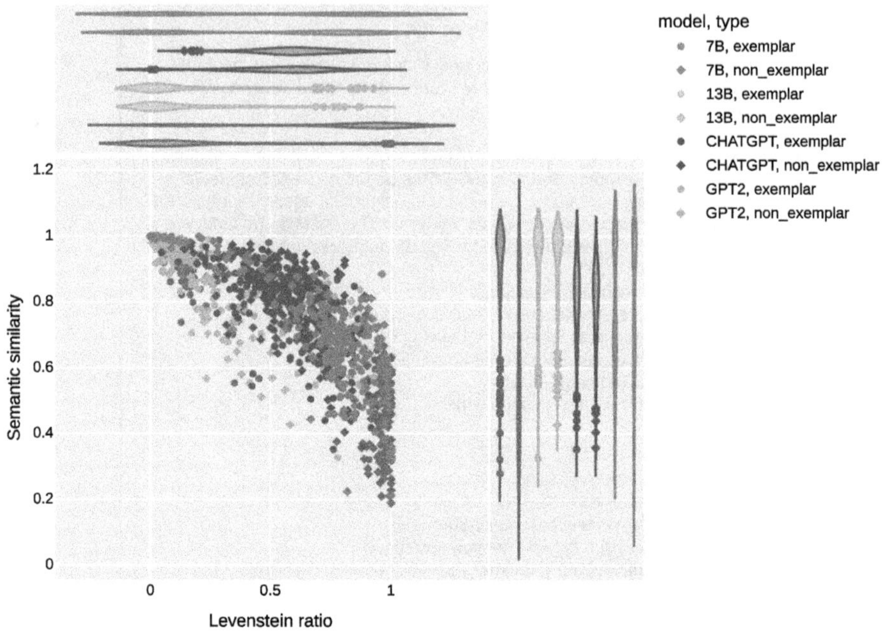

Fig. 2. Semantic similarity against edit distance for paraphrases. ChatGPT shows a promising balance, while Llama-based models tend to repeat the text verbatim. Using exemplars shows performance improvements for the Llama-2 7B model.

that an LLM that understands a comment will be able to paraphrase it. Comments are used for paraphrasing in two scenarios: (i) instruction-based paraphrasing, and (ii) exemplar-based paraphrasing. Instruction-based paraphrasing consists of an Alpaca format prompt where the instruction asks the language model to paraphrase, the input is the text to paraphrase, and the response is generated by the model.

4.1 Similarity Analysis

We use four main metrics to evaluate the quality of generations: (i) BLEU score, (ii) ROGUE score, (iii) Jaccard similarity, and, (iv) Semantic similarity. One of the quirks with generation is that in many cases, the LLMs reproduce the provided text verbatim. This leads to an increase in most of the similarity metrics. Thus, we also report the Levenshtein ratio for each strata. BLEU score measures the correspondence between the model's output and the reference paraphrase based on n-gram overlap. ROGUE score also evaluates the overlap of n-grams, but with a focus on recall, thus capturing the extent to which the reference n-grams are present in the generated text. Jaccard similarity assesses the similarity and diversity of the sets of n-grams between the generated and reference texts. More specifically, it is the ratio of the intersection and union of the n-grams in the two texts. Semantic similarity explores the likeness in meaning between the texts. In this instance, it is extracted from the encodings of the BERT (base) model.

Table 1. Comparison of similarity metrics and the corresponding edit distance across various language models when paraphrases generated on social media comments are compared with the original comment. Ideal paraphrasing should vary from the original comment but keep the meaning intact.

		Similarity Metrics						Edit Distance
Model	Type	BLEU	ROUGE-1	ROUGE-2	ROUGE-L	Jaccard	Semantic	Levenshtein
13B	exemplar	**0.725 (0.352)**	**0.865 (0.219)**	**0.801 (0.299)**	**0.859 (0.230)**	**0.784 (0.285)**	**0.917 (0.124)**	0.157 (0.228)
13B	non_exemplar	0.687 (0.383)	0.844 (0.248)	0.777 (0.322)	0.838 (0.256)	0.757 (0.310)	0.915 (0.128)	0.175 (0.260)
7B	exemplar	0.654 (0.378)	0.768 (0.352)	0.734 (0.373)	0.763 (0.355)	0.678 (0.354)	0.888 (0.161)	0.239 (0.329)
7B	non_exemplar	0.182 (0.310)	0.270 (0.396)	0.239 (0.393)	0.265 (0.396)	0.211 (0.313)	0.604 (0.233)	**0.733 (0.351)**
CHATGPT	exemplar	0.187 (0.258)	0.527 (0.219)	0.305 (0.264)	0.493 (0.225)	0.345 (0.233)	0.805 (0.127)	0.521 (0.211)
CHATGPT	non_exemplar	0.084 (0.094)	0.456 (0.172)	0.197 (0.183)	0.410 (0.165)	0.264 (0.137)	0.795 (0.119)	0.599 (0.145)
GPT2	exemplar	0.401 (0.465)	0.489 (0.427)	0.426 (0.469)	0.479 (0.433)	0.451 (0.433)	0.788 (0.189)	0.488 (0.387)
GPT2	non_exemplar	0.453 (0.466)	0.532 (0.439)	0.480 (0.463)	0.526 (0.443)	0.522 (0.418)	0.800 (0.213)	0.467 (0.419)

The previous metrics show the similarity between different text pairs. However, we observed that in many instances, the generations repeated the input verbatim leading to high similarity. Figure 2 shows the distribution of similarities in the form of violin plots. To differentiate verbatim or close-to-verbatim generations from generations of higher quality, we also calculate the Levenshtein distance (edit distance) between the generation and the original text. Because this distance varies depending on the length of the text, in practice, we rely on the Levenshtein ratio:

$$\text{Levenshtein Ratio}(s_1, s_2) = \frac{\text{Levenshtein Distance}(s_1, s_2)}{max(|s_1|, |s_2|)} \tag{1}$$

4.2 Comparison of Social Understanding Among LLMs

We compare different LLMs to understand their ability to comprehend social conversations. Specifically, GPT-2, Llama-2 7B/13B, and ChatGPT are used in our comprehension analysis. Some of the comments include language of toxic nature that ChatGPT refuses to answer due to ethical constraints. For some of our analysis, we used ChatGPT (GPT 3.5) generations near the timeframe of its release when the restrictions were comparatively lax. Unfortunately, current versions of ChatGPT do not allow most of these prompts. We compare generations with and without exemplars to gauge any improvement in understanding. For the social media content, we use a dateset of Instagram sessions [18]. The results are tabulated in Table 1.

Common LLM Mis-Generations. We identify three common types of generation mistakes made by the models: (i) verbatim generation; (ii) repetition of exemplars; and (iii) gibberish generation. When a model generates responses verbatim, the Levenshein distance is 0. In some cases, exemplars serve as a double-edged sword, as the model reproduces an example instead of producing novel paraphrasing. In many cases, LLMs produce gibberish unrelated to the query comment or the provided exemplars. In such cases, we observe a low similarity score and a high edit distance since the generated text is semantically dissimilar to the origin text and contains few common sub-sequences.

Significance of Exemplars. Adding exemplars to the prompt is a common technique to guide generations toward a specific goal. However, in our analysis, we only see a significant change in generation statistics for the 7B variant of Llama-2. For this case, the Levenshtein distance also decreases, hinting at verbatim repetitions of the input. These changes are evident in the distributions represented in the violin plots in Fig. 2.

Importance of Distributional Analysis. Although the semantic scores in Table 1 appear to favor Llama-2 13B, even compared to ChatGPT, qualitative analysis and the distributional stratification in Fig. 2 better contextualises the tradeoffs between textual and semantic similarity. ChatGPT generations produce text that is consistently semantically similar ($similarity \rightarrow 1$) to the input, but avoids being verbatim repetition (Levenshtein ratio $\rightarrow$ 0) or gibberish (Levenshtein ratio $\rightarrow$ 1). Thus, ChatGPT shows a robust understanding of social context, while GPT-2 tends to produce gibberish. The Llama-2 based models tend to produce a mixture of repetition, valid responses, and gibberish. A possible explanation for this phenomenon could be the pre-training corpus, which includes mostly academic text in the case of Llama-2. The results also suggest that exemplars may be insufficient to improve language model understanding, as they only guide the generation process rather than improving the model's inherent understanding (encoded in the weights).

For social understanding, we conclude that *even though common metrics such and BLEU and ROGUE allude to language understanding, we find problems with regurgitation and verbatim reproductions. We also find that multi-faceted evaluation paints a more nuanced picture.*

5 Can Language Models Understand Directionality in Social Contexts?

Many social platforms enable their users to express whom they target with their comments. A majority of social media platforms provide some variation of the 'mention' functionality. In the context of this work, we are specifically interested in the ability of LLMs of follow the directionality pattern.

5.1 Language Models and Directionality Detection

The primary objective of training an LLM for next-word prediction is to learn the conditional probability distribution of the next word given the context of preceding words. Formally, let $w_1, w_2, \ldots, w_{n-1}$ be a sequence of $n-1$ words, and w_n be the target word to predict. The goal is to learn the probability $P(w_n|w_1, w_2, \ldots, w_{n-1})$, which captures the likelihood of different words occurring next in the sequence. To train LLMs, large text corpora are used as datasets. These corpora contain a wide variety of sentences and documents, providing diverse linguistic contexts for the model to learn from. Common sources of data include books, articles, websites, and other textual sources.

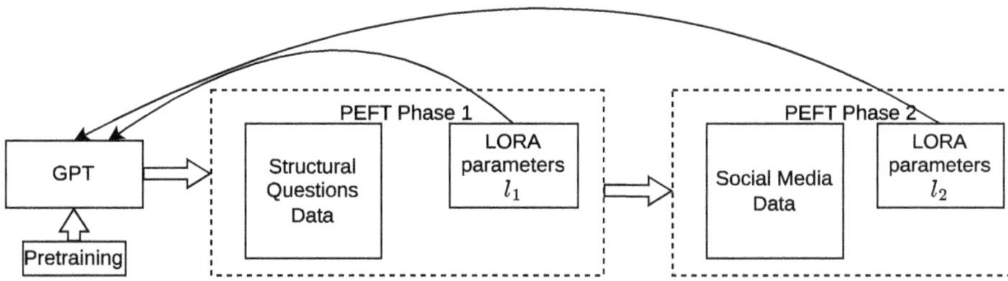

Fig. 3. We divide our fine-tuning analysis in two phases. The first phase adds structural understanding, while the second phase adds social understanding to the under-privileged models.

LLMs Capabilities Under Fine-Tuning Because prompt tuning is insufficient for models to develop social understanding, as noted in Sect. 4, we use a fine-tuning process consisting of two phases to add knowledge to the LLM (illustrated in Fig. 3). Both phases of fine-tuning use Parameter Efficient Fine-Tuning (PEFT)[1]. In our problem setting, rather than just predicting the next words, we aim to gain an understanding of the relation between different comments. For instance, a comment in a session may target the previous comment, the original post that spawned the session, or some comment in the middle of the discourse. To glean insight into the target of the comment in terms of its context, reasoning about the structure of the conversation is critical. Unfortunately, the LLM pre-training does not consider these relationships specifically and there is no public data related to reasoning at the comment level in social media discourse. Thus, we rely on other general purpose structured data to act as a surrogate to learn structure and reasoning. We use the WikiTableQuestions [30] dataset to infuse structural intelligence into the model. This dataset consists of a large variety of independent tables, questions based on one of the tables, and a corresponding answer. To answer a question, it is vital to use the data in the table.

Phase 2 of the PEFT process aims to improve the social understanding of under-privileged LLMs. ChatGPT shows improved social language understanding compared to competitors[2]. Even though, the advantage of ChatGPT may be attributed to a more generalized and customized training corpus, this phase of PEFT seeks to study the transfer learning abilities of competitor models in the confines of our problem setting.

5.2 Can LLMs Emergent Abilities Improve Directional Understanding?

For the directionality analysis task, we utilized a corpus of 4chan threads [29] and efficient fine-tuning using JORA [34]. Since 4chan permits its users to tag the individuals to whom they are replying, we employ this data as the ground truth for directionality information. Our objective is to determine whether our designed PEFT phases (i) enhance the model's ability to identify the post being targeted for behavior comprehension, and (ii) improve the model's capacity to discern the individual being targeted by the poster. Given that 4chan users can reference multiple comments as the target of their

[1] See Sect. 3.3.
[2] See Sect. 4.2.

Table 2. We find that the veracity of the directionality identification significantly improves with the proposed PEFT fine-tuning phases compared to the base models.

	Target Post	Reply Post	p(Reply—Target)
7B	0.082	0.153	0.643
13B	0.159	0.200	0.815
7B-PEFT	0.865	0.541	0.558
13B-PEFT	**0.971**	**0.847**	**0.855**

replies, we consider the model successful in this analysis if it correctly identifies any one of the multiple mentioned comments.

We stratify our analysis based on the number of model parameters and whether the models were fine-tuned. Table 2 shows the accuracy of directionality identification for the target post and the reply post. The PEFT models significantly outperform the base models. In general, the greater the number of model parameters, the better the veracity of identification. We observe this for the base models as well as the fine-tuned ones. Surprisingly, we find that the base models perform better with reply identification than with target post identification. This behavior is the result of the model relying on randomly mentioning the tags to the other posts in the comments. Because not all posts tag other comments, the probability of returning a valid comment that is tagged at random is greater than the probability of selecting the target comment at random. In the conversation data, the number of comments being replied to is a subset of the total comments. Because PEFT models rely on more educated guesses, the veracity of reply identification is less than the veracity of target identification.

Finally, we examine whether each model can capture the reply direction given that it was able to identify the target. Our results indicate that even the base models yield promising findings for identifying directionality in such a scenario. However, because the probability of identifying the target post is low, this may represent a very small sample of easy examples not representative of the bigger picture. For PEFT models, we see a minor increase in $p(reply_{\text{predicted}} = reply_{\text{ground truth}}|target)$, suggesting that these models have learned the relation between the target and the reply during the fine-tuning phase. In conclusion, *LLMs show promising prospects for learning directionality when they are trained to do so by fine-tuning structural and social intelligence. This result is aligned to recent findings about the emergent abilities of LLMs* [40].

6 Can LLM's Detect Instances of Cyberbullying and Anti-bullying?

To further validate our findings with respect to understanding social context and to analyze the veracity of behaviour classification as cyberbullying or anti-bullying, we design our experiment on a dataset of Instagram sessions [16] which tests both language and directional understanding. We use 100 labeled sessions. Each session consists of a post and a set of comments. This dataset serves two purposes: (i) Phase 2 of our PEFT is based on 4chan threads. Using a separate dataset allows us to avoid any potential data

Table 3. Both base and PEFT models show random performance for identification of cyberbullying and anti-bullying comments.

	7B	7B-PEFT	13B	13B-PEFT
accuracy	0.500	0.481	0.494	0.513
precision	0.500	0.480	0.496	0.514
recall	0.897	0.462	0.769	0.462
f1	0.642	0.471	0.603	0.486

leakage and measures the generalization capabilities of the model, and (ii) Instagram comments are short compared to 4chan comments. This allows us to add additional context in the input of the prompt. In this experiment, we stratify the evaluation based on the number of model parameters and whether PEFT was applied.

We partition a random sample of the comments into two sets. The first set consists of comments labeled as cyberbullying and the second set consists of comments labeled as anti-bullying. Comments that fall into neither of the two categories are ignored. This simplifies the analysis to a binary classification problem, which is easier to interpret. Different variations of the LLMs are prompted and allowed to respond in a constrained manner as illustrated in Fig. 1. Table 3 summarizes the results of our experiment. In terms of accuracy, all variations of the models appear to perform close to random chance. Further investigation of the responses, including the summary, revealed that the models were incapable of understanding the social commentary at a semantic level. Non-PEFT models display a higher recall (and f1) score since they tend to positively label both cyberbullying and anti-bullying cases. The results are consistent with the findings in Sect. 4 where we show that these models struggle to understand the informal language used in social networks. We were unable to compare with ChatGPT for two reasons. First, ChatGPT is a closed source model and thus does not allow constrained generation. Second, because cyberbullying and anti-bullying comments contain content or context of a toxic nature, safeguards prevent the generation of responses. Given that ChatGPT showed promising ability in informal language understanding, we hypothesize that it would have displayed improved performance in this particular experiment. This highlights the need for, and importance of, a social language comprehension dataset, analogous to SQuAD [31], and more robust mechanisms to perform academic evaluations of popular LLMs.

7 Discussion

Making inferences on social data is a multi-faceted problem. In this work, we study three dimensions relevant to exploiting LLMs for social behaviour analysis. Foundational models pretrained on large corpora of text, such as Llama, have enabled a larger audience to enjoy their benefits by tuning them for specific use-cases. First, we compare the ability of different models to understand social context by measuring how well they can generate paraphrases. Second, we study whether LLMs can understand directionality in social media. Third, we measure the effectiveness of LLMs in a classification task that uses a combination of directional and social understanding. Our experiments converged in the finding that the weakest link in the considered problem setting is the lack of semantic understanding of the language used in social media. This finding is corroborated by the results in Sects. 4 and 6. In addition, our results show promising ability of LLMs in the domain of directional understanding. Because LLMs are trained on large corpus of semi-formal and formal text, our results show a need for large scale informal training corpuses, preferably focused on language comprehension.

8 Conclusion, Limitations, and Future Work

LLMs have seen a spike in popularity recently due to their astounding ability in zero-shot response generation for general-purpose prompts. There have been various prior studies that either question [28,37] or praise [7,20] their abilities in different domains. In this work, we focus on the subdomain of social analysis through the lens of social media discourse. More specifically, we considered the detection of cyberbullying and anti-bullying behavior. We highlight strengths and limitations throughout a taxonomy of learning in language models through prompt engineering, fine-tuning, and constrained generation. We apply relevant parts of this taxonomy to answer research questions related to social context understanding, directionality understanding, and identification of cyberbullying and anti-bullying activity, which requires both social and directional understanding. We believe future breakthroughs in this domain will require larger coded social datasets as well as new model features that better capture the complex nature of social interaction semantics.

Acknowledgment. This work was supported by NSF Awards #2227488 and #1719722 and a Google Award for Inclusion Research. Cheng was supported by a Cisco Faculty Research Award.

References

1. Araci, D.: Finbert: financial sentiment analysis with pre-trained language models. arXiv preprint arXiv:1908.10063 (2019)
2. Atske, S.: The state of online harassment (2021). https://www.pewresearch.org/internet/2021/01/13/the-state-of-online-harassment/
3. Basile, V., et al.: Semeval-2019 task 5: multilingual detection of hate speech against immigrants and women in twitter. In: Proceedings of the 13th International Workshop on Semantic Evaluation, pp. 54–63 (2019)
4. Bubeck, S., et al.: Sparks of artificial general intelligence: early experiments with GPT-4 (2023). https://doi.org/10.48550/arXiv.2303.12712
5. Calderón, F.H., Cheng, L.K., Lin, M.J., Huang, Y.H., Chen, Y.S.: Content-based echo chamber detection on social media platforms. In: Proceedings of the 2019 IEEE/ACM International Conference on Advances in Social Networks Analysis and Mining (ASONAM 2019), p. 597600. Association for Computing Machinery, New York (2020). https://doi.org/10.1145/3341161.3343689
6. Chatterjee, A., Narahari, K.N., Joshi, M., Agrawal, P.: Semeval-2019 task 3: emocontext contextual emotion detection in text. In: Proceedings of the 13th International Workshop on Semantic Evaluation, pp. 39–48 (2019)

7. Chen, C., Shu, K.: Can llm-generated misinformation be detected? arXiv preprint arXiv:2309.13788 (2023)
8. Cheng, L., Guo, R., Shu, K., Liu, H.: Causal understanding of fake news dissemination on social media. In: Proceedings of the 27th ACM SIGKDD Conference on Knowledge Discovery and Data Mining (KDD 2021), p. 148157. Association for Computing Machinery, New York (2021). https://doi.org/10.1145/3447548.3467321
9. Davidson, T., Bhattacharya, D., Weber, I.: Racial bias in hate speech and abusive language detection datasets. arXiv preprint arXiv:1905.12516 (2019)
10. Devlin, J., Chang, M.W., Lee, K., Toutanova, K.: Bert: pre-training of deep bidirectional transformers for language understanding. arXiv preprint arXiv:1810.04805 (2018)
11. Duzen, Z., Riveni, M., Aktas, M.S.: Misinformation detection in social networks: a systematic literature review. In: Gervasi, O., Murgante, B., Misra, S., Rocha, A.M.A.C., Garau, C. (eds.) Computational Science and Its Applications ICCSA 2022 Workshops, pp. 57–74. Springer (2022)
12. Felbo, B., Mislove, A., Søgaard, A., Rahwan, I., Lehmann, S.: Using millions of emoji occurrences to learn any-domain representations for detecting sentiment, emotion and sarcasm. arXiv preprint arXiv:1708.00524 (2017)
13. Gaind, B., Syal, V., Padgalwar, S.: Emotion detection and analysis on social media. arXiv preprint arXiv:1901.08458 (2019)
14. Gillani, N., Yuan, A., Saveski, M., Vosoughi, S., Roy, D.: Me, my echo chamber, and I: introspection on social media polarization. In: Proceedings of the 2018 World Wide Web Conference (WWW 2018), p. 823831. International World Wide Web Conferences Steering Committee, Republic and Canton of Geneva, CHE (2018). https://doi.org/10.1145/3178876.3186130
15. Guerra, G.A., et al.: GPT-4 artificial intelligence model outperforms ChatGPT, medical students, and neurosurgery residents on neurosurgery written board-like questions (2023). https://doi.org/10.1016/j.wneu.2023.08.042
16. Hamlett, M., Powell, G., Silva, Y.N., Hall, D.: A labeled dataset for investigating cyberbullying content patterns in Instagram. In: Proceedings of the International AAAI Conference on Web and Social Media, vol. 16, pp. 1251–1258 (2022)
17. He, X., et al.: AnnoLLM: making large language models to be better crowdsourced annotators (2023). https://doi.org/10.48550/arXiv.2303.16854
18. Hosseinmardi, H., Mattson, S.A., Ibn Rafiq, R., Han, R., Lv, Q., Mishra, S.: Presented at the Analyzing Labeled Cyberbullying Incidents on the Instagram Social Network (2015). https://doi.org/10.1007/978-3-319-27433-1_4
19. Hu, E.J., et al.: Lora: low-rank adaptation of large language models. arXiv preprint arXiv:2106.09685 (2021)
20. Kıcıman, E., Ness, R., Sharma, A., Tan, C.: Causal reasoning and large language models: opening a new frontier for causality. arXiv preprint arXiv:2305.00050 (2023)
21. Lan, M., Law, N., Pan, Q.: Effectiveness of anti-cyberbullying educational programs: a socio-ecologically grounded systematic review and meta-analysis. Comput. Hum. Behav. **130**, 107200 (2022)
22. Liu, P., Yuan, W., Fu, J., Jiang, Z., Hayashi, H., Neubig, G.: Pre-train, prompt, and predict: a systematic survey of prompting methods in natural language processing (2021)
23. Liu, Y., Wu, Y.F.: Early detection of fake news on social media through propagation path classification with recurrent and convolutional networks. **32**(1), 1 (2018). https://doi.org/10.1609/aaai.v32i1.11268
24. Malmasi, S., Zampieri, M.: Challenges in discriminating profanity from hate speech. J. Exp. Theor. Artif. Intell. **30**(2), 187–202 (2018)
25. Min, S., Lewis, M., Zettlemoyer, L., Hajishirzi, H.: Metaicl: learning to learn in context. arXiv preprint arXiv:2110.15943 (2021)

26. Minici, M., Cinus, F., Monti, C., Bonchi, F., Manco, G.: Cascade-based echo chamber detection. In: Proceedings of the 31st ACM International Conference on Information and Knowledge Management (CIKM 2022), p. 15111520. Association for Computing Machinery, New York (2022). https://doi.org/10.1145/3511808.3557253
27. Nguyen, D.Q., Vu, T., Nguyen, A.T.: Bertweet: a pre-trained language model for English Tweets. arXiv preprint arXiv:2005.10200 (2020)
28. Onoe, Y., Zhang, M.J., Padmanabhan, S., Durrett, G., Choi, E.: Can LMS learn new entities from descriptions? Challenges in propagating injected knowledge. arXiv preprint arXiv:2305.01651 (2023)
29. Papasavva, A., Zannettou, S., De Cristofaro, E., Stringhini, G., Blackburn, J.: Raiders of the lost kek: 3.5 years of augmented 4chan posts from the politically incorrect board. In: Proceedings of the International AAAI Conference on Web and Social Media, vol. 14, pp. 885–894 (2020)
30. Pasupat, P., Liang, P.: Compositional semantic parsing on semi-structured tables. arXiv preprint arXiv:1508.00305 (2015)
31. Rajpurkar, P., Jia, R., Liang, P.: Know what you don't know: Unanswerable questions for squad. arXiv preprint arXiv:1806.03822 (2018)
32. Risch, J., Stoll, A., Wilms, L., Wiegand, M.: Overview of the germeval 2021 shared task on the identification of toxic, engaging, and fact-claiming comments. In: Proceedings of the GermEval 2021 Shared Task on the Identification of Toxic, Engaging, and Fact-Claiming Comments, pp. 1–12 (2021)
33. Rosa, H., et al.: Automatic cyberbullying detection: a systematic review. Comput. Hum. Behav. **93**, 333–345 (2019). https://doi.org/10.1016/j.chb.2018.12.021
34. Tahir, A., Cheng, L., Liu, H.: Jora: Jax tensor-parallel lora library for retrieval augmented fine-tuning. arXiv preprint arXiv:2403.11366 (2024)
35. Tamkin, A., Brundage, M., Clark, J., Ganguli, D.: Understanding the capabilities, limitations, and societal impact of large language models (2021). https://doi.org/10.48550/arXiv.2102.02503
36. Touvron, H., et al.: LLaMA: open and efficient foundation language models (2023). https://doi.org/10.48550/arXiv.2302.13971, version: 1
37. Valmeekam, K., Sreedharan, S., Marquez, M., Olmo, A., Kambhampati, S.: On the planning abilities of large language models (a critical investigation with a proposed benchmark). arXiv preprint arXiv:2302.06706 (2023)
38. Vaswani, A., et al.: Attention is all you need. Adv. Neural Inf. Process. Syst. **30** (2017)
39. Wang, S.: Standing up or standing by: bystander intervention in cyberbullying on social media. New Media Soc. **23**(6), 1379–1397 (2021). https://doi.org/10.1177/1461444820902541
40. Wei, J., et al.: Emergent abilities of large language models. arXiv preprint arXiv:2206.07682 (2022)
41. Wei, J., et al.: Chain-of-thought prompting elicits reasoning in large language models. Adv. Neural. Inf. Process. Syst. **35**, 24824–24837 (2022)
42. Wiedemann, G., Yimam, S.M., Biemann, C.: Uhh-lt at semeval-2020 task 12: fine-tuning of pre-trained transformer networks for offensive language detection. arXiv preprint arXiv:2004.11493 (2020)
43. Ye, H., Liu, T., Zhang, A., Hua, W., Jia, W.: Cognitive mirage: a review of hallucinations in large language model (2023). https://doi.org/10.48550/arXiv.2309.06794
44. Zotova, E., Agerri, R., Rigau, G.: Semi-automatic generation of multilingual datasets for stance detection in twitter. Expert Syst. Appl. **170**, 114547 (2021)

Two-Stage Stance Labeling: User-Hashtag Heuristics with Graph Neural Networks

Joshua Melton(✉), Shannon Reid, Gabriel Terejanu, and Siddharth Krishnan

University of North Carolina at Charlotte, Charlotte, NC 28223, USA
{jmelto30,sreid33,gterejan,skrishnan}@charlotte.edu

Abstract. The high volume and rapid evolution of content on social media present major challenges for studying the stance of social media users. In this work, we develop a two stage stance labeling method that utilizes the user-hashtag bipartite graph and the user-user interaction graph. In the first stage, a simple and efficient heuristic for stance labeling uses the user-hashtag bipartite graph to iteratively update the stance association of user and hashtag nodes via a label propagation mechanism. This set of soft labels is then integrated with the user-user interaction graph to train a graph neural network (GNN) model using semi-supervised learning. We evaluate this method on two large-scale datasets containing tweets related to climate change from June 2021 to June 2022 and gun control from January 2022 to January 2023. Our experiments demonstrate that enriching text-based embeddings of users with network information from the user interaction graph using our semi-supervised GNN method outperforms both classifiers trained on user textual embeddings and zero-shot classification using LLMs such as GPT4. We discuss the need for integrating nuanced understanding from social science with the scalability of computational methods to better understand how polarization on social media occurs for divisive issues such as climate change and gun control.

Keywords: graph neural networks · polarization · social network analysis · stance labeling

1 Introduction

In today's digital society, online social media have become the main platform for dissemination of partisan and political information as well as the main forum for public discussion of political and social issues. The immense scale and interconnected nature of social media has meant that the scope of public discussion online permits varied and diverse viewpoints about a broad range of topics. Despite the potential for the co-existence of diverse perspectives online, research has shown that online communities are increasingly *polarized*, especially with respect to contentious issues [7,21,23,28]. The increasingly polarized nature of

L. M. Aiello et al. (Eds.): ASONAM 2024, LNCS 15213, pp. 245–261, 2025.
https://doi.org/10.1007/978-3-031-78548-1_19

online communities poses a real challenge to many democracies in today's world. Studies have shown that over the last 30 years, both Democrats and Republicans have become more negative in their views towards the opposition party [7], and these negative views have affected outcomes in areas such as scholarship fund allocation, mate selection, employment decisions, and acceptance of public health measures [9,20]. Over the last several years, social media platforms like Twitter, Reddit, YouTube, and others have amassed millions of users who engage both with mainstream and fringe media outlets as well as engaging directly with other users on numerous political and social issues. As such, large social media platforms like Twitter present an excellent opportunity to study web-scale user behavior and polarization among differing viewpoints on contentious issues such as climate change and gun control.

The study of polarization online has broadly been divided into two main categories. The majority of computational social science research has focused on analyzing *interactional polarization*, which occurs when users interact to a large extent exclusively with like-minded individuals in highly balkanized communities and are only minimally exposed to users with opposing viewpoint. Such *echo chambers* isolate users in an in-group community that reinforces users' existing beliefs and skews user perspectives towards a particular issue [12]. Another important form of polarization that is less well-studied is known as *affective polarization*. Affective polarization refers to highly negative sentiments expressed by users towards out-groups of users with opposing viewpoints [23,28]. Prior work on affective polarization has been limited in part due to reliance on manually annotated datasets and challenges in identifying user stances and group memberships in a computationally efficient manner. With recent advances in language modeling, stance detection, sentiment analysis, and network science-based approaches to community detection, researchers have begun to study affective polarization using large, unannotated corpora and interaction networks. Despite this recent progress, much work on affective polarization remains *ad hoc* and incommensurate between studies.

Stance labeling remains a particularly challenging task due to the informal and noisy nature of social media language, the presence of sarcasm and irony, and the potential biases introduced by human annotators and automated models. To further facilitate research into polarization online, we develop a stance labeling method that incorporates both textual content and user social interactions in a scalable two-stage pipeline. In the first stage, we apply a reciprocal label propagation algorithm to the user-hashtag bipartite graph that generates a soft-labeled set of users. In the second stage, we construct a signed, weighted, and attributed user-user interaction network and use transformer language models to represent the content of user posts and to determine the sentiment of interactions between users. Using the soft labels generated from the label propagation stage, we employ semi-supervised training to train a graph neural network classifier to classify user stances.

Our stance labeling approach utilizes an understanding about linguistic and network homophily. Linguistic homophily states that people who share similar ideologies also share similar textual content, not only through similar keyword and hashtag usages but also through sentiment and linguistic features [15,27], and the inverse phenomenon—that users with dissimilar views use different language—has also been investigated by researchers [21]. Likewise, the concept of network homophily, which describes the phenomenon that similar nodes tend to be connected to one another has long been studied in graph theory [14,17] and informs the success of graph neural networks [8,13,26]. By combining the content-based and network-based approaches into a single, scalable pipeline for stance labeling, we facilitate further large-scale analysis of both interactional and affective polarization online.

In sum, the contributions of this work are:

- We develop an approach for estimating user stance labels that incorporates both textual features and social network interactions.
- A two-stage pipeline allows our approach to scale to large datasets with soft labeling and semi-supervised training.
- We discuss challenges and potential biases in stance determination of users when relying on short text content as commonly found on social media.

2 Related Work

As social media has become the primary forum for discussion on many topics, including political issues, there has been growing interest in user stance detection [2,6,10]. Many existing methods can be divided into two broad categories: textual-based methods that use the content of messages, and network-based methods that use social or conversational structure. Textual-based methods use the content of tweets, user profiles, and hashtags to determine a user's stance on a particular topic. Several studies have used supervised classification of user stance labels where classifiers are trained using a set of features encoding tweet text, hashtags, and user profile information. These techniques have been applied to determine user political ideology [1,3]. These approaches all generally rely on manually annotated sets of users to train the classifier models, and such methods are often limited by the size and diversity of the training content as manual annotation is both time-consuming and expensive. Furthermore, these methods can be susceptible to annotator bias and poor generalizability due to the limited distribution of training examples.

Topic modeling and user clustering are two unsupervised techniques that have been employed for stance labeling. These unsupervised methods generally embed text or user profile information in a latent feature space and then apply a variety of topic modeling or cluster algorithms to identify texts of similar topics or clusters of similar users. While such techniques avoid the issues with manual annotation for supervised training, they suffer from the *curse of dimensionality*—the search space for a solution grows exponentially with the increase in dimensionality—as there are many more possible patterns than in lower-dimensional subspaces. Selecting useful and informative features, and the computation and memory complexity of clustering methods on high-dimensional data create problems for unsupervised clustering-based methods. For high-dimensional data, many clustering techniques fail to produce meaningful clusters, but on the other hand, such techniques can be very efficient on low-dimensional features spaces [6]. Several studies have therefore employed dimensionality reduction techniques prior to clustering to first project content embeddings to a low-dimensional space, and such techniques can be robust to class imbalance between different stance groups.

Network-based approaches use the structure of social interaction networks to determine user ideology or stance. Such methods employ techniques such as label propagation, min-cut, and node embedding techniques to leverage structural information from retweet [10] and user follower [4] networks. Xiao et al. [25] generate a multi-relational network using retweets, mentions, likes, and follows for binary classification of user ideologies.

Recent works have combined both textual content and network features in socially-infused text mining. Li and Goldwasser [16] combine user interactions and user sharing of news media to detect bias in news articles, and Johnson, Jin, and Goldwasser [11] combine multiple representations from lexical features and social interactions to categorize tweets by politicians on a variety of political topics. Pan et al. [19] use the network structure and node content to learn node representations for classifying the category of scientific publications. Such works combining textual and network information have demonstrated improvements in user stance detection compared with methods that rely solely on textual content or network structure alone.

Our approach is likewise informed by socially-infused text mining that combines information from text and network relationships for user stance labeling, but the approach we present in this paper is unique from the methods described above in that we i) combine language features from state-of-the-art transformer language models with the structure and sentiment of social interactions using graph neural networks, and ii) our two-stage process allows our method to scale to large datasets with minimal supervision.

3 Methods

This section describes our proposed method to estimate the stance of Twitter users on political issues. We first describe the collection of data for the gun control and climate changes datasets. Then we characterize the hashtag-based heuristic method to generate soft labels for two stance groups of users, which are used as seed users for training stance labeling methods. We then introduce several baseline methods followed by our GNN-based approach for stance labeling.

3.1 Data Collection

We collect tweets related to two prominent and contentious political issues—climate change and gun control. The keywords listed in Table 1 are used to scrape the initial set of relevant tweets related to each topic. To fully capture conversation cascades and user interactions, we then recursively collect all referenced tweets from each tweet in the initially matched set. This ensures that all available tweets from relevant conversations are included in our analysis, even if particular tweets do not explicitly use any of the keywords. The climate change dataset consists of tweets published between June 1st, 2021 and May 31st, 2022. In all, the climate change dataset consists of a total of 46M tweets authored by 4.8M unique users and contains 726,378 conversation threads of at least three tweets. The gun control dataset consists of tweets published between January 1st, 2022 and December 31st, 2022. In all, the gun control dataset consists of a total of 14.4M tweets from 2.66M unique users containing 335,000 conversation threads of at least three tweets.

Table 1. Set of keywords used to collect tweets related to Climate Change and Gun Control from Twitter.

Climate Change		Gun Control	
climate change	global warming	gun control	gun rights
climate hoax	global warming hoax	second amendment	2nd amendment
global cooling	#ActOnClimate	#guncontrol	#guncontrolnow
#ClimateChange	#climatechangehoax	#gunreform	#gunviolence
#globalwarminghoax	#globalcooling	#endgunviolence	#2a
#globalwarmingisahoax	#climatehoax	#nra	#gunrights
		#secondamendement	#shallnotbeinfringed
		#righttobeararms	

These issues are good models for understanding the differences in behavior on social media for topics that are highly event driven, like gun control, and those that are less focused on particular events, like climate change. Figure 1 illustrates that conversation threads related to climate change are started at a fairly consistent rate over time with some spikes in conversation activity connected with specific events. But compared with the conversation activity on Twitter surrounding gun control, illustrated in Fig. 2, the frequency of discussions surrounding gun control spikes sharply in response to mass shooting events. The summer of 2022 saw several deadly and highly publicized mass shootings in Buffalo, Uvalde, and Chicago that generated large amounts of discourse surrounding gun control and gun violence on social media, illustrated by the sharp spikes from the end of May through late July. Shortly after mass shooting events though, the level of discourse abates, returning to near baseline levels between early August until November, when a shooting event at UVA caused a small spike in conversation activity. Event driven issues such as gun control exhibit different posting behavior and user interactions that we hypothesize lead to differences in polarization and require different techniques for intervention and mitigation of polarization on social media. Developing approaches for stance labeling that are both topic independent and robust to different types of discourse online is an important step for facilitating further analysis of polarization online.

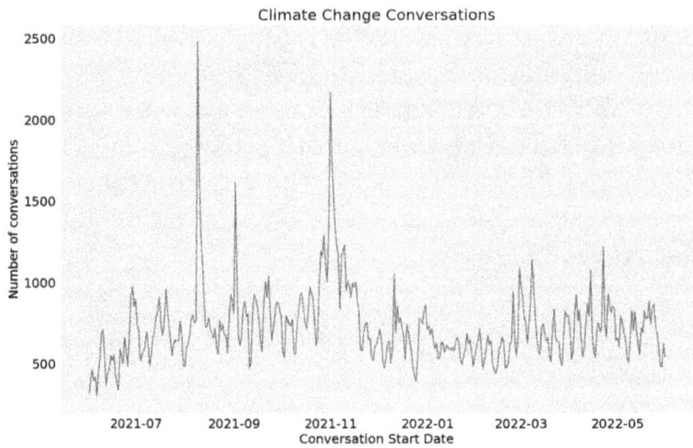

Fig. 1. Timeline of the number of Twitter conversations about climate change started each day.

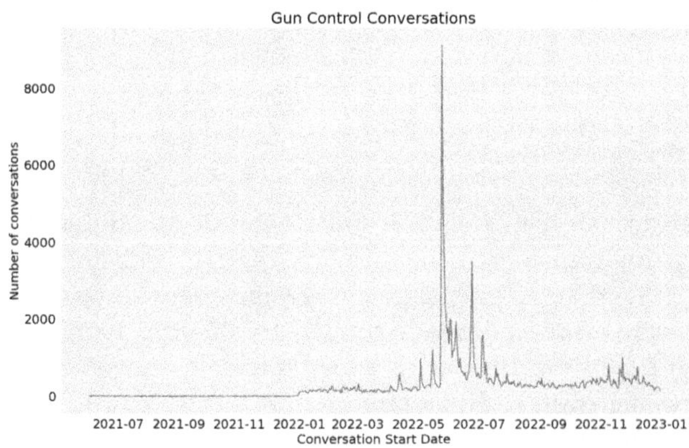

Fig. 2. Timeline of the number of Twitter conversations about gun control started each day.

3.2 User-Hashtag Stance Labeling

From the set of collected tweets for a given topic, we first construct a bipartite graph connecting users with the hashtags that each user posted. The user-hashtag bipartite graph $\mathcal{G} = (V, E)$ consists of two disjoint node sets containing the users and the hashtags posted in tweets by the user set. An edge $(u, h) \in E$

indicates that a user u posted a tweet containing hashtag h. The weight $w(u, h)$ represents the number of times the author posted a particular hashtag across all of their authored tweets. Using the bipartite user-hashtag graph, we then develop a reciprocal label propagation-based method for automatically assigning stance labels to users and hashtags (Table 2).

Table 2. Sets of seed hashtags used for heuristic stance labeling of users in the climate change and gun control datasets.

Climate Change		Gun Control	
Believe	Disbelieve	Pro	Anti
actonclimate	climatechangehoax	guncontrolnow	shallnotbeinfringed
climatecrisis	globalwarminghoax	endgunviolence	righttobeararms
climateaction	globalcooling	gunreform	gunrights
climateemergency	globalwarmingisahoax		
climateactionnow	climatehoax		

Based on a small (3–5) set of seed hashtags associated with each stance, we first propagate the stance associated with each seed hashtag across the user-hashtag bipartite graph to the set of users who posted a given labeled hashtag. The importance of a hashtag's stance to a particular destination user is weighted by the number of times the user posted each hashtag, as described in Algorithm 2. Users are then assigned a label based on the number of hashtag usages for each stance group. User labels are then propagated back to the hashtags used by each user, as described in Algorithm 3. Each hashtag is then scored based on the difference between the normalized count of usages from users of each stance group. Hashtag scores are normalized via min-max normalization, and the mean and standard deviation of hashtag scores is computed. Hashtags that score one or more standard deviations above or below the mean are assigned the corresponding stance label. We then repeat alternating label propagation until the model converges, i.e. the set of labeled users is the same from the prior iteration to the next.

Algorithm 1. Bipartite stance label propagation algorithm

Require: Bipartite user-hashtag graph G, seed hashtag set H^0, s stance label
$H^0 \leftarrow \{(h, s) \in \text{Set of labeled seed hashtags}\}$
$U^0 \leftarrow \emptyset$
while $U^{(k-1)} \neq U^k$ **do**
  $U^k \leftarrow propagateTagsToUsers\left(G, H^{(k-1)}\right)$
  $H^k \leftarrow propagateUsersToTags\left(G, U^k\right)$
end while
return U^K, H^K

Algorithm 2. Propagate hashtag labels to users

Require: Bipartite user-hashtag graph G, labeled hashtag set H, s stance label, $\mathcal{N}$ neighbor set

user_counts = {}
for $h, s \in H$ **do**
  for $u, w \in \mathcal{N}(h)$ **do**
    user_counts[u][s] += w
  end for
end for
for $u \in$ user_counts **do**
  $U \leftarrow (u, \max_s(\text{user\_counts[u]}))$
end for
return U

Algorithm 3. Propagate user labels to hashtags

Require: Bipartite user-hashtag graph G, labeled user set U, s stance label, $\mathcal{N}$ neighbor set

tag_counts = {}
for $u, s \in U$ **do**
  for $h \in \mathcal{N}(u)$ **do**
    tag_counts[h][s] += 1
  end for
end for
for $h \in$ tag_counts **do**
  $h_{score} = \frac{\text{tag\_counts}[h][s_2]}{|U_{s2}|} - \frac{\text{tag\_counts}[h][s_1]}{|U_{s1}|}$
end for
$\mu = \text{mean}(h_{score} \mid h \in \text{tag\_counts})$
$\sigma = \text{stdev}(h_{score} \mid h \in \text{tag\_counts})$
for $h \in$ tag_counts **do**
  if $h_{score} \geq \mu + \sigma$ **then**
    $H \leftarrow (h, s_1)$
  else if $h_{score} \leq \mu - \sigma$ **then**
    $H \leftarrow (h, s_2)$
  end if
end for
return H

While the user-hashtag bipartite label propagation method can quickly label a number of users with high precision. It suffers from the limitation that a large proportion of users do not use hashtags, and while it is well suited for determining hashtags that are highly associated with one particular stance group as compared with the other, hashtags that are used across both stances albeit in different contexts will remain unlabeled.

3.3 GNN Stance Labeling

The user-hashtag stance labeling method can efficiently determine the stance group of users and discover important hashtags associated with each stance group, but it is limited by the fact that not all users post hashtags and that hashtags consist of only a small part of the overall textual content contained in tweets. Furthermore, the method does not take into account any social interactions between users. The hashtag-based heuristic has high precision but relatively low recall due to the limited breadth of users covered by the stance labeling method. To develop a scalable model capable of labeling users in large-scale datasets, we employ the user-hashtag bipartite method as a means to generate a soft-labeled set of users from which we train a graph neural network classifier over the attributed user-user interaction graph.

We define a user interaction as a retweet, mention, reply, or quote from the author of the tweet towards another user. We then construct a weighted, signed, and attributed user-user interaction network $G = (V, E, X)$, where V is the set of users and a directed edge $(u, v)_i \in E$ connects the author user to the target user of their interaction as defined above. The edge weight $w(u, v)_i$ corresponds to the sentiment of the tweet, scored by a BERTweet-based sentiment analysis model [18]. To characterize the overall interactions between users, we consolidate individual tweets between users by combining each directed edge $(u, v)_i$ into a single edge (u, v) with the composite edge weight $w(u, v)$ computed as the mean sentiment score across all tweets between two users. Each user $u \in V$ is also associated with a feature vector $x_u \in X$ corresponding to the element-wise mean of the BERTweet embeddings of all tweets authored by the user. We train a graph neural network model as a stance label classifier using semi-supervised learning with the set of seed users labeled by our hashtag-based heuristic.

For the GNN, we compare two implementations from the prominent families of graph neural networks—convolutional and attentional. For convolutional GNNs, the neighbor coefficients are fixed weights; whereas, in attentional GNNs these weights are computed implicitly by an attention mechanism. Specifically, we evaluate graphSAGE [8] as an example of a convolutional GNN and graph attention network (GAT) [24] as an example attentional GNN.

GraphSAGE defines a message passing function as:

$$\hat{\boldsymbol{h}}_v^{(k)} = \sigma\left(\mathbf{W^k} \sum_{u \in \tilde{\mathcal{N}}(v)} \frac{1}{|\tilde{\mathcal{N}}|} \boldsymbol{h}_u^{(k-1)}\right) \tag{1}$$

where $\tilde{\mathcal{N}}(v)$ is the local neighborhood of node v with added self-loop, $\mathbf{W}$ is a trainable weight matrix, and σ is a non-linear activation function. We use the mean aggregator to aggregate the neighboring node features.
We also consider a GAT layer operator, defined as:

$$\hat{\boldsymbol{h}}_v^{(k)} = \sigma\left(\mathbf{W^k} \sum_{u \in \tilde{\mathcal{N}}(v)} a_{vu} \boldsymbol{h}_u^{(k-1)}\right) \tag{2}$$

where aggregation coefficient a_{vu} is computed dynamically in a node-dependent manner via edge softmax attention. In this case, the GAT function is a first-order approximation of a convolution applied over the weighted adjacency matrix, where the edge weights are computed via the attention mechanism (Table 3).

Table 3. Models utilized for user stance labeling and indication whether the models incorporate textual based information and/or network structural information.

Model Type	Model	Text Content	Network Structure
Random	Weighted random	✗	✗
Transformer	BERT-base-uncased	✓	✗
	BERT-large-uncased	✓	✗
	RoBERTa-base	✓	✗
	RoBERTa-large	✓	✗
	BERTweet	✓	✗
	GPT-4	✓	✗
GNN	GraphSAGE	✓	✓
	GAT	✓	✓

3.4 Baselines

We compare our stance labeling method against six transformer language models that leverage the textual content of tweets to represent users. With each language model, we compute a user embedding by computing the element-wise mean of the embedding of all the users' tweets from BERT, RoBERTa, and BERTweet models. Using the set of labeled users from our heuristic method, we train an MLP classifier to predict user stance label.

We evaluate classification performance of the stance labeling methods using a set of manually annotated users and a set of users annotated via GPT-4 using zero-shot classification. Two social science experts independently annotated a set of 250 users for the climate change dataset and 350 users for the gun control dataset by analyzing the top 20 most popular tweets from each user. We provide the same set of tweets to GPT-4 and prompt the LLM to provide the sentiment of content with respect to gun control or climate change. For the gun control dataset, the function call prompt is: "The stance of the content. Is the message pro gun control, anti gun control, or neutral?" with three classification options of "pro", "anti", or "neutral". For the climate change dataset, "The stance of the content. Does the message capture the user's belief in climate change, disbelief, or is it neutral?" with classification options of "belief", "disbelief", or "neutral". We then aggregate the stance label for each individual tweet to determine a user's overall stance. Users who have a majority of tweets belonging to one stance are labeled accordingly. Users whose tweets are all labeled as neutral or have an equivalent number of tweets labeled as both stances are classified as undetermined.

Using the set of manual and GPT-4 annotated users for the two datasets, we compare the classification performance of both text-based methods and our GNN-based method informed by socially infused text mining. We additionally compare the performance of GPT-4 for zero-shot stance labeling of users against the manual annotation by domain experts.

4 Results

We report the macro averaged precision, recall, and F1 score for each model, and we use the F1 score as the primary metric due to the class imbalance in the datasets. Metrics reported are averaged across five trials for all models.

Table 4. Results for stance labeling classification on the `Gun Control` dataset ($N = 350$). The best scores for each model type are shown in bold and the best overall scores are underlined.

Model	`GPT4 Labels`			`Manual Labels`		
	Prec.	Recall	F1	Prec.	Recall	F1
Weighted random	47.36	49.99	39.45	40.76	49.92	38.11
BERT-base-uncased	**84.25**	85.38	**84.66**	83.08	82.75	82.88
BERT-large-uncased	80.47	82.16	79.96	81.44	81.71	81.13
RoBERTa-base	83.33	83.94	83.60	84.79	83.99	84.25
RoBERTa-large	81.39	82.69	79.52	84.41	84.18	83.14
BERTweet	83.59	**85.43**	83.97	85.45	**85.76**	85.13
GPT-4	–	–	–	**89.81**	85.17	**87.05**
GraphSAGE	**87.58**	88.75	**88.02**	91.09	90.46	90.70
GAT	87.17	**88.99**	87.56	**91.28**	**91.46**	**91.36**

4.1 Gun Control Dataset

The classification results for the stance labeling models on the gun control dataset are shown in Table 4. On the gun control dataset, BERT and RoBERTa transformer methods achieve a macro averaged F1 score of between 82%–84% when evaluated on the set of manually annotated users. BERTweet, having been finetuned on tweets, surpasses the performance of the other transformer models achieving an F1 score of 85.13%. GPT-4 zero-shot classification outperforms the trained BERTweet classifier by two percent on F1 score and a four percent increase in precision while maintaining an equivalent recall performance.

For the gun control dataset, the GNN-based models that incorporate network structure information with the content-based representations of users produced by BERTweet, outperform both the trained text-based classifiers and zero-shot classification with GPT-4. Using graphSAGE and GAT to enrich users' content-based representations results in an improvement in stance labeling classification compared with the unenriched BERTweet embeddings. With GAT, incorporating users' local neighborhood information with their text-based embeddings improves classification performance by four percent in F1 score. Compared with zero-shot classification using GPT-4, the BERTweet + GAT model performs 3%–4% better in F1 score on classifying user stance labels for the manually labeled user set. Compared with graphSAGE, which takes the mean representation of a node's neighborhood effectively weighting each incident edge on the node equally, GAT utilizes edge softmax attention to dynamically compute edge weights when aggregating node neighborhood information. The dynamic computation of edge importance for users results in a marginal improvement in classification performance.

Table 5. Results for stance labeling classification on the `Climate Change` dataset ($N = 250$). The best scores for each model type are shown in bold and the best overall scores are underlined.

Model	`GPT4 Labels`			`Manual Labels`		
	Prec.	Recall	F1	Prec.	Recall	F1
Weighted random	47.36	49.99	39.45	45.26	47.55	42.87
BERT-base-uncased	76.74	79.82	77.36	78.60	82.54	79.36
BERT-large-uncased	75.63	79.14	73.86	76.11	80.28	74.17
RoBERTa-base	76.30	79.92	76.21	76.63	80.92	76.50
RoBERTa-large	75.72	79.36	75.16	76.91	81.33	76.29
BERTweet	**77.26**	**80.78**	**77.64**	78.35	82.66	78.78
GPT-4	–	–	–	<u>**85.36**</u>	<u>**86.08**</u>	<u>**85.71**</u>
GraphSAGE	<u>**84.62**</u>	86.06	<u>**85.25**</u>	**83.80**	85.91	**84.67**
GAT	83.73	<u>**86.43**</u>	84.70	83.47	**87.02**	84.62

4.2 Climate Change Dataset

The classification results for the stance labeling models on the climate change dataset are shown in Table 5. Compared with the gun control dataset, where pro gun control users outnumber anti gun control users by roughly a ratio of 2:1 in the heuristic labeled user set, the imbalance in the climate change dataset is significant with climate change believers outnumbering disbelievers by a ratio of nearly 10:1. We also observe that climate change related conversations are less topic-focused than gun control discussions on Twitter. Due in part to the severe class imbalance in the dataset in addition to the less focused of climate change conversation activity on Twitter, stance determination of users in the climate change dataset is more difficult than in the gun control dataset.

Of the BERT and RoBERTa based classifier models, BERT-base-uncased performs the best with an F1 score of 79.36%, narrowly outperforming BERTweet with an F1 score of 78.78%. Enriching the BERTweet user embeddings with structural information from the user interaction graph improves the classification performance by six percent to 84.67% F1 score. Despite the performance uplift from the user social network, for the climate change dataset zero-shot classification with GPT-4 outperforms all other models with an F1 score of 85.71%, outperforming other transformer methods by six percent and GNN-based approaches by one percent.

In all, enriching text-based representations of users with structural information from the user-interaction graph results in a significant improvement in classification performance compared with the text-based embeddings alone. Zero-shot classification of user stances using GPT-4 consistently outperforms BERT and RoBERTa-based classifier models, but incorporating social network information surpasses GPT-4 classification performance in the gun control dataset and significantly narrows the gap in the climate change dataset.

5 Discussion

When researching online polarization, serious divides exist in how questions regarding user stance are structured, and this divide is also reflected in how different researchers integrate the literature into future work. Oftentimes, this split can be seen in how social science researchers integrate computer science research into their studies and vice versa. This methodological gap can inhibit the understanding of the phenomena of online polarization and radicalization. Big data research can provide scalability not feasible with traditional qualitative social science research, while social science expertise provides understanding of the subtleties of these interactions on human behavior. As seen in the aftermath of events like radicalized mass shooters, the online behavior of these individuals is not well understood, and we need to understand how and why these processes take place online. The integration of subject-matter expertise with AI is necessary for the application of big data methodology into human-level polarization research and policy work, requiring a balance of the nuance provided by social science subject-matter experts with the speed and volume provided by computation experts. Qualitative research offers tools like narrative analysis to analyze the cultural and historical contingency of the terms, beliefs, and issues narrators address in their writings [22]. The small narratives created from tweets can provide insights into the range of ongoing societal arguments that create collective identity online [5]. The wealth of this data adds critical information to understanding how people discuss their thoughts and opinions about polarizing topics like gun rights or climate change, and the ability to combine these two approaches can greatly benefit the understanding of online polarization.

One area of difficulty in automating stance detection on topics like gun control or climate change is the need to dichotomize beliefs that really exist on a spectrum. The extreme ends of these beliefs are straight-forward to categorize, but the majority of people exist in the space in-between. A better goal is to qualitatively code the tipping point as people's stance moves between stances on issues. Automated stance analysis can mislabel users because of difficulties in discerning written cues like tone, connotation, coded words, or pictures. Some of these coding errors are due to a purposeful misdirection of a tweet that is clarified by the picture. For example, user 2945287090 said "I'm pro-choice are you? https://t.co/WtYW6zohSB" but the picture has them wearing a shirt that says pro-choice with an array of guns to "choose" from. Since the picture cannot be assessed by many stance labeling algorithms, these misdirections would be potentially miscoded. There are also users who frame their arguments in neutral language or the language of their opposition. For example, user 66811907 states "Gun control advocates do not have a monopoly on outrage or sorrow. They do not have a monopoly on ideas for addressing gun violence. They do not have a monopoly on wanting to fix the problem." The larger context of this user's tweets balance being pro-Second Amendment and being critical to the responses to mass shooting events. As a whole, this user falls more strongly on the side of being anti-gun control but certainly has criticism of open market gun purchasing. The neutral framing can be difficult to for automated approaches to parse since the full narrative of the users' tweets need to be taken into account.

However, coding errors are not only on the side of computational approaches. One of the key limitations to using only human-based qualitative coding is that the deep understanding of stances is highly time-consuming and delayed—especially in online space where posting moves rapidly. This means that qualitative researchers are forced to limit their focus to a subset of tweets or users, and balancing efficiency and accuracy means that there will be miscoded information that could potentially be corrected by looking at all the available data. The ability to integrate big data and computational methods can be a force multiplier for qualitative researchers as a mixed methods approach greatly improves the scope and generalizability of work on online polarization.

As mentioned previously, these stances exist on a spectrum and the middle ground can often be coded either way by a researcher making a qualitative decision on which end of spectrum the user exists. For example, is a person who owns guns for hunting but pushes for very strong hand gun control measures, pro-gun control or anti-gun control? Does wanting strong background checks or other restrictions inherently make someone pro/anti gun? Even in the context of a large body of tweets, it may not be fully clear how this stance should be coded. One example is user 66811907 who has a large body of tweets that different coders could code differently "This is your friendly reminder that New York already bans "assault weapons", that the shooter procured his NY-compliant "non-assault weapon" legally, and that the cops who stopped him showed up with their "assault weapon" version of the same gun to combat a civilian threat." and "Rifles of any kind account for fewer annual homicides than hands and feet. Mass shooters don't need pistol grips, etc., to cause horrific carnage, but on the flip side those features can be quite useful for lawful owners using them in self-defense.

That's why cops carry them, too." This example is just one that highlights that neither AI nor human coding is without issues, error, or disagreement and the reporting of inter-rater reliability could be a useful measure to add for both human annotators and automated approaches.

6 Conclusion

In this work, we take a step towards a broader understanding of user stance labeling through socially infused text mining—incorporating both textual content and network interactions using language models and graph neural networks. Both gun control and climate change are issues that have a range of historical contexts, political connotations, and intersections with race and gender. It is in this space that subject-matter expertise can provide the contextual information to inform the work of computational approaches. Social science insights provide the ground truths necessary to properly contextualize computational approaches to complex human stances to ensure that the models are acknowledging the real-world context, minimizing errors, and maximizing the relevance of the generated outputs to policymakers and to the broader research community.

Acknowledgments. The research was sponsored by the Army Research Office and was accomplished under Grant Number W911NF-22-1-0035. The views and conclusions contained in this document are those of the authors and should not be interpreted as representing the official policies, either expressed or implied, of the Army Research Office or the U.S. Government. The U.S. Government is authorized to reproduce and distribute reprints for Government purposes notwithstanding any copyright notation herein.

Disclosure of Interests. The authors have no competing interests to declare that are relevant to the content of this article.

References

1. Addawood, A., Badawy, A., Lerman, K., Ferrara, E.: Linguistic cues to deception: identifying political trolls on social media. In: ICWSM, vol. 13, pp. 15–25 (2019)
2. ALDayel, A., Magdy, W.: Stance detection on social media: state of the art and trends. Inf. Process. Manage. **58**(4), 102597 (2021). https://doi.org/10.1016/j.ipm.2021.102597
3. Badawy, A., Ferrara, E., Lerman, K.: Analyzing the digital traces of political manipulation: the 2016 Russian interference Twitter campaign. In: ASONAM, pp. 258–265 (2018). https://doi.org/10.1109/ASONAM.2018.8508646
4. Barberá, P., Jost, J.T., Nagler, J., Tucker, J.A., Bonneau, R.: Tweeting from left to right: is online political communication more than an echo chamber? Psychol. Sci. **26**(10), 1531–1542 (2015). https://doi.org/10.1177/0956797615594620
5. Coupland, C., Brown, A.D.: Constructing organizational identities on the web: a case study of Royal Dutch/Shell. J. Manage. Stud. **41**(8), 1325–1347 (2004)

6. Darwish, K., Stefanov, P., Aupetit, M., Nakov, P.: Unsupervised user stance detection on Twitter. ICWSM **14**(1), 141–152 (2020). https://doi.org/10.1609/icwsm.v14i1.7286
7. Garimella, K., Morales, G.D.F., Gionis, A., Mathioudakis, M.: Quantifying controversy in social media. CoRR abs/1507.05224 (2015)
8. Hamilton, W., Ying, Z., Leskovec, J.: Inductive representation learning on large graphs. In: NeurIPS, pp. 1024–1034 (2017)
9. Jiang, J., Ren, X., Ferrara, E.: Social media polarization and echo chambers in the context of Covid-19: case study. JMIRx Med. **2**(3), e29570 (2021)
10. Jiang, J., Ren, X., Ferrara, E.: Retweet-BERT: political leaning detection using language features and information diffusion on social networks. In: ICWSM, vol. 17, pp. 459–469 (2023)
11. Johnson, K., Jin, D., Goldwasser, D.: Modeling of political discourse framing on Twitter. In: ICWSM, vol. 11, pp. 556–559 (2017)
12. Karlsen, R., Steen-Johnsen, K., Wollebaek, D., Enjolras, B.: Echo chamber and trench warfare dynamics in online debates. Eur. J. Commun. **32**(3), 257–273 (2017). https://doi.org/10.1177/0267323117695734
13. Kipf, T.N., Welling, M.: Semi-supervised classification with graph convolutional networks. In: ICLR (2017)
14. Kossinets, G., Watts, D.J.: Origins of homophily in an evolving social network. Am. J. Sociol. **115**(2), 405–450 (2009)
15. Kovacs, B., Kleinbaum, A.M.: Language-style similarity and social networks. Psychol. Sci. **31**(2), 202–213 (2020). https://doi.org/10.1177/0956797619894557
16. Li, C., Goldwasser, D.: Encoding social information with graph convolutional networks for political perspective detection in news media. In: ACL, pp. 2594–2604 (2019)
17. McPherson, M., Smith-Lovin, L., Cook, J.M.: Birds of a feather: homophily in social networks. Ann. Rev. Sociol. **27**(1), 415–444 (2001)
18. Nguyen, D.Q., Vu, T., Tuan Nguyen, A.: BERTweet: a pre-trained language model for English tweets. In: Liu, Q., Schlangen, D. (eds.) EMNLP, pp. 9–14. ACL, October 2020. https://doi.org/10.18653/v1/2020.emnlp-demos.2
19. Pan, S., Wu, J., Zhu, X., Zhang, C., Wang, Y.: Tri-party deep network representation. In: IJCAI, pp. 1895–1901. AAAI (2016)
20. Peretti-Watel, P., et al.: A future vaccination campaign against Covid-19 at risk of vaccine hesitancy and politicisation. Lancet. Infect. Dis. **20**(7), 769–770 (2020)
21. KhudaBukhsh, A.R., Sarkar, R., Kamlet, M.S., Mitchell, T.: We don't speak the same language: Interpreting polarization through machine translation. AAAI **35**(17), 14893–14901 (2021). https://doi.org/10.1609/aaai.v35i17.17748
22. Rosenwald, G.C., Ochberg, R.L.: Storied Lives: The Cultural Politics of Self-understanding. Yale University Press (1992)
23. Tyagi, A., Uyheng, J., Carley, K.M.: Affective polarization in online climate change discourse on Twitter. In: ASONAM, pp. 443–447 (2020). https://doi.org/10.1109/ASONAM49781.2020.9381419
24. Veličković, P., Cucurull, G., Casanova, A., Romero, A., Lió, P., Bengio, Y.: Graph attention networks. In: ICLR (2018)
25. Xiao, Z., Song, W., Xu, H., Ren, Z., Sun, Y.: TIMME: Twitter ideology-detection via multi-task multi-relational embedding. In: SIGKDD, pp. 2258–2268 (2020)
26. Xu, K., Hu, W., Leskovec, J., Jegelka, S.: How powerful are graph neural networks? In: ICLR (2019)

27. Yang, Y., Eisenstein, J.: Overcoming language variation in sentiment analysis with social attention. Trans. Assoc. Comput. Linguist. **5**, 295–307 (2017)
28. Yarchi, M., Baden, C., Kligler-Vilenchik, N.: Political polarization on the digital sphere: a cross-platform, over-time analysis of interactional, positional, and affective polarization on social media. Polit. Commun. **38**(1–2), 98–139 (2021). https://doi.org/10.1080/10584609.2020.1785067

Enhancing School Success Prediction with FRC and Merged GNN

Melike Yildiz Aktas[1(✉)], Aadyant Khatri[1], Mariam Almutairi[1], Lulwah Alkulaib[2], and Chang-Tien Lu[1]

[1] Department of Computer Science, Virginia Tech, Falls Church, VA 22043, USA
{melike,aadyant,malmutair,lalkulaib,clu}@vt.edu
[2] Department of Computer Science, Kuwait University, Kuwait City, Kuwait

Abstract. This study presents an innovative framework for predicting school success, leveraging Large Language Models (LLMs) to define ground truth labels based on comprehensive school information with Factor-Reasoning-Classification (FRC) prompting. In this article, we conceptualize school data as a complex social network and create two different graphs where schools are defined as nodes. Similarity Graph captures school similarities, integrating factors such as graduation rates, ACT scores, socioeconomic conditions, crime rates, and community resources. Geographic Proximity Graph models spatial relationships among schools using geographical coordinates. We define Merged GNN that enhances prediction accuracy by incorporating both similarity-based and spatial proximity-based information. Our approach leverages Graph Neural Networks (GNN) to predict the most probable labels the LLM model identifies. Experimental results on the school success dataset not only demonstrate the superior predictive performance of our methodology over baseline models but also highlight the importance of integrating diverse sources of information for accurate prediction and analysis.

Keywords: Data labeling · School success prediction · Graph neural network · Classification · Spatial analysis

1 Introduction

The choice of high school profoundly impacts students' futures including college admissions and career trajectories. Yet, public school quality is often affected by external factors like geographic location and community resources [1]. Given the significant impact of these factors on educational outcomes, understanding school performance is essential. Prioritizing school achievement prediction empowers parents, urban planners, and policymakers to make informed decisions for societal benefit. A school's success hinges on several factors, which can be grouped into four categories. First, school climate like teaching environment and safety of students are pivotal [2]. Second, crime rates can directly impact the educational environment [3]. Third, social amenities such as libraries and parks

L. M. Aiello et al. (Eds.): ASONAM 2024, LNCS 15213, pp. 262–277, 2025.
https://doi.org/10.1007/978-3-031-78548-1_20

enrich the learning experience [4]. Finally, socioeconomic factors like employment rates, income distribution, and community health can influence the educational landscape. Understanding these socioeconomic dynamics is crucial for enhancing educational equity and opportunity [5].

Students' prior academic achievements, demographic background, and psychological traits are widely recognized as the primary factors influencing the prediction of academic success [6,7]. Predicting school and student success can be determined by factors such as residential crowding [8], crime rates [9,10], socioeconomic background [11,12], and mental [13] and physical health [14] of the community or individuals.

In our study, the prediction of school success rather than individuals is highlighted. We present an innovative approach for predicting school success by integrating two distinct graph structures. The first graph is constructed based on the similarity of schools, utilizing various parameters such as graduation rate, educational attainment levels, socioeconomic conditions, crime rates, and the availability of community resources such as parks and libraries. This graph encapsulates the multidimensional aspects of school environments and their surrounding communities, comprehensively representing school characteristics. The second graph is built using geographical coordinates (latitude and longitude) to capture the spatial proximity between schools. By exploiting the geographic proximity of different locations, this graph allows us to model the spatial dependencies, reflecting their geographical relationships.

For ground truth labeling of school success, we present LLM-based Factor-Reasoning-Classification prompting. In our method, critical factors that might impact school success are asked, encompassing both positive and negative influences, and is provided a brief explanation of the diagnostic reasoning process, explaining the rationale behind the classification of each school's success level. Drawing upon the identified factors and diagnostic reasoning, the model assigns success classes (high, medium, or low) to each school. Merged GNN is designed to predicts the success levels that our prompt model assigns. Through this comprehensive approach, our methodology aims to provide a robust framework for assessing and understanding school performance.

Experimental evaluations conducted on real-world school success dataset demonstrates the effectiveness of our proposed approach. Comparative analyses against baseline models showcase the superior predictive performance achieved by integrating multiple graph structures. Our findings underscore the importance of incorporating both similarity-based and spatial proximity-based information for accurate school success prediction, highlighting the potential of graph-based techniques in education analytics.

We summarize our contribution as follows:

- **Data labeling prompting:** We introduce a novel data labeling prompting strategy utilizing Large Language Models (LLMs), which enables the automatic definition of school success classes based on comprehensive information that reduces the need for human labeling.

- **Predictive modeling (Merged GNN) with MergeGraph:** We advance predictive modeling techniques by employing different types of message passing of GNN, a cutting-edge approach for analyzing graph-structured data.
- **Analyzing comprehensive school success datasets:** We introduce an approach of combining heteregenous data sources into a single framework, enabling a nuanced understanding of school success rather than solely focusing on student success.
- **Conducting extensive experiment:** We investigate an experiment on a real-world Chicago school dataset to demonstrate that our model is capable of capturing the shape of success classification.

2 Related Work

In this section, three groups of related work are discussed: 1) School and student success, 2) LLM-Based data classification and labeling, and 3) Node classification with graph neural networks.

2.1 School and Student Success

The discovery of factors affecting academic success is an ongoing field of study as the definition of success evolves over time. Emotional and social competencies play a pivotal role in self-awareness, higher educational attainment, improved mental and emotional health, and a decrease in engagement with criminal activities [13]. Conversely, it's evident that reductions in crime rates and improvements in overall welfare also contribute to the pathways to success. Woehr and Newman [15] demonstrated that the presence of dogs can alleviate stress among elementary and high school students, shedding light on innovative methods to support student well-being. Bastedo et al. [16] concluded that both raw and contextualized measures of high school GPA, along with ACT scores, serve as robust predictors for underrepresented students, influencing outcomes such as freshman college GPA, retention rates, and graduation success. Additionally, providing a positive school environment has been identified as a powerful intervention to improve academic performance and promote holistic student well-being by Daily et al. [2]. Rodriguez et al. [17] delved into the interaction between socioeconomic status and academic achievement, finding a positive correlation, highlighting social and economic factors in shaping educational outcomes.

Hashim et al. [7] compared the performances of several supervised machine learning algorithms, such as Decision Tree, Naïve Bayes, Logistic Regression, Support Vector Machine, K-Nearest Neighbour, Sequential Minimal Optimisation, and Neural Network for forecasting student success in bachelor study programmes. Malik and Jothimani [18] proposed a framework to choose important features that impact student quality and reduce dropout rates. Natek and Zwilling [19] conducted a study focusing on specific students, analyzing their academic records and family backgrounds to predict performance, utilizing models such as Rep Tree, J48, and M5P. Khalaf et al. [20] forecasted exam outcomes leveraging student-filled questionnaires on social activities. In the context of predicting student dropout, Yukselturk et al. [21] explored various models, including K-Nearest Neighbour, Decision Tree, Naïve Bayes, and Neural Network.

2.2 LLM-Based Data Classification and Labeling

Recently, large language models (LLMs) have drawn significant attention for tasks like text classification, data annotation, and content creation. Trained on vast text and multi-modal data, these models excel in identifying patterns and connections, enabling them to tackle various NLP tasks with exceptional adaptability and generalization [22–24].

One of the most significant benefits of LLMs is their ability to reduce the need for extensive data annotations, which can be especially difficult to obtain for complex tasks. Sushil et al. [22] investigated the use of LLMs in breast cancer pathology classification, showcasing their potential to alleviate the burden of extensive data labeling. Bansal et al. [23] and Clavié et al. [24] also emphasized the importance of LLMs in improving the generalization of NLP models through innovative annotation techniques. Sun et al. [25] propose clue and reasoning prompting (CARP) for the task of text classification. Gilardi et al. [26] found that ChatGPT outperformed crowd workers in tasks such as relevance, stance, topic, and frame detection, with a zero-shot accuracy approximately 25% points higher than that of human annotators across different datasets. This highlights not only the efficiency of LLMs in text classification but also their ability to handle tasks with high accuracy and reliability.

Lastly, Wang et al. [27] addressed the scalability of text classification by proposing a set of best practices for deploying LLMs to classify large volumes of comments efficiently and cost-effectively. These practices underscore the practical implications of LLMs in real-world settings, where they can offer substantial improvements over traditional methods.

2.3 Node Classification with Graph Neural Networks

Node classification, the task of predicting a category for a node in a network, is an active area of research with various approaches being developed. Among these, semi-supervised learning methods play a significant role in leveraging both labeled and unlabeled data to improve classification accuracy. Abu-El-Haija et al. [28] introduced N-GCN, employing multi-scale graph convolution to capture information from nodes at varying distances in the network. Wang et al. [29] proposed NodeAug, which utilizes data augmentation techniques tailored for graphs to enhance model generalizability and classification performance. Yang et al. [30] developed MGCN, employing contrastive learning to effectively leverage information from labeled and unlabeled nodes, resulting in improved classification accuracy in semi-supervised settings. Xu et al. [31] introduced LC-GNN, emphasizing the importance of label consistency and enlarging the receptive field of nodes to improve classification accuracy.

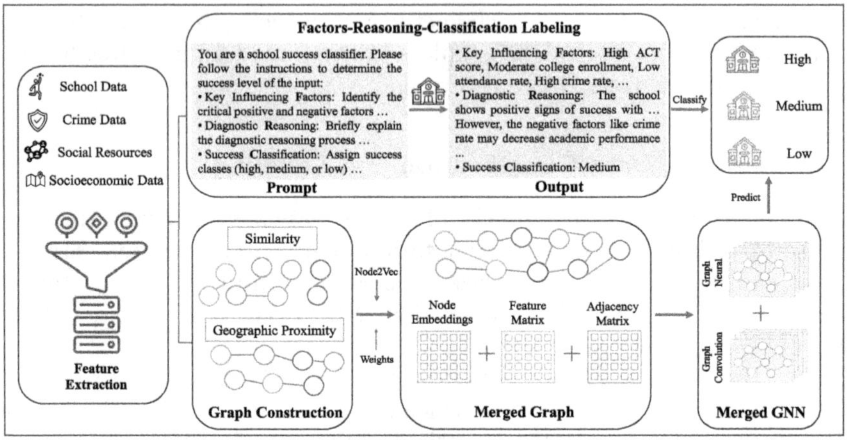

Fig. 1. School Success Prediction with FRC and Merged GNN

In multi-label classification, Xu et al. [32] introduce AdaNN, leveraging gated recurrent units (GRUs) to combine node attributes and network topology at different time points, facilitating adaptation to dynamic graph changes. Zhou et al. [33] propose LANC, employing a label-attentive convolution mechanism to focus on informative neighboring nodes for each label, enhancing classification accuracy in multi-label tasks. Zan et al. Zhang et al. [34] present a novel multi-label relational classifier that identifies similar nodes based on local network structure performs clustering on nodes with known labels.

3 Methodology

In this study, we aim to precisely define and predict the success levels of schools by analyzing a comprehensive set of features. Our methodology is structured around three core components: robust feature extraction, innovative FRC (Factors-Reasoning-Classification) prompting, and strategic Merged Graph integration with Merged GNN, each contributing uniquely to the predictive accuracy of our model. We showed our framework in Fig. 1.

3.1 Preprocessing and Feature Extraction

In the feature extraction phase of our study, we utilized a comprehensive approach that integrates multiple datasets, enabling us to effectively capture the intricate dynamics of school environments and their communities. Our dataset included comprehensive details about the spatial distribution of key amenities such as parks and libraries within a 2 km radius of each school, enhancing our analysis with critical insights into the accessibility of important recreational and educational resources. To further enrich our feature set, we grouped schools using the K-means clustering into four different feature categories: school-related metrics, socioeconomic indicators, social amenities, and safety statistics. We determined the optimal number of clusters for each category using the elbow method, enabling precise capture and analysis of the underlying patterns and interrelationships within the data.

Additionally, to mitigate issues with missing data, we implemented an imputation strategy that involves replacing missing values with the mean of available corresponding features for each school. For example, if attainment data were missing for a given year, we calculated the mean from the data of adjacent years to ensure our dataset's continuity and integrity.

3.2 Factors-Reasoning-Classification Labeling

To determine the success levels of schools, we developed a specialized prompting method tailored for Large Language Models (LLMs). This innovative approach fosters a gradual decision-making environment.

The method entails a systematic approach consisting of three steps:

- **Key Influencing <u>Factors</u> Identification:** The model prompts to identify critical factors that influence school success, both positively and negatively. This initial step is instrumental in explaining the multifaceted aspects contributing to a school's overall performance.
- **Diagnostic <u>Reasoning</u> Process:** Model provides a concise explanation of the diagnostic reasoning process, justifying the classification of each school's success level. Model takes the positive and negative influencing factors that might affect school success. This step enables the expression of the thought process behind their classification decisions, enhancing transparency and interpretability.
- **Success <u>Classification</u>:** Based on the identified factors and diagnostic reasoning, the model assigns success classes (high, medium, or low) to each school. This final step synthesizes the input data and user rationale to produce actionable insights regarding school success levels.

This prompting model processes by ingesting the individual features of each school as inputs for detailed analysis. Once the analysis is complete, it assigns a specific class label high, medium, or low to each school based on the insights gained. These class labels are compared to human evaluation results and then utilized as the ground truth within our Merged GNN methodology. This integration significantly enhances both the accuracy and reliability of our overall approach.

3.3 Graph Construction

We conceptualize schools as nodes within a network, with their connections representing various interrelationships. This setup forms the basis of a node classification problem, where schools are systematically categorized into three distinct success levels: high, medium, and low.

In the graph construction phase, we define two distinct graph structures, denoted as $G_m = (V, E_m)$, where m specifies the graph (either Similarity Graph (1) or Geographic Proximity Graph (2)), and each graph contains a set of n nodes $V = \{v_1, v_2, \ldots, v_n\}$. The adjacency matrix for each graph is represented as $A_m \in \mathbb{R}^{n \times n}$, and the node feature matrix is indicated by $X \in \mathbb{R}^{n \times F}$, where F represents the number of features for each node. For Similarity Graph, edges are defined based on the cosine similarity between nodes; an edge e_{ij}^m between nodes v_i and v_j is established if their cosine similarity exceeds a predetermined threshold (62%), with the edge weight equal to this similarity value. For Geographic Proximity Graph, edges are determined by the haversine distance between nodes, where an edge is formed if the distance is below another specified threshold (5 km). These thresholds ensure that each node is connected to at least one other node in the graph. The ultimate objective is to use these graph structures to predict success levels, represented by $Y = \{Y_1, Y_2, \ldots, Y_n\}$, where Y_i is labeled as 2 for high, 1 for medium, and 0 for low success.

We employed the Node2Vec algorithm [35] to generate node embeddings that effectively capture the topological proximity of nodes within our graphs. Node2Vec operates by taking a graph $G(V, E)$ as input where V and E represent nodes and edges respectively and it learns an n-dimensional embedding for each node. The learned embeddings are then represented as follows:

$$\text{node2vec}(G = (V, E)) \rightarrow \mathbb{R}^{n \times c} \quad (1)$$

where c indicates the number of embedding dimensions per node. These embeddings are integrated with the original node features $X \in \mathbb{R}^{n \times F}$, resulting in an enriched node feature matrix $X \in \mathbb{R}^{n \times (F+c)}$, which combines both the original features and the new embeddings, thus updating the node features for enhanced graph analysis.

3.4 Merged GNN

To leverage the distinct topological information embedded in the two types of graphs representing our school network, we implement a Merged Graph Neural Network (Merged GNN) described in Algorithm 1. This approach combines the node embeddings generated from both graphs using the node2vec algorithm, along with the base node features common to each graph. We called Similarity Graph as G_1 and Geographic Proximity GraphG_2 in the formula. The resultant new node feature matrix is formalized for use in Merged GNN as follows:

$$X_{\text{new}} = \text{Concat}(\text{node2vec}(G_1), \text{node2vec}(G_2), X) \quad (2)$$

We create a combined adjacency matrix by summing the adjacency matrices of the two graphs, each weighted appropriately to optimize the model's performance. We denote the weights for Similarity Graph as γ which is a learnable parameter and for Geographic Proximity Graph as $1-\gamma$. The new adjacency matrix is updated as:

$$A_{\text{new}} = (\gamma \times A_1 + (1-\gamma) \times A_2) \tag{3}$$

In our node classification approach, we employ both Graph Convolutional Network (GCN) [36] and Graph Neural Network (GNN) [37] techniques on the merged graph structure. Given the merged graph G with node feature matrix $X_{\text{new}} \in \mathbb{R}^{n\times(F+2c)}$ and the combined adjacency matrix A_{new}, the GCN operates as follows:

$$H = \widetilde{A}_{\text{new}} X_{\text{new}} \Theta \tag{4}$$

where $\widetilde{A}_{\text{new}}$ is the normalized version of A_{new}, calculated by:

$$\widetilde{A}_{\text{new}} = D^{-\frac{1}{2}} A_{\text{new}} D^{-\frac{1}{2}} + I \tag{5}$$

with D representing the diagonal node degree matrix and I the identity matrix. Each GCN layer i updates as:

$$H^i = \widetilde{A}_{\text{new}} H^{i-1} \Theta \tag{6}$$

and the output from the final layer is computed by:

$$H^{\text{final}} = \sigma(H^l \Theta) \tag{7}$$

Here, σ represents the ReLU activation function, Θ learnable parameter, and l the number of hidden layers.

The GNN component begins by defining the neighborhood N_i for each node v_i, consisting of nodes v_j directly connected to v_i:

$$N_i = \{j : e_{ij} \in E\} \tag{8}$$

Message Passing in GNNs involves transforming node features using a function F, aggregating these transformations from neighbors:

$$F(x_j) = W_j \cdot x_j + b \tag{9}$$

$$\bar{m}_i = G\left(\{W_j \cdot x_j : j \in N_i\}\right) \tag{10}$$

Node features are updated by combining the node's current feature vector with the aggregated messages:

$$h_i = \sigma(K(H(x_i), \bar{m}_i)) \tag{11}$$

where K is a projection neural network (MLP) that integrates the features into a new dimension.

By integrating the outputs from both GCN and GNN layers, we derive the final predictions for our node classification task, effectively capturing both local and global structural information. In Algorithm 1, linear refers to a linear transformation applied to combine features, and softmax is used to normalize the output logits into probabilities.

Algorithm 1. Merged Graph Neural Network (Merged GNN) for School Success Prediction

Require: Graph adjacency matrix A, node features X, number of classes C
Ensure: Predicted class labels for each node
Step 1: Initialize Graph Convolutional Network Layers
Setup transformation matrices $\Theta^{(1)}, \Theta^{(2)}$ for GCN layers
Setup weight matrix W for GNN layer
Step 2: Apply Graph Convolutional Network Layers
for $i = 1$ to 2 **do**
    if $i == 1$ **then**
        $H^{(1)} \leftarrow \sigma(\widetilde{A}X\Theta^{(1)})$ ▷ First GCN layer
    else
        $H^{(2)} \leftarrow \sigma(\widetilde{A}H^{(1)}\Theta^{(2)})$ ▷ Second GCN layer
    end if
end for
Step 3: Apply Graph Neural Network Operations
$H^{(GNN)} \leftarrow \sigma(A(XW))$ ▷ GNN operation using learned weights
Step 4: Integrate GCN and GNN Outputs
$H^{(combined)} \leftarrow H^{(2)} + H^{(GNN)}$
Step 5: Class Prediction
$Z \leftarrow \text{softmax}(\text{Linear}(H^{(combined)}))$ ▷ Compute class probabilities
    return Z ▷ Return the class probabilities

4 Experiments

To assess the performance of our proposed model, FRC, and MergeGraph, we conducted extensive experiments to validate its effectiveness.

4.1 Datasets

We are utilizing various datasets sourced from the City of Chicago and the Chicago Metropolitan Agency for Planning (CMAP) [38] for our research. These datasets include information on crime, parks, libraries, boundaries, demographic, and economic data as shown in Table 1. For the Chicago School Dataset, we analyze essential metrics such as graduation rate, college enrollment percentage, and attendance rate for each school over the years 2014, 2015, and 2016. The Chicago Crime Dataset provides comprehensive crime data from the Chicago Data Portal [39]. We focus on specific crime types robbery, arson, homicide, and offense and analyze their correlation with school success. Crime rates around each school within a 2 km radius are aggregated for each distinct year.

Incorporating region-specific public health data [40] enhances our dataset with crucial health statistics related to each school's geographic location. This enrichment includes features such as teen birth rate, poverty level, housing conditions, and per capita income sourced from the Community Data Snapshots (CDS) project. We merge data on Chicago Public Parks [39] and libraries [41] with school data based on latitude and longitude features. Our analysis extends to parks conducive to teenage engagement, focusing on amenities suitable for teenagers. Additionally, we consider the count of libraries near a school, acknowledging their impact on learning opportunities and access to resources like computers, books, and the internet within a 2 km radius.

Table 1. Dataset Description

Node Features	Types
School Dataset	Graduation Rate, Attainment, ACT Scores, College Enrollment, Attendance Rate
Crime Dataset	Robbery, Offense, Arson, Homicide
Socioeconomic Dataset	Race, Age, Employment, and Income Distributions, Community Health Data
Social Resource Dataset	Parks and Libraries

4.2 Baselines

We compared our proposed method with five baselines that can be categorized into two groups: graph-based methods and traditional machine-learning models. To compare traditional machine learning models, we used node embeddings and node features as tabular data.

Graph-based methods:

- GNN [37]: Predicts school success labels by learning node representations based on the underlying graph structure.
- GCN [36]: Classifies school nodes by leveraging their spatial and temporal features along with the features of their neighboring nodes.
- Graph Sage [42]: Aggregates information from neighboring nodes and learns representations based on school features and the collective information of neighboring schools and their features.

Traditional machine learning methods:

- XGBoost [43]: Classifies schools by identifying complex patterns and relationships among the provided features.
- Random Forest [44]: Captures the relationships between input variables and the target by constructing decision trees based on subsets of school features.

4.3 Evaluation Metrics and Experiment Settings

We compared the results obtained with the learned weights in the final adjacency matrix required for our proposed model with the results obtained by trying various weight configurations. Additionally, we conducted a comparative analysis between our model and baseline methods. The performance of our model was assessed using key metrics such as F1-Score and accuracy.

For FRC prompting, we leveraged the Chat-GPT 3.5 Turbo model due to its advanced capabilities, which offered several benefits to our study. With its vast language understanding and generation capabilities, this model enabled more precise predictions. To facilitate manageable input sizes and enhance the computational workflow, we imposed a maximum token limit of 256.

5 Results

5.1 LLM-Based Data Labeling Results

To assess the efficacy of the FRC prompting method, we conducted a comparative analysis against human-labeled data. Three individuals assigned success labels to 50 schools based on the subset of the dataset that is provided to the LLM. We determined the label for each school by selecting the most frequently assigned category. For instance, if two evaluators labeled a school as "medium" while another labeled it as "high", we assigned the label "medium" to that school as human-labeled data and compared it to prompt labels. Figure 2 shows the confusion matrix of labeling. While our prompt model doesn't suggest that a school commonly perceived as "high" by humans is actually "low", only one of the schools identified as "high" by the model is deemed "low" by humans. Regarding "medium" schools, both human evaluation and our model generally yield similar assessments. Notably, there are instances where schools classified as "high" by the model are labeled as "medium" by humans.

We compared human-labels and prompting result to reach F1-score, precision, and recall. Table 2 presents the convergence of our model's scores with those assigned by humans across various evaluation metrics. Remarkably, our model demonstrated comparable performance in all aspects. On average, there's a 65% similarity between human evaluations and prompting results. Consequently, the LLM enabled us to discern all labels without the need for additional human labeling efforts, showcasing its efficiency and accuracy in predictive tasks.

Table 2. FRC and Human Evaluation Comparison

Class	F1-Score	Precision	Recall
High	0.6316	0.5455	0.7500
Medium	0.6522	0.7143	0.6000
Low	0.6857	0.6667	0.7059

5.2 Merged GNN Performance with Different Weights

In our study, we put together two distinct graph structures: one defining the interconnectivity of schools based on node similarity (Similarity Graph), while

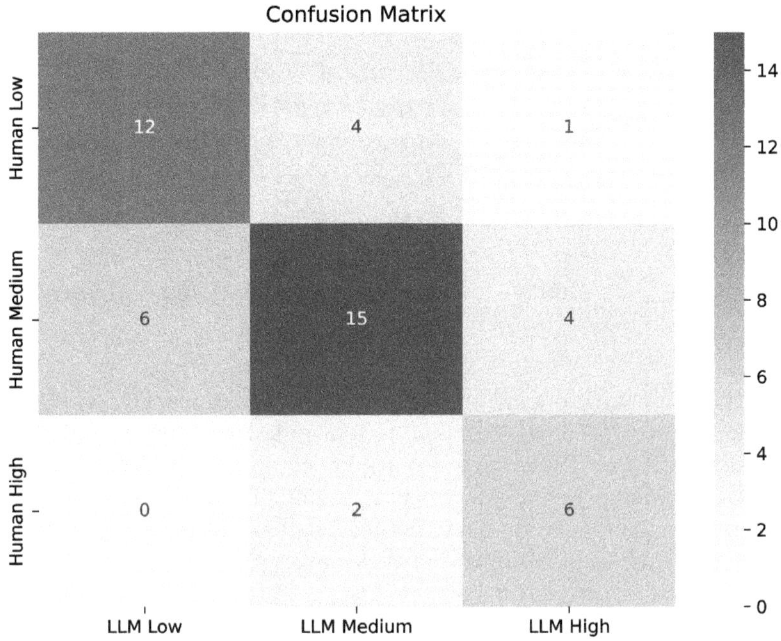

Fig. 2. Confusion Matrix of Human-Labels and LLM-Labels

the other encapsulated the geographical proximity between schools (Geographic Proximity Graph). As detailed in Table 3, we systematically evaluated the impact of different weight distributions on model performance and compared the results of these weights with the weights learned by our model. The Merged GNN assigns weights of 0.3063 and 0.6937 to the Similarity Graph and Geographic Proximity Graph, respectively. The near balance between these weights at (0.3, 0.7) leads to highly competitive results. This distribution outperforms alternative weight assignments. This strategic weighting strategy optimally leverages the unique characteristics of each graph, enhancing the predictive capabilities of our model.

5.3 School Success Prediction Results

As shown in Table 4, our proposed methodology, FRC and Merged GNN, demonstrates notable superiority when compared to existing graph neural network models, XGBoost, and Random Forest classifier. Our proposed model achieves an accuracy of 0.897 and an F1 score of 0.8928, indicating its efficacy in school success prediction task. Comparative analysis reveals the individual performance of graph neural network (GNN), graph convolutional network (GCN), and GraphSAGE on the constituent graphs. Specifically, while GNN achieves better results for Geographic Proximity Graph (G2) with 0.5297, GCN attains 0.7335 and 0.5925 for Similarity Graph (G1) and Geographic Proximity Graph (G2), respectively. Graph SAGE gives similar results with random forest in terms of Similar-

Table 3. Weights For Merge Graph

Weight for G1	Weight for G2	Accuracy	F1 Score	Precision	Recall
0.9	0.1	0.682	0.6711	0.7069	0.6753
0.7	0.3	0.7568	0.7486	0.7858	0.7486
0.5	0.5	0.7872	0.7817	0.8056	0.7906
0.3	0.7	0.897	0.8928	0.8931	0.893
0.1	0.9	0.738	0.7299	0.7578	0.7377
0.3063	**0.6937**	**0.8951**	**0.8914**	**0.892**	**0.8915**

ity Graph (G1) but fails to learn from Geographic Proximity Graph (G2). Moreover, Random Forest predicts classes in Geographic Proximity Graph (G2) with 70% accuracy, whereas XGBoost's results hover around 0.50 for both graphs. These findings underscore the effectiveness of our proposed model in leveraging the complementary information from merged graphs, offering a robust framework for predictive modeling in complex network data.

Table 4. Comparison

Models	Graphs	Accuracy	F1 Score	Precision	Recall
GNN	G1	0.4053	0.3944	0.3946	0.3952
	G2	0.5297	0.4949	0.5642	0.5148
GCN	G1	0.7335	0.729	0.7545	0.7301
	G2	0.5925	0.5552	0.6266	0.577
Graph SAGE	G1	0.5741	0.4771	0.4887	0.5494
	G2	0.2556	0.1357	0.0852	0.3333
XGBoost	G1	0.4815	0.4618	0.4588	0.4815
	G2	0.5926	0.5956	0.6005	0.5926
Random Forest	G1	0.5556	0.5302	0.5377	0.5556
	G2	0.7037	0.6956	0.7007	0.7037
Merged GNN	**Merged Graph**	**0.897**	**0.8928**	**0.8931**	**0.893**

6 Conclusion

As a consequence, our investigation attempts to forecast school achievement by explaining the multifaceted influences. We introduce a novel methodology by leveraging Factor-Reasoning-Classification (FRC) prompting and Merged GNN integration. The FRC prompting method facilitates the automated identification of school achievement labels through comprehensive knowledge and diagnostic reasoning. Simultaneously, the Merged GNN technique combines two distinct graph structures, one based on school similarity and the other on geographic proximity, to provide a comprehension of school attributes and interconnections. Through extensive experiments on real-world datasets, our methodology shows significant improvements over baseline models. Our approach achieves superior performance in school achievement prediction by effectively integrating multiple graph structures and leveraging advanced predictive modeling techniques.

Acknowledgment. Melike Yildiz Aktas is financially supported by the Turkish Ministry of National Education for her PhD research. Many thanks to Dr. Taoran Ji, Dr. Abdulaziz Alhamadani, and Shailik Sarkar for their invaluable feedback and Priya Nitin Pitre for her help in collecting some datasets used in this study.

References

1. Wenger, M., Gärtner, H., Brunner, M.: To what extent are characteristics of a school's student body, instructional quality, school quality, and school achievement interrelated? Sch. Eff. Sch. Improv. **31**(4), 548–575 (2020)
2. Daily, S.M., Mann, M.J., Kristjansson, A.L., Smith, M.L., Zullig, K.J.: School climate and academic achievement in middle and high school students. J. Sch. Health **89**(3), 173–180 (2019)
3. Torrats-Espinosa, G.: Crime and inequality in academic achievement across school districts in the united states. Demography **57**(1), 123–145 (2020)
4. Githaka, P.W., Gachahi, M.W., Mwaruvie, J.: Social amenities and academic performance in primary schools in gichugu constituency, Kenya. Eur. J. STEM Educ. **4**(1), 4 (2019)
5. Dunga, H., Mafini, C.: Socio-economic factors influencing the education of the girl child in Zomba, Malawi. Int. J. Soc. Sci. Humanit. Stud. **11**(2), 20–38 (2019)
6. Alyahyan, E., Düştegör, D.: Predicting academic success in higher education: literature review and best practices. Int. J. Educ. Technol. High. Educ. **17**(1), 1–21 (2020). https://doi.org/10.1186/s41239-020-0177-7
7. Hashim, A.S., Awadh, W.A., Hamoud, A.K.: Student performance prediction model based on supervised machine learning algorithms. In: IOP Conference Series: Materials Science and Engineering, vol. 928, no. 3, p. 032019. IOP Publishing (2020)
8. McCartney, S., Rosenvasser, X.: Not your parents' dorm room: changes in universities' residential housing privacy levels and impacts on student success. SAGE Open **13**(2), 21582440231178540 (2023)
9. Bell, B., Costa, R., Machin, S.: Why does education reduce crime? J. Polit. Econ. **130**(3), 732–765 (2022)
10. Lochner, L.: Education and crime. In: The Economics of Education, pp. 109–117. Elsevier (2020)
11. Jin, R., Wu, R., Xia, Y., Zhao, M.: What cultural values determine student self-efficacy? An empirical study for 42 countries and economies. Front. Psychol. **14**, 1177415 (2023)
12. Souchon, N., Kermarec, G., Trouilloud, D., Bardin, B.: Influence of teachers' political orientation and values on their success prediction toward students from different socioeconomic background. Eur. Rev. Appl. Psychol. **70**(5), 100553 (2020)

13. Mahoney, J.L., et al.: Systemic social and emotional learning: promoting educational success for all preschool to high school students. Am. Psychol. **76**(7), 1128 (2021)
14. Gueldner, B.A., Feuerborn, L.L., Merrell, K.W.: Social and Emotional Learning in the Classroom: Promoting Mental Health and Academic Success. Guilford Publications (2020)
15. Woehr, A., Newman, K.: Effects of dog ownership and academic success: a correlational study. J. Student Res. **9**(1) (2020)
16. Bastedo, M.N., Umbricht, M., Bausch, E., Byun, B.K., Bai, Y.: How well do contextualized admissions measures predict success for low-income students, women, and underrepresented students of color? J. Divers. High. Educ. (2023)
17. Rodríguez-Hernández, C.F., Cascallar, E., Kyndt, E.: Socio-economic status and academic performance in higher education: a systematic review. Educ. Res. Rev. **29**, 100305 (2020)
18. Malik, S., Jothimani, K.: Enhancing student success prediction with featurex: a fusion voting classifier algorithm with hybrid feature selection. Educ. Inf. Technol. 1–51 (2023)
19. Natek, S., Zwilling, M.: Student data mining solution-knowledge management system related to higher education institutions. Expert Syst. Appl. **41**(14), 6400–6407 (2014)
20. Hamoud, A., Hashim, A.S., Awadh, W.A.: Predicting student performance in higher education institutions using decision tree analysis. Int. J. Interact. Multimedia Artif. Intell. **5**, 26–31 (2018)
21. Yukselturk, E., Ozekes, S., Türel, Y.K.: Predicting dropout student: an application of data mining methods in an online education program. Eur. J. Open Distance e-Learn. **17**(1), 118–133 (2014)
22. Sushil, M., et al.: A comparative study of zero-shot inference with large language models and supervised modeling in breast cancer pathology classification (2024)
23. Bansal, P., Sharma, A.: Large language models as annotators: enhancing generalization of NLP models at minimal cost. arXiv preprint arXiv:2306.15766 (2023)
24. Clavié, B., Soulié, G.: Large language models as batteries-included zero-shot esco skills matchers. arXiv preprint arXiv:2307.03539 (2023)
25. Sun, X., et al.: Text classification via large language models. arXiv preprint arXiv:2305.08377 (2023)
26. Gilardi, F., Alizadeh, M., Kubli, M.: Chatgpt outperforms crowd workers for text-annotation tasks. Proc. Natl. Acad. Sci. **120**(30) (2023). https://doi.org/10.1073/pnas.2305016120
27. Wang, R.E., Wirawarn, P., Goodman, N., Demszky, D.: Sight: a large annotated dataset on student insights gathered from higher education transcripts. arXiv preprint arXiv:2306.09343 (2023)
28. Abu-El-Haija, S., Kapoor, A., Perozzi, B., Lee, J.: N-GCN: multi-scale graph convolution for semi-supervised node classification. In: Uncertainty in Artificial Intelligence, pp. 841–851. PMLR (2020)
29. Wang, Y., Wang, W., Liang, Y., Cai, Y., Liu, J., Hooi, B.: Nodeaug: semi-supervised node classification with data augmentation. In: Proceedings of the 26th ACM SIGKDD International Conference on Knowledge Discovery & Data Mining, pp. 207–217 (2020)
30. Yang, X., et al.: Mixed graph contrastive network for semi-supervised node classification. ACM Trans. Knowl. Discov. Data (2022)

31. Xu, B., Huang, J., Hou, L., Shen, H., Gao, J., Cheng, X.: Label-consistency based graph neural networks for semi-supervised node classification. In: Proceedings of the 43rd International ACM SIGIR Conference on Research and Development in Information Retrieval, pp. 1897–1900 (2020)
32. Xu, D., et al.: Adaptive neural network for node classification in dynamic networks. In: 2019 IEEE International Conference on Data Mining (ICDM), pp. 1402–1407. IEEE (2019)
33. Zhou, C., Chen, H., Zhang, J., Li, Q., Hu, D., Sheng, V.S.: Multi-label graph node classification with label attentive neighborhood convolution. Expert Syst. Appl. **180**, 115063 (2021)
34. Zhang, Z., Wang, H., Liu, L., Li, J.: Multi-label relational classification via node and label correlation. Neurocomputing **292**, 72–81 (2018)
35. Grover, A., Leskovec, J.: node2vec: scalable feature learning for networks. In: Proceedings of the 22nd ACM SIGKDD International Conference on Knowledge Discovery and Data Mining, pp. 855–864 (2016)
36. Kipf, T.N., Welling, M.: Semi-supervised classification with graph convolutional networks. arXiv preprint arXiv:1609.02907 (2016)
37. Scarselli, F., Gori, M., Tsoi, A.C., Hagenbuchner, M., Monfardini, G.: The graph neural network model. IEEE Trans. Neural Networks **20**(1), 61–80 (2008)
38. Chicago Metropolitan Agency for Planning: CMAP data hub (2023). https://datahub.cmap.illinois.gov
39. Chicago Police Department. Chicago data portal (2023). https://data.cityofchicago.org
40. P. C. a. U. Chicago Department of Public Health and Metopio. Chicago health atlas (2023). https://chicagohealthatlas.org/
41. Data.gov. Libraries - 2022 circulation by location (2023). https://catalog.data.gov/dataset/libraries-2022-circulation-by-location
42. Liu, J., Ong, G.P., Chen, X.: Graphsage-based traffic speed forecasting for segment network with sparse data. IEEE Trans. Intell. Transp. Syst. **23**(3), 1755–1766 (2020)
43. Chen, T., Guestrin, C.: Xgboost: a scalable tree boosting system. In: Proceedings of the 22nd ACM SIGKDD International Conference on Knowledge Discovery and Data Mining, pp. 785–794 (2016)
44. Breiman, L.: Random forests. Mach. Learn. **45**, 5–32 (2001)

Balancing Efficiency and Quality in LLM-Based Entity Resolution on Structured Data

Navapat Nananukul(✉) and Mayank Kekriwal

Information Sciences Institute, University of Southern California, Los Angeles, CA 90292, USA
{nananuku,kejriwal}@isi.edu

Abstract. Entity Resolution (ER) is the problem of automatically determining when two or more entities refer to the same underlying entity. ER has been researched for over fifty years across multiple domains (including healthcare, e-commerce, and census data). In graph-based applications, such as deduplicating identities across (or even within) social media platforms, as well as knowledge graphs, ER can be particularly important. Traditionally, ER was a difficult problem both within Artificial Intelligence (AI) and in databases, owing to the quadratic $O(n^2)$ complexity of comparing n entities to each other, given one or more graphs with n total nodes. However, recent emergence of large language models (LLMs) allow us to address the challenges of ER as an AI problem, but a clear framework for applying LLMs in a cost-effective way remains an open issue. In this paper, we present such a framework and validate it through early experiments on real-world ER benchmarks. The framework is LLM-agnostic and is premised on assumptions that resemble pragmatic real-world requirements.

Keywords: Entity resolution · identity matching · knowledge graphs · blocking · efficiency · large language models

1 Background

Entity resolution (ER) is the algorithmic problem of determining when two or more entities refer to the same underlying entity. With increased growth of social media and user data on the Web, and with users typically having accounts (sometimes, more than one) across multiple internet platforms, ER has become central in applications ranging from customer analytics to data integration. Figure 1 provides an intuitive illustration of the problem. Many classes of ER solutions [4,5,7,11,14] have been proposed over the decades, with deep learning [6,16], and most recently, large language models (LLMs) yielding improved performance on challenging benchmarks [18,19,21].

ER is a problem with inherently quadratic complexity in the worst case. The simplest case (and one that is theoretically sufficient for analyzing other variants)

L. M. Aiello et al. (Eds.): ASONAM 2024, LNCS 15213, pp. 278–293, 2025.
https://doi.org/10.1007/978-3-031-78548-1_21

arises when a single network with n nodes needs to be *deduplicated.* Given a vertex-set V with n nodes, suppose that a perfect constant-time oracle $f: V \times V \rightarrow \{True, False\}$ exists that, given two nodes from V, outputs a decision on whether two input entities are semantic duplicates[1] (True) or not (False). Even given such a *similarity* function, which in practice is imperfect and based on some form of machine learning, all pairs of entities would have to be compared in the worse case, yielding $O(n^2)$ complexity. Conducting so many comparisons is wasteful in real-world scenarios because most nodes either don't have a duplicate, and those that do tend to have some small number of duplicates. For this reason, ER has previously been referred to as a *micro-clustering* problem [3].

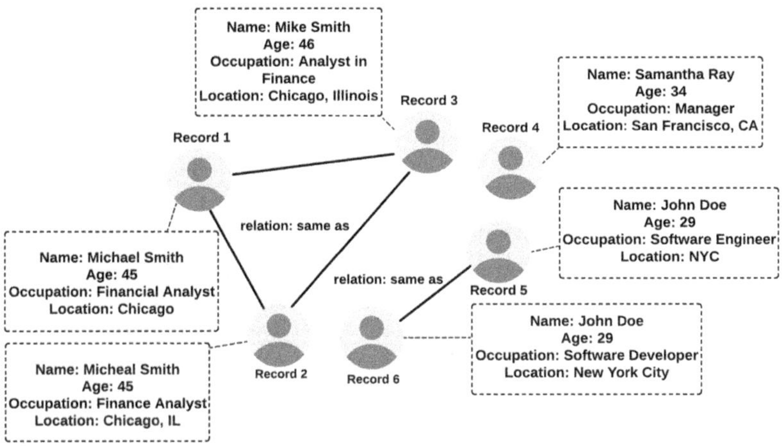

Fig. 1. Illustration of the ER problem where the same user (indicated by the *same as* edge) has multiple records with different semantic representation.

To mitigate the quadratic $O(n^2)$ complexity of comparing n entities to each other, a typical ER architecture comprises two steps: *blocking* and *similarity.* Blocking is the process of using an inexpensive algorithm to cluster entities into 'blocks', such that entities sharing a block have higher probability of being duplicates. Figure 2 illustrates a simple blocking scenario where blocks are created based on matching last names. Following blocking, the similarity step only compares pairs of entities within the *same* blocks. As we noted earlier, LLMs like GPT-3.5 and GPT-4 have emerged as a viable similarity step, but are too expensive to invoke for hundreds of thousands of entity pairs, which is not uncommon (even post-blocking) for reasonably-sized datasets.

[1] We clarify, as also suggested by the figure, that the entities are not *syntactic* duplicates, otherwise the problem would neither require AI, nor a sophisticated approach to reduce quadratic computation. Deduplication and ER are *semantic* processes at their core i.e., entities that are actually the same 'look' different.

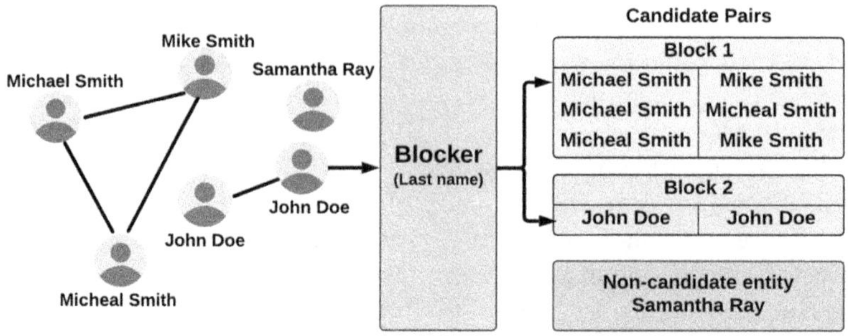

Fig. 2. An illustration of the blocking process in entity resolution, where blocks are created based on matching last names. Entities are grouped into blocks according to their last names, resulting in Block 1 containing a permutation of candidate pairs with the last name "Smith" and Block 2 containing candidate pairs with "Doe." The entity "Samantha Ray," which does not share a last name with any other entity, is classified as a non-candidate entity and is excluded from further comparisons.

Motivated by this challenge, we propose a novel cost-sensitive approach for balancing efficiency and quality in LLM-based ER on structured data. An important aspect of the framework is that, rather than only consider the LLM as the similarity function, we consider using *two* similarity functions: one high-quality but expensive (like an LLM, or in the extreme case, an oracle), and the other, moderate quality but much cheaper. The question now becomes: which candidate pairs (obtained after blocking) should we prioritize sending into the former, rather than accepting the output of the 'default' latter method? Intuitively, the second case will apply to the vast majority of pairs in limited-budget settings, which any prioritization function must be sensitive to.

The proposed framework is designed to take a budget as input and, rather than treat all blocks equally, uses an *entropy-based* methodology for block prioritization. Next, the feedback from the LLM is used to further refine the prioritization score to better navigate the cost-benefit tradeoff. We conduct detailed experiments across three ER benchmarks, using two blocking baselines, and show that the framework significantly outperforms two competitive prioritization baselines by achieving competitive ER performance 30% faster than traditional blocking methods. Incorporating feedback from the LLM is further found to boost efficiency by 5% compared to the normal entropy-based approach on some ER benchmarks.

2 Related Work

Blocking is a crucial step in reducing the quadratic complexity of entity resolution (ER) and has been approached through various methodologies [27]. Originally developed in the late 1960s [9], blocking workflows involve block building [9,24,42], block cleaning [23,25], and comparison cleaning [26], organizing records into blocks based on derived signatures. However, ER datasets' growing complexity and size have introduced new challenges and limitations [10] where budget constraints have become important. This leads to a new entity resolution technique called *progressive ER*. Progressive ER integrates additional components into existing blocking techniques to enhance early recall performance [30]. Recent works [1,2] explored methods for handling large datasets with limited time and computational resources. [31] proposed a system to minimize missed duplicates by estimating the matching likelihood of attributes in unresolved candidate pairs. These studies have opened new avenues for improving ER performance within constrained budgets.

The recent emergence of large language models (LLMs) has demonstrated state-of-the-art performance in many fields [8,15,35]. LLMs such as ChatGPT represent a significant milestone in the field. Recently, several notable LLMs have been developed and released, including OpenAI's ChatGPT [22], Meta's LLaMA [36], and Google's Gemini [34]. LLMs have been applied in various domains, including natural language understanding, machine translation, question answering, and text summarization [20,40,41]. They are also being used in specialized applications such as legal document analysis, medical research, and educational tools [13,33].

In ER, evidence suggests that LLMs can offer promising results in unsupervised ER. LLMs can determine whether two records refer to the same entity [28] using their pre-trained knowledge of products across the internet. This has led to a growing body of research integrating LLMs' capability with ER [18]. Moreover, prompt engineering [12,38,43] also shows that LLMs' performance is significantly influenced by how different information or examples are given in the prompt. Recent work [21] illustrated that prompt engineering techniques influence ER performance at different costs depending on the prompts' complexity. However, one obvious trade-off when using state-of-the-art commercial LLMs still persists. Creating or resolving thousands of blocks can be expensive and impractical in a research setup. Existing works related to using LLMs-ER field focus on finding methods to integrate LLMs with ER or maximize ER performance. To our knowledge, no existing work has explored cost-efficient methods of employing ER with LLMs for entity matching. Our proposed methods aim to achieve cost-efficient ER by combining a progressive ER technique with a token-based entropy sorting approach to maximize ER performance under a budget.

3 Framework

Figure 3 illustrates the schematic of the entropy-based block prioritization approach. We integrate the entropy-based approach into the traditional blocking process to have high recall early when having a limited entity-matching budget. First, this approach tokenized attributes in an ER dataset, yielding a probability distribution of tokens in the ER dataset. Second, entropy scores are calculated at an entity level using the probability distribution from the first step. Third, candidate pairs are generated and sorted based on their entropy scores, from lowest to highest, to prioritize more homogeneous pairs, expected to have a higher

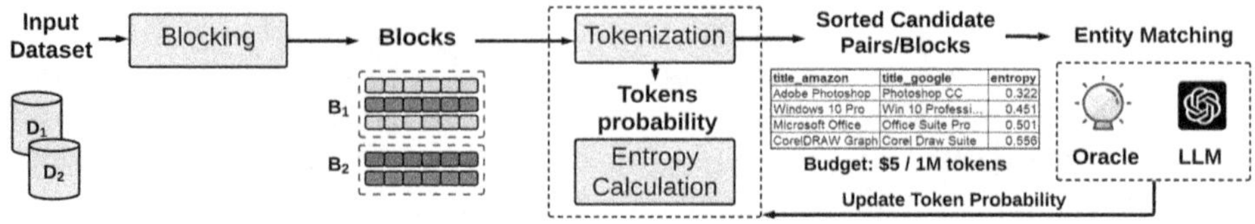

Fig. 3. An illustrative example of how the entropy-based block prioritization approach fits in an ER workflow. This approach begins by tokenizing the attributes of entities within blocks to establish a probability distribution. Next, it calculates entropy scores for all entities, sorting them from lowest to highest. Using this sorted list of entities, candidate pairs are formed according to a predetermined blocking method. After all entropy scores are assigned to all entities, blocks are sorted using the average scores of the candidate pairs. Lastly, candidate pairs are sent to a theoretically constructed Oracle or an LLM to resolve. The token probability pool is then updated based on the remaining candidate pairs.

potential for matching. Lastly, the probability distribution can be updated while entity matching is performed. We implemented a feedback loop to recalculate the probability distribution and the entropy scores using the remaining candidate pairs. Next, we provide specific details, using a practical example.

3.1 Prioritization Using Entropy

As a first step, entropy scores are computed for all entities in blocks to quantify the 'information' in each entity in the block. We use the attributes of entities (e.g., names and descriptions) to calculate entropy scores. The entropy score E is based on the token probability distributions and is given by the formula:

$$E = - \sum_{\text{token} \in \text{tokens}} p(\text{token}) \log_2 p(\text{token})$$

where $p(\text{token})$ is the probability of occurrence of each token, calculated as the frequency of the token divided by the total token count across inside a block.

The entropy scores are subsequently utilized to reorder the entities within each block. Additionally, they serve as a key variable for traditional blocking methods, such as Progressive Sorted Neighborhood (PSN) and Hierarchical Record Partition (HRP), in generating the final list of candidate pairs. The following subsection provides a detailed explanation of these selected blocking methods and illustrates how entropy scores are integrated using a practical example.

3.2 Blocking Methods

We selected established blocking techniques commonly utilized in progressive entity resolution frameworks. Notable methods, including the Progressive Sorted Neighborhood (PSN) and Hierarchical Record Partition (HRP), systematically reorder candidate pairs using sorting keys derived from dataset attributes. This makes them excellent blocking benchmarks for integrating the entropy score and using it as a sorting key.

Progressive Sorted Neighborhood (PSN). [37,39] initially sorts the data using a predefined sorting key and subsequently compares records that fall within a specified sliding window. The underlying rationale for this method is that records positioned closely in the sorted order are more likely to be duplicates than those that are distantly placed.

Hierarchical Record Partition (HRP). [37] organizes records into a structured hierarchy of partitions, where each level represents groups of potentially matching entities at varying degrees of granularity. The most granular partition at the hierarchy's base applies the strictest criteria for grouping, resulting in tightly clustered entities. The closer to the root of the hierarchy, the coarser each partition becomes. For example, at level 1, entities might be grouped based on a single matching token; at level 2, the criteria might require two matching tokens to align entities as potentially similar.

Entropy Score Application. To integrate entropy scores into a blocking method, we illustrate the application using PSN as an example. Assuming we have a block $X = \{e_1, e_2, e_3, e_4\}$, where the entities must be progressively resolved. Initially, entities within X are not sorted and are randomly placed within a block. Suppose the entities, after being sorted by entropy, are listed as $[e_3, e_2, e_4, e_1]$ with corresponding entropy values indicating $[e_3, e_2]$ as having lower entropy, suggesting closer similarity or higher match potential. The $\text{Rank}(e_i)$ of each entity e_i in the sorted list is determined, and the distance between any two entities e_i and e_j, $\text{distance}(e_i, e_j)$, is calculated as $|\text{Rank}(e_i) - \text{Rank}(e_j)|$. For example, if $\text{Rank}(e_2) = 2$ and $\text{Rank}(e_4) = 3$, then $\text{distance}(e_2, e_4) = |2 - 3| = 1$.

The sorting mechanism starts resolving the pairs in X in an ascending order based on their distance measures within window size w. For instance, it will resolve the pair (e_3, e_2) before (e_3, e_4) because the former pair has a smaller distance and potentially lower combined entropy, implying a higher likelihood of being duplicates. This process continues until a stopping condition is met, such as when the rate of newly discovered duplicate pairs drops below a certain threshold or until all pairs whose distance value is less than a specified window size w have been compared.

3.3 Feedback Mechanism

We developed a feedback mechanism that dynamically updates entropy scores during the entity resolution (ER) process. Initial entropy scores are computed using all tokens from unresolved candidate pairs at the start of the ER step. Subsequently, Oracle or LLMs perform entity matching based on the entropy-sorted list of candidate pairs. After resolving a certain number of pairs, entropy scores are recalibrated based on the updated token pool, excluding resolved pairs. The number of pairs resolved before recalibration is a hyperparameter that can be adjusted in our implementation, with the default value set to every 10% of the total number of pairs in a block. This recalibration aims to reorder candidate pairs using the current token pool from unresolved entities, ensuring more accurate and efficient matching. The setup and results are reported separately for (1) entropy without feedback and (2) entropy with feedback in Sect. 5.

4 Experiments

In this section, we detail the experimental setup used to evaluate the efficacy of the entropy-based prioritization approach. We describe the datasets, Large Language Models (LLMs), baselines, and experimental outcomes, including insights from the results. All experiments, including blocking, tokenization, entropy calculation, and subsequent evaluations, were conducted using Python version 3.9.13.

We used three ER datasets to evaluate our approach. Specifically, we used two e-commerce datasets—WDC Products and Amazon-Google—and an academic dataset, DBLP-ACM.

1. **Amazon-Google:** [17] This dataset primarily contains software product listings featuring various versions of the Windows operating system alongside popular image and video editing applications. Compiled from both Amazon and Google, the dataset provides a rich environment for entity resolution with its overlap of product types and differing naming standards.
2. **WDC Products:** [29] This dataset contains various products offered across categories such as electronics and software, clothing, and tools. We chose the WDC computer products (small) dataset to conduct the experiment.
3. **DBLP-ACM:** [32] This dataset contains bibliographic entries from two prominent academic sources, DBLP and ACM. It includes titles, authors, publication venues, and other metadata. Although generally considered less challenging than the other two datasets, DBLP-ACM presents unique challenges due to the diversity in conventions and formatting styles between the sources.

We used GPT-3.5 turbo (specifically, the gpt-3.5-turbo-0125 model) developed by OpenAI as the expensive entity matching method. It is a commercial LLM that is known to have excellent performance on a range of tasks.

Blocking Benchmarks and Baselines: We compared the results of candidate pairs ordered using an entropy-based approach with standard progressive entity resolution techniques. As explained in Sect. 3.2, we used two blocking methods commonly utilized in progressive entity resolution frameworks: PSN and HRP. Each blocking method is applied to each dataset before serving as input for the sorting step, where the baselines are compared against the entropy-based approach. We report the dataset statistics and the Pair Completeness (PC) of the initial blocking step for each dataset in Table 1. PC measures the proportion of

Table 1. Dataset size and pair completeness (PC) for PSN and HRP blocking Methods across three datasets

Dataset	# Rows	PC (PSN)	PC (HRP)
Amazon-Google	1363, 3226	92.3%	93.6%
WDC Products	1100, 1100	89.3%	89.7%
DBLP-ACM	2216, 2294	87.9%	88.4%

true matches retained after blocking, which is the maximum recall performance we can get for each dataset (in the overall ER pipeline). The details on how we compare the entropy-based approach with the baseline are provided below.

PSN Baseline. The Progressive Sorted Neighborhood (PSN) method employs heuristic sorting keys and a fixed window slide value to generate candidate pairs with a high probability of matching. In the Amazon-Google and WDC Product datasets, we utilized *product_ title* as the baseline sorting key with a window size of 3. For the DBLP-ACM dataset, *biography_ title* was used as the sorting key. We compared the baseline performance against two entropy-based approaches: (1) PSN using entropy as the sorting key without any feedback update and (2) PSN with entropy as the sorting key with feedback update.

HRP Baseline. We used traditional HRP to generate a hierarchy of blocks where we set the level of hierarchy = 3 and resolve blocks from lowest level to highest. Each individual block at a different level of hierarchy is sorted using the same baseline sorting key we used in PSN, which are *product_ title* for Amazon-Google and WDC Product datasets, and *biography_ title* for DBLM-ACM dataset. The baseline sorted blocks are then compared with (1) HRP using entropy as the sorting key without any feedback update and (2) PSN with entropy as the sorting key with feedback update.

5 Results and Discussion

We evaluate the effectiveness of the entropy-based approach by measuring progressive recall relative to token costs. Specifically, we analyze two variations: entropy without feedback and entropy with feedback. This analysis demonstrates how entropy-based strategies can effectively prioritize candidate pairs with a high probability of matching within a constrained budget. Additionally, to ensure a comprehensive assessment of entity resolution (ER) performance at any budget level, we implement a cost-effective method using TF-IDF and cosine similarity with a defined threshold of 0.8. This cost-effectiveness similarity method is designed to resolve any remaining pairs at specific budget points, allowing us to gauge the overall ER performance even when the budget does not extend to all candidate pairs. For instance, consider a scenario with 1000 total candidate pairs and a limited budget that only allows for resolving 500 pairs. In this case, we prioritize the top 500 pairs based on entropy using language model lookups

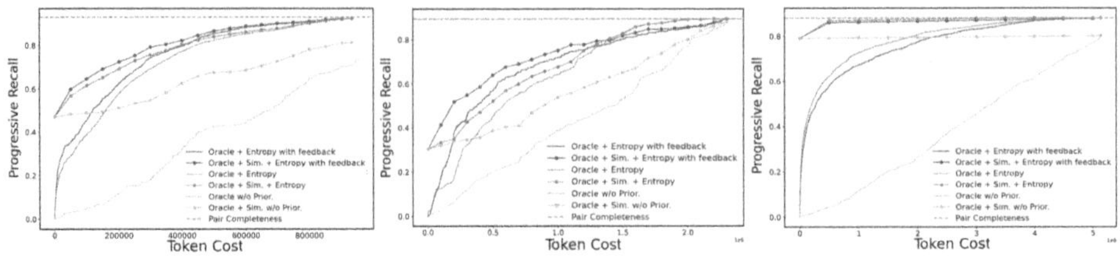

Fig. 4. Progressive recall performance of PSN blocking from three datasets: Amazon-Google (left), WDC Products (middle), and DBLP-ACM (right) when using Oracle. Baselines are presented using red lines, while entropy-based variations are presented in green (without feedback) and blue (with feedback). (Color figure online)

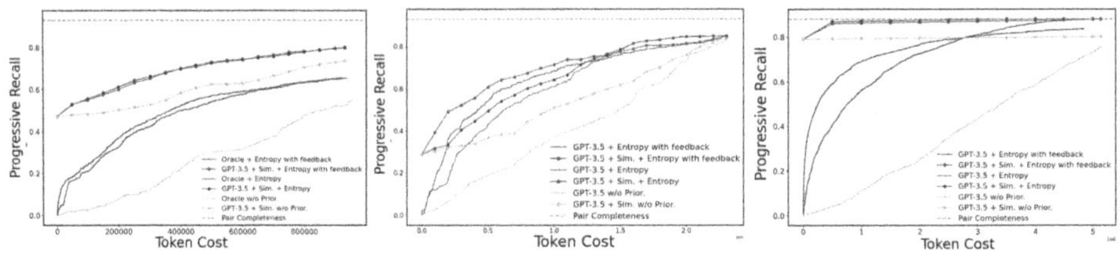

Fig. 5. Progressive recall performance of PSN blocking from three datasets: Amazon-Google (left), WDC Products (middle), and DBLP-ACM (right) when using GPT-3.5. Baselines are presented using red lines, while entropy-based variations are presented in green (without feedback) and blue (with feedback). (Color figure online)

(LLMs) and address the remaining 500 pairs using a more economical similarity-based approach. This dual strategy enables us to maintain performance insights across the full dataset under budget constraints.

Performance Using PSN as Blocking Method: Figures 4 and 5 show progressive recall performance when Oracle and GPT-3.5 are used for entity matching, respectively. We found that entropy-based approaches outperformed all default baselines. We further analyze the token cost cutoff where entropy-based approaches (with and without similarity method) performed relatively well. As shown in Fig. 4 (left), the performance from Amazon-Google shows that both entropy-based approaches outperformed the default PSN. When employing entropy scores as the sorting key within the PSN framework, recall performance significantly improves, exceeding 0.5 from token cost between 0 to 200k (25% of total cost). Table 2 (Oracle columns) also shows that at 44.2% and 42.6% of the total cost, both variations of entropy-based approaches achieved a recall of 0.8. Moreover, the maximum performance of both variations of the entropy-based approaches is higher than the default approach, where both entropy-based methods beat the maximum baseline performance at 44.8% and 43.0% of the total cost. This is due to the different order of initial candidate pairs within blocks resulting from different sorting keys. PSN uses a fixed sliding window to generate the final candidate pairs. Therefore, different sorting keys make the list of

Table 2. The percentage out of total cost required to achieve competitive entity resolution (ER) performance using the entropy-based approach with the PSN blocking method. We report two scenarios from the ER results: (1) percentage out of the total cost to achieve 80% recall and (2) percentage out of the total cost to achieve the maximum recall of the baseline approach. The dash (-%) indicates that the ER performance did not reach 0.8 recall. The notation "E" refers to the normal entropy approach, while "E+F" refers to the entropy with feedback approach.

Metric	Amazon-Google		WDC Products		DBLP-ACM	
	Oracle	GPT-3.5	Oracle	GPT-3.5	Oracle	GPT-3.5
% of total cost achieving 0.8 recall (E)	44.2%	-%	50.6%	59.2%	45.2%	51.9%
% of total cost achieving 0.8 recall (E+F)	42.6%	-%	53.7%	68.5%	35.6%	55.7%
% of total cost achieving maximum baseline perf. (E)	44.8%	44.6%	100%	100%	45.2%	35.7%
% of total cost achieving maximum baseline perf. (E+F)	43.0%	54.3%	100%	100%	35.6%	42.3%

candidate pairs different when using entropy scores as a sorting key compared to the default PSN.

For the WDC Products dataset, as depicted in Fig. 4 (middle), the recall plot illustrates that entropy-based approaches consistently outperform the default Progressive Sorted Neighborhood (PSN). Specifically, the recall for entropy-based methods increased approximately 25% faster than the default PSN over the initial 1M token cost range. Towards the end of this range, the recall rates of both entropy-based approaches begin to converge, ultimately reaching the same maximum recall. As shown in Table 2 (Oracle columns), the entropy-based approaches used 50.6% and 53.7% of the total cost to achieve 0.8 recall. However, they did not exceed the maximum performance of the baseline, as the recall rates converged to the same value in the end.

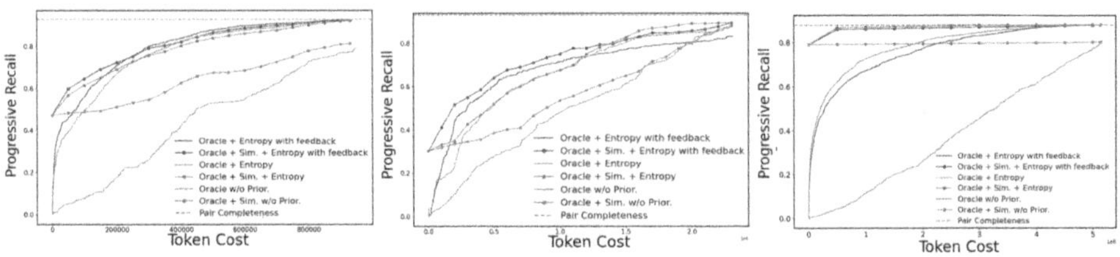

Fig. 6. Progressive recall performance of HRP blocking from three datasets: Amazon-Google (left), WDC Products (middle), and DBLP-ACM (right) when using Oracle. Baselines are presented using red lines, while entropy-based variations are presented in green (without feedback) and blue (with feedback). (Color figure online)

For the DBLP-ACM dataset, which is the simplest among the three, even basic similarity methods achieve high performance without the need for prioritization. However, when employing Oracle as an entity matcher, entropy-based approaches still outperform the default Progressive Sorted Neighborhood (PSN),

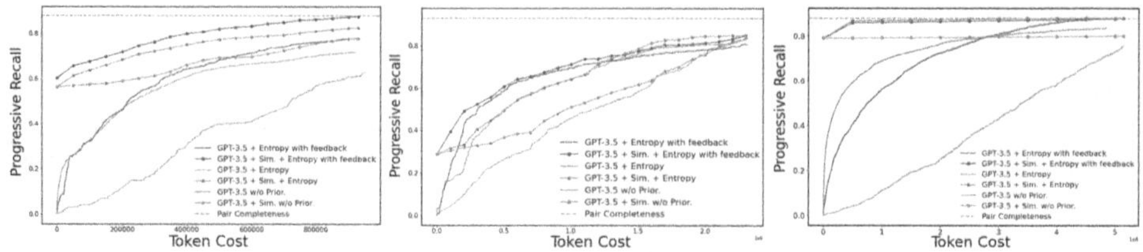

Fig. 7. Progressive recall performance of HRP blocking from three datasets: Amazon-Google (left), WDC Products (middle), and DBLP-ACM (right) when using GPT-3.5. Baselines are presented using red lines, while entropy-based variations are presented in green (without feedback) and blue (with feedback). (Color figure online)

Table 3. The percentage out of total cost required to achieve competitive entity resolution (ER) performance using the entropy-based approach with the HRP blocking method. We report two scenarios from the ER results: (1) percentage out of the total cost to achieve 80% recall and (2) percentage out of the total cost to achieve the maximum recall of the baseline approach. The dash (-%) indicates that the ER performance did not reach 0.8 recall. The notation "E" refers to the normal entropy approach, while "E+F" refers to the entropy with feedback approach.

Metric	Amazon-Google		WDC Products		DBLP-ACM	
	Oracle	GPT-3.5	Oracle	GPT-3.5	Oracle	GPT-3.5
% of total cost achieving 0.8 recall (E)	32.6%	-%	55.2%	68.5%	45.7%	52.0%
% of total cost achieving 0.8 recall (E+F)	33.6%	-%	64.8%	92.6%	35.8%	55.8%
% of total cost achieving maximum baseline perf. (E)	33.1%	45.6%	100%	100%	46.7%	35.9%
% of total cost achieving maximum baseline perf. (E+F)	34.2%	41.3%	100%	100%	35.9%	42.4%

as shown in Fig. 4 (right). This advantage is particularly noticeable in the early stages of entity matching, where entropy-based methods achieve a recall exceeding 0.8 at approximately 45.2% and 35.6% of the total token cost. Since the baseline method did not reach 0.8 recall, both entropy-based methods beat the baseline approach at the same budget points.

When using GPT-3.5 as an entity matcher, the performance is shown in Fig. 5. The overall trend of recall performance mirrors that of using Oracle, with both entropy-based variations outperforming the baseline at the same rate. However, the average recall performance decreased by approximately 10–20% due to GPT-3.5 incorrectly resolving some true positive (TP) pairs as negatives. Consequently, the budget percentage required to achieve 80% recall increased by 10–20%, as shown in Table 2 (GPT-3.5 columns). This outcome aligns more closely with typical entity-matching scenarios, where errors are expected.

Performance Using HRP as Blocking Method: Figures 6 and 7 illustrate the recall performance with Oracle and GPT-3.5 as entity matchers, respectively. The overall performance is similar to that obtained with PSN as the blocking method across all datasets and entity matchers. Both entropy-based methods consistently outperformed the baseline HRP method, accelerating early recall

Table 4. Qualitative Examples from the ground truth of three ER datasets including the Amazon-Google, WDC Products, and DBLP-ACM Datasets.

Dataset	Source	Title
Amazon-Google	A	adobe premiere pro cs3 upgrade
	G	premiere pro cs3 mac upgrade
	A	mia's math adventure: just in time
	G	kutoka interactive 61208 mia's math adventure: just in time!
	A	noah's ark activity center (jewel case ages 3–8)
	G	the beginners bible: noah's ark activity center: activity center
WDC Products	P1	Benq ZOWIE RL2455 24 Full HD TN Grey computer monitor 4718755065736) Product data Benq monitor monitors
	P2	Zowie RL2455 E-Sports 24 Full HD LED Monitor 24 inch 1ms Monitor - HDMI 9H.LF4LB.DBE
	P1	Matrox G550 Low Profile Graphics Card
	P2	MATROX G55MDDE32LPDF 32MB Millenium G550 LP PCI-E Video Card with Cable
	P1	WD Black 2TB Performance Desktop Hard Disk Drive - 7200 RPM SATA 6 Gb/s 64MB Cache 3.5 in.
	P2	HDD Black 2TB 3.5 SATA 6Gbs 64MB@es Cables
DBLP-ACM	DBLP	Database Systems: The Complete Book
	ACM	Database Systems: The Complete Book (2nd Edition)
	DBLP	Introduction to Algorithms
	ACM	Introduction to Algorithms, 3rd Edition
	DBLP	Artificial Intelligence: A Modern Approach
	ACM	Artificial Intelligence: A Modern Approach (3rd Edition)

performance across all datasets. Notably, the early recall acceleration rate is slightly higher with HRP than with PSN for the Amazon-Google and DBLP-ACM datasets.

For the Amazon-Google dataset, HRP with entropy-based approaches achieved 80% recall using only 32.6% and 33.6% of the total budget, as shown in Table 3 (Oracle). This indicates that HRP with entropy-based methods accelerates early recall performance by about 10–11% faster than PSN for this dataset. A similar improvement is seen in Table 3 (GPT-3.5) when using GPT-3.5 as the entity matcher where GPT-3.5 used about 7–12% more budget to achieve the same result as Oracle.

However, for the WDC Products dataset, HRP with entropy-based approaches achieved 80% recall using 55.2% and 64.8% of the total budget, which is 5–11% slower than PSN. Even with a slower pace compared to PSN, the entropy-based approaches still consistently outperformed the baselines by accelerating approximately 25% faster than the default HRP. This trend is also observed when using GPT-3.5 as an entity matcher.

Due to the simplicity of the DBLP-ACM dataset, the results from both Oracle and GPT-3.5 with HRP are almost identical to those with PSN across all sorting methods (baseline and entropy-based approaches). Both entropy-based approaches consistently outperformed the baseline at similar rates and budget points. The maximum recall performance is also influenced by the difficulty of the dataset. As shown in Table 4, the DBLP-ACM dataset is the easiest, as illustrated by the candidate pair "Database Systems: The Complete Book" from DBLP and "Database Systems: The Complete Book (2nd Edition)" from ACM, where the differences are minimal and they share multiple identical tokens. This makes entity matching straightforward for GPT-3.5. In contrast, the Amazon-Google dataset is more complex, exemplified by the pair "mia's math adventure: just in time" versus "kutoka interactive 61208 mia's math adventure: just in time!" which includes differences in formatting and additional model information, making entity resolution challenging. The WDC Products dataset also shows high difficulty, exemplified by pairs "WD Black 2TB Performance Desktop Hard Disk Drive" versus "HDD Black 2TB 3.5 SATA," where the brand name and description are omitted in one of the entities. These variations led to mismatches and performance drop when using GPT-3.5 as an entity matcher.

6 Conclusion and Future Work

Entropy-based block prioritization methods are able to effectively prioritize high-probability candidate pairs, achieving superior performance at a reduced cost. On average, our approach accelerates the expected ER performance by 30% compared to traditional methods on both Oracle and GPT-3.5. This demonstrates that our approach effectively prioritizes pairs that LLMs can resolve efficiently. The cost of resolving candidate pairs is correlated with the size of the datasets. For example, two datasets with 2,000 records each would typically require evaluating over 4 million candidate pairs without blocking. With blocking, this number might still exceed 40,000 pairs, assuming a 99% reduction ratio. The proposed method mitigates this cost by easily integrating with existing blocking techniques and significantly enhances early matching performance through budgeted use of LLMs. Future work should expand the scope of datasets to include ER benchmarks from different fields, such as the Cora, FEBRL, and MusicBrainz20K datasets, to further validate the robustness of entropy-based approaches. Additionally, improving feedback mechanisms through different methods for dynamically updating entropy scores could enhance performance. Finally, as large language models evolve, it is crucial to compare our entropy-based approaches with more LLMs, such as Gemini, LLaMA, GPT-4, and beyond, to ensure competitive results across multiple LLMs.

References

1. Altowim, Y., Kalashnikov, D.V., Mehrotra, S.: Progresser: adaptive progressive approach to relational entity resolution. ACM Trans. Knowl. Discov. Data (TKDD) **12**(3), 1–45 (2018)
2. Altowim, Y., Mehrotra, S.: Parallel progressive approach to entity resolution using mapreduce. In: 2017 IEEE 33rd International Conference on Data Engineering (ICDE), pp. 909–920. IEEE (2017)
3. Betancourt, B., Zanella, G., Miller, J.W., Wallach, H., Zaidi, A., Steorts, R.C.: Flexible models for microclustering with application to entity resolution. In: Advances in Neural Information Processing Systems, vol. 29 (2016)
4. Christophides, V., Efthymiou, V., Palpanas, T., Papadakis, G., Stefanidis, K.: An overview of end-to-end entity resolution for big data. ACM Comput. Surv. (CSUR) **53**(6), 1–42 (2020)
5. Christophides, V., Efthymiou, V., Stefanidis, K.: Entity Resolution in the Web of Data, vol. 5. Springer (2015)
6. Ebraheem, M., Thirumuruganathan, S., Joty, S., Ouzzani, M., Tang, N.: Deeper–deep entity resolution. arXiv preprint arXiv:1710.00597 (2017)
7. Elmagarmid, A.K., Ipeirotis, P.G., Verykios, V.S.: Duplicate record detection: a survey. IEEE Trans. Knowl. Data Eng. **19**(1), 1–16 (2006)
8. Fan, L., Li, L., Ma, Z., Lee, S., Yu, H., Hemphill, L.: A bibliometric review of large language models research from 2017 to 2023 (2023)
9. Fellegi, I.P., Sunter, A.B.: A theory for record linkage. J. Am. Stat. Assoc. **64**(328), 1183–1210 (1969)
10. Galhotra, S., Firmani, D., Saha, B., Srivastava, D.: Efficient and effective ER with progressive blocking. VLDB J. **30**(4), 537–557 (2021)
11. Getoor, L., Machanavajjhala, A.: Entity resolution for big data. In: Proceedings of the 19th ACM SIGKDD International Conference on Knowledge Discovery and Data Mining, pp. 1527–1527 (2013)
12. Giray, L.: Prompt engineering with chatgpt: a guide for academic writers. Ann. Biomed. Eng. **51**(12), 2629–2633 (2023)
13. Kasneci, E., et al.: Chatgpt for good? On opportunities and challenges of large language models for education. Learn. Individ. Differ. **103**, 102274 (2023)
14. Kejriwal, M.: Populating entity name systems for big data integration. In: Mika, P., et al. (eds.) ISWC 2014. LNCS, vol. 8797, pp. 521–528. Springer, Cham (2014). https://doi.org/10.1007/978-3-319-11915-1_34
15. Kojima, T., Gu, S.S., Reid, M., Matsuo, Y., Iwasawa, Y.: Large language models are zero-shot reasoners. In: Koyejo, S., Mohamed, S., Agarwal, A., Belgrave, D., Cho, K., Oh, A. (eds.) Advances in Neural Information Processing Systems, vol. 35, pp. 22199–22213. Curran Associates, Inc. (2022). https://proceedings.neurips.cc/paper_files/paper/2022/file/8bb0d291acd4acf06ef112099c16f326-Paper-Conference.pdf
16. Kooli, N., Allesiardo, R., Pigneul, E.: Deep learning based approach for entity resolution in databases. In: Asian Conference on Intelligent Information and Database Systems, pp. 3–12. Springer (2018)
17. Köpcke, H., Thor, A., Rahm, E.: Evaluation of entity resolution approaches on real-world match problems (2010). https://doi.org/10.14778/1920841.1920904
18. Li, H., et al.: On leveraging large language models for enhancing entity resolution. arXiv preprint arXiv:2401.03426 (2024)

19. Li, H., Li, S., Hao, F., Zhang, C.J., Song, Y., Chen, L.: Booster: leveraging large language models for enhancing entity resolution. arXiv preprint arXiv:2403.06434 (2024)
20. Lyu, C., Xu, J., Wang, L.: New trends in machine translation using large language models: case examples with chatgpt. arXiv preprint arXiv:2305.01181 (2023)
21. Nananukul, N., Sisaengsuwanchai, K., Kejriwal, M.: Cost-efficient prompt engineering for unsupervised entity resolution (2024)
22. OpenAI: GPT-4 technical report (2024)
23. Papadakis, G., Alexiou, G., Papastefanatos, G., Koutrika, G.: Schema-agnostic vs schema-based configurations for blocking methods on homogeneous data. Proc. VLDB Endow. **9**(4), 312–323 (2015)
24. Papadakis, G., Ioannou, E., Niederée, C., Palpanas, T., Nejdl, W.: Eliminating the redundancy in blocking-based entity resolution methods. In: Proceedings of the 11th Annual International ACM/IEEE Joint Conference on Digital Libraries, pp. 85–94 (2011)
25. Papadakis, G., Ioannou, E., Palpanas, T., Niederée, C., Nejdl, W.: A blocking framework for entity resolution in highly heterogeneous information spaces. IEEE Trans. Knowl. Data Eng. **25**(12), 2665–2682 (2012)
26. Papadakis, G., Koutrika, G., Palpanas, T., Nejdl, W.: Meta-blocking: taking entity resolutionto the next level. IEEE Trans. Knowl. Data Eng. **26**(8), 1946–1960 (2013)
27. Papadakis, G., Skoutas, D., Thanos, E., Palpanas, T.: Blocking and filtering techniques for entity resolution: a survey. ACM Comput. Surv. **53**(2) (2020). https://doi.org/10.1145/3377455
28. Peeters, R., Bizer, C.: Using chatgpt for entity matching. In: European Conference on Advances in Databases and Information Systems, pp. 221–230. Springer (2023)
29. Primpeli, A., Peeters, R., Bizer, C.: The WDC training dataset and gold standard for large-scale product matching (2019)
30. Simonini, G., Papadakis, G., Palpanas, T., Bergamaschi, S.: Schema-agnostic progressive entity resolution. IEEE Trans. Knowl. Data Eng. **31**(6), 1208–1221 (2019). https://doi.org/10.1109/tkde.2018.2852763
31. Stefanidis, K., Christophides, V., Efthymiou, V.: Web-scale blocking, iterative and progressive entity resolution. In: 2017 IEEE 33rd International Conference on Data Engineering (ICDE), pp. 1459–1462 (2017). https://doi.org/10.1109/ICDE.2017.214
32. Tang, J., Zhang, J., Yao, L., Li, J., Zhang, L., Su, Z.: Arnetminer: extraction and mining of academic social networks. In: Proceedings of the 14th ACM SIGKDD International Conference on Knowledge Discovery and Data Mining, pp. 990–998 (2008)
33. Tang, L., et al.: Evaluating large language models on medical evidence summarization. NPJ Digit. Med. **6**(1), 158 (2023)
34. Team, G.: Gemini: a family of highly capable multimodal models (2023)
35. Thirunavukarasu, A.J., Ting, D.S.J., Elangovan, K., Gutierrez, L., Tan, T.F., Ting, D.S.W.: Large language models in medicine. Nat. Med. **29**(8), 1930–1940 (2023)
36. Touvron, H., et al.: Llama: open and efficient foundation language models. arXiv preprint arXiv:2302.13971 (2023)
37. Whang, S.E., Marmaros, D., Garcia-Molina, H.: Pay-as-you-go entity resolution. IEEE Trans. Knowl. Data Eng. **25**(5), 1111–1124 (2012)
38. White, J., et al.: A prompt pattern catalog to enhance prompt engineering with chatgpt. arXiv preprint arXiv:2302.11382 (2023)

39. Yan, S., Lee, D., Kan, M.Y., Giles, L.C.: Adaptive sorted neighborhood methods for efficient record linkage. In: Proceedings of the 7th ACM/IEEE-CS Joint Conference on Digital Libraries, pp. 185–194 (2007)
40. Yasunaga, M., Ren, H., Bosselut, A., Liang, P., Leskovec, J.: QA-GNN: reasoning with language models and knowledge graphs for question answering. arXiv preprint arXiv:2104.06378 (2021)
41. Zhang, T., Ladhak, F., Durmus, E., Liang, P., McKeown, K., Hashimoto, T.B.: Benchmarking large language models for news summarization. Trans. Assoc. Comput. Linguist. **12**, 39–57 (2024)
42. Zhang, W., Wei, H., Sisman, B., Dong, X.L., Faloutsos, C., Page, D.: Autoblock: a hands-off blocking framework for entity matching. In: Proceedings of the 13th International Conference on Web Search and Data Mining, pp. 744–752 (2020)
43. Zhou, Y., et al.: Large language models are human-level prompt engineers. arXiv preprint arXiv:2211.01910 (2022)

Analyzing Characteristics of Nontrivial Information Diffusion via Implicit Links on Social Media

Yuto Tamura[1(✉)], Sho Tsugawa[1], and Kohei Watabe[2]

[1] University of Tsukuba, Tsukuba, Japan
yuto.tamura@snlab.cs.tsukuba.ac.jp, s-tugawa@cs.tsukuba.ac.jp
[2] Saitama University, Saitama, Japan
kwatabe@mail.saitama-u.ac.jp

Abstract. On social media, information spreads primarily through explicit relationships, such as follow connections, among users. However, information diffusion isn't limited to explicit relationships alone. Users may come across posts from others via external websites or searches and subsequently repost them. The characteristics of such information diffusion, which bypass explicit social relationships, have not been thoroughly examined to date. In this paper, we examine the characteristics of nontrivial information diffusion facilitated by implicit-links among social media users. We define information diffusion that occurs without explicit social relationships as dissemination through implicit links. Utilizing three Twitter datasets, we investigate how users engage in information diffusion through implicit-links. Our findings suggest that users situated farther from the source are more inclined to involve in information diffusion via implicit-links, contrasting with those in closer proximity. This implies that implicit-links originate from sources beyond direct social connections among users. Additionally, we observe that while implicit-links have a weaker influence on diffusion size compared to explicit-links, they tend to distribute information across diverse communities distinct from the source user.

Keywords: Social networks · Social media · Information diffusion · Implicit links

1 Introduction

Social media platforms are utilized by a large number of users across diverse contexts. The number of users on social media such as Twitter (currently X), Facebook, and Instagram has been increasing [9,24]. Social media platforms play crucial roles to disseminate important information for people under several contexts such as political elections [10], natural disasters [16], and pandemic of infectious viruses [23].

L. M. Aiello et al. (Eds.): ASONAM 2024, LNCS 15213, pp. 294–306, 2025.
https://doi.org/10.1007/978-3-031-78548-1_22

Analyzing and understanding the characteristics of information diffusion on social media has been an important research topic for facilitating various applications, such as limiting the spread of misinformation, and optimizing viral marketing strategies. Information posted by companies or individuals on social media has the potential to spread extensively. Consequently, social media is utilized as a platform for implementing viral marketing strategies that leverage word-of-mouth [12]. In contrast, the ease of information dissemination inherent in social media also leads to the proliferation of fake news, posing significant challenges [23]. Therefore, there is a growing body of research aimed at understanding the characteristics of information diffusion on social media and intervening to either enhance or suppress information dissemination by leveraging these characteristics. For instance, there exist several studies on identifying influencers [1,14], and adding or removing links within social networks [7,19] to control the scale of information diffusion.

Existing studies assume that information is disseminated through explicit relationships (i.e., follow relationships) among social media users, and use a cascade graph for representing the paths of information diffusion on social media [9]. In a cascade graph, nodes represent social media users, and link (u, v) represents that information is disseminated from user u to user v. Most social media platforms allow users to explicitly specify receiving information from other users by following them. If user u follows user v and user u reposts user v's post, it is natural to consider that information is disseminated from user v to user u. With such assumptions, cascade graphs are constructed. Figure 1 shows an example of a follow network and a cascade graph that represents information diffusion on the follow network.

In contrast, information diffusion does not necessarily occur only among users in a explicit follow relationship. For instance, users may encounter posts from other users through external websites or searches and may repost those posts. In such cases, it is difficult to estimate the paths of information diffusion solely from follow relationships, which produces a disconnected cascade graph. Figure 2 illustrates an instance of such a disconnected cascade graph. In this example, despite node A having no direct paths to node D, it reposts node D's post. In such cases, it is not obvious how node A came to know about node D's post. We hypothesize that in such follow networks, where paths are not readily apparent, information diffusion occurs via implicit links. Existing studies have shown that when constructing cascade graphs from the history of information diffusion on social media and the social graph among users, the resulting cascade graph often becomes disconnected [13,15,17,18]. This observation suggests that real information diffusion cascades comprise nontrivial information diffusion paths via implicit links that cannot be directly inferred from a follow network representing explicit relationships among users.

However, the characteristics of such nontrivial information diffusion via implicit links have not been examined before. Understanding the characteristics of nontrivial diffusion, which are prevalent in many real information diffusion cascades, is valuable for constructing realistic information diffusion models.

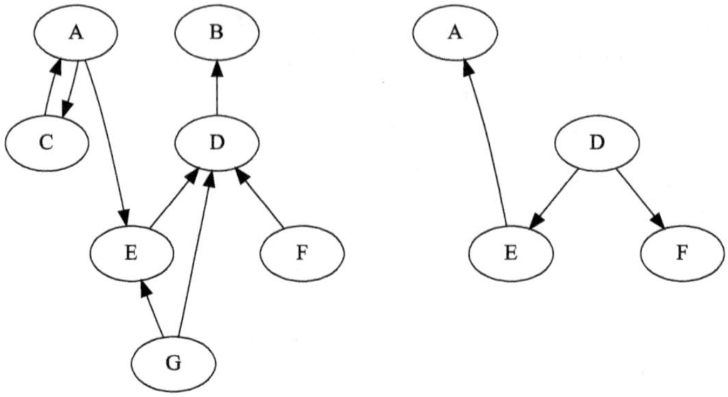

(a) Follow Network (b) Cascade Graph

Fig. 1. Example of a follow network and cascade graph: (a) An illustration of a follow network where each directed link represents a follow relationship. (b) An example of a cascade graph when a post by node D is reposted by nodes A, E, and F. Given that nodes E and F follow node D (as depicted in Fig. 1(a)), the post is considered to be disseminated from node D to nodes E and F, thus resulting in the cascade graph having links (D, E) and (D, F). Similarly, since node A follows node E, the post is considered to be disseminated from node E to A, and the cascade graph has link (E, A).

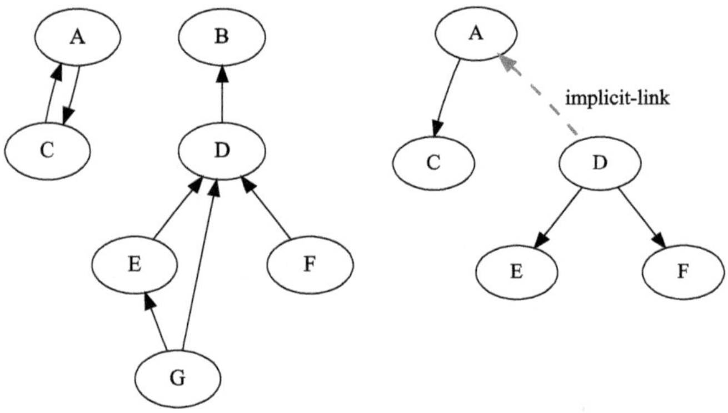

(a) Follow Network (b) Cascade Graph

Fig. 2. Example of a follow network and a cascade graph demonstrating nontrivial information diffusion via an implicit link: (a) An illustration of a follow network where each directed link represents a follow relationship. (b) An example of a cascade graph when a post by node D is reposted by nodes A, C, E, and F. Given that nodes E and F follow node D, the post is considered to be disseminated from node D to nodes E and F, resulting in the cascade graph containing links (D, E) and (D, F). However, since node A lacks an explicit path to node D, the route through which node D's post reaches node A is not readily apparent. Such nontrivial pathways of information diffusion are inferred to have occurred via implicit links.

Most existing information diffusion models assume that information spreads only through explicit relationships between users. In other words, existing models of information diffusion do not account for the existence of nontrivial diffusion paths. Understanding the characteristics of nontrivial diffusion is expected to contribute to the development of more realistic information diffusion models that incorporate nontrivial diffusion paths.

In this paper, we investigate the characteristics of nontrivial information diffusion paths, which cannot be directly inferred from a social graph, utilizing datasets of retweet cascades from Twitter. Specifically, we address the following research questions:

- **(RQ1)**: How does nontrivial information diffusion, which does not rely on explicit relationships between users, occur on social media?
- **(RQ2)**: To what extent does nontrivial information diffusion affect the size of the diffusion cascade?
- **(RQ3)**: Which users receive information through nontrivial information diffusion?

To address these research questions, we analyze the characteristics of nontrivial information diffusion using multiple datasets related to information diffusion on Twitter [1,4,20]. Using datasets of retweet cascades among users constituting the social network, which represents the follow relationships among Twitter users, we analyze the characteristics of nontrivial diffusion. We investigate the distance between the source of information and users engaging in nontrivial reposting to analyze how nontrivial information diffusion occurs (**RQ1**). Furthermore, we examine how the subsequent sizes of information diffusion differ after trivial reposts through the follow network and nontrivial reposts that cannot be inferred from follow relationships, aiming to analyze the impact of nontrivial information diffusion on diffusion sizes (**RQ2**). Finally, by examining the communities of users who originated posts and those who reposted them, we analyze which users receive information through trivial and nontrivial reposts (**RQ3**).

The rest of this paper is structured as follows. Section 2 introduces related work on information diffusion on social media. In Sect. 3, we define the terms used in this paper and describe the datasets. Section 4 shows the results, and Sect. 5 discusses their implications. Finally, Sect. 6 concludes this paper.

2 Related Work

Researchers have been actively investigating the dynamics of information diffusion on social media [1,4,5,10,14,16,21,22]. For instance, Vosoughi et al. [22] found that false information often spreads more extensively than accurate information. Similarly, Tsugawa and Ohsaki [21] revealed that negative posts have a tendency to disseminate more rapidly and widely compared to positive or neutral content on social media. Moreover, researchers have investigated the characteristics of information diffusion within specific contexts. For example, Domenico et al. [4] examined how information about the discovery of the Higgs boson propagated. Additionally, investigations have focused on understanding information diffusion patterns during significant events such as elections, hurricanes [10], earthquake disasters [16], and the COVID-19 pandemic [23].

Insights into information diffusion on social media have led to practical applications, including methods for identifying influential users and strategies for controlling information diffusion. The identification of influencers, who can spread information to many users, has been a hot research topic. Techniques for influencer identification utilizing the network structure of follow relationships [14] and leveraging the historical patterns of information diffusion [1] have been proposed. Tong et al. [19] proposed a method to limit the size of information diffusion by manipulating network connections. Furukawa and Tsugawa [7] evaluated the effectiveness of link deletion strategies in limiting the size of information diffusion using social media data. Their findings suggest that link deletion strategies are not effective in limiting the sizes of actual information diffusion cascades. Moreover, their findings suggest that one contributing factor to the ineffectiveness of the link deletion strategies is the prevalence of nontrivial information diffusion paths that do not rely on social ties within the follow network.

While there has been limited research on nontrivial information diffusion paths that do not rely on explicit social ties, a few pioneering studies have addressed this issue. Numann and Fischer [15] highlighted that constraints imposed by the API used for data acquisition and user privacy settings may influence the observation of nontrivial information diffusion. Taxidou et al. [18] found that approximately 50% of information diffusion can be attributed to reposts by users with explicit social ties, with an additional 13% explained by other forms of interaction, such as quotes. Myers et al. [13] reported that 29% of diffusion was influenced by diffusion outside the follow network. Shioda and Nakajima [17] defined users involved in information diffusion through nontrivial paths as untraceable users and demonstrated a negative correlation between the number of untraceable users involved in information diffusion and the number of followers of the source of that information.

While most existing studies have primarily reported the occurrence frequency of nontrivial information diffusion, this study goes beyond by analyzing the impact of nontrivial diffusion on the dynamics of information diffusion through a comparison with conventional information diffusion facilitated by social ties. Additionally, we explore the implications of these findings for the mechanisms underlying nontrivial diffusion.

3 Preliminaries

3.1 Terminologies and Notations

A follow network $G = (V, E)$ is a directed network, where V denotes a set of nodes representing social media users and E denotes a set of links indicating follow relationships between users. The set of users followed by user u is denoted as $\Gamma(u)$. The author of post t is represented as $a(t)$, and a repost of post t by user v is denoted as $r(t, v)$. Each original post t and its subsequent reposts form a diffusion cascade.

A diffusion graph representing the diffusion paths of a cascade of post t is denoted as $H_t = (R_t, E_t)$. The set of nodes in H_t comprises users who posted or reposted post t. The link $(u, v) \in E_t$ in H_t represents the spread of post t from user u to user v. In the diffusion graph, $(u, v) \in E_t$ if user v follows user u ($u \in \Gamma(v)$) and the timing of $r(t, v)$ precedes the timing of repost $r(t, u)$ or the author of the tweet t, $a(t)$ is user u.

To differentiate between reposts facilitated by explicit social ties and those occurring through implicit connections, we introduce the terms "explicit-link" and "implicit-link". When the diffusion path of a repost $r(t, u)$ explicitly exists in the diffusion graph H_t (i.e., when node $u \in V_t$ has an incoming link, denoted as $(v, u) \in E_t$, where node v exists), we refer to the spread of repost $r(t, u)$ as being facilitated by an explicit-link. Conversely, when node $u \in V_t$ does not have an incoming link (i.e., when there is no node v such that $(v, u) \in E_t$), we consider the spread of repost $r(t, u)$ as being facilitated by an implicit-link.

3.2 Datasets and Methodology

We employed Twitter datasets comprising follow networks representing user relationships, along with users' posts and their reposts, for our analyses. Each repost in the dataset was categorized as disseminated either by an implicit-link or an explicit-link, and differences between these categories were examined. We determined whether a repost is disseminated by an implicit-link using information regarding who follows whom and who reposted which post and when.

The three Twitter datasets utilized in this study are referred to as Higgs [4], Nepal [1], and Ordinary [20] datasets. The Higgs dataset was collected following the announcement of the discovery of the Higgs boson, the Nepal dataset was gathered after the 2015 Nepal earthquake, and the Ordinary dataset comprises randomly selected English reposts from 2018. Basic statistics of these datasets are presented in Table 1. The datasets contain approximately 10,000 to 50,000 posts and associated reposts, along with follow relationships between users who posted and reposted them.

Firstly, we examine the distance between the source and reposting users on the follow network to analyze the characteristics of users who reposted through implicit-links (Sect. 4.1). Secondly, we investigate the impact of reposts through implicit-links on the sizes of diffusion cascades by comparing the number of reposts following a repost through an implicit-link and an explicit-link (Sect. 4.2). Finally, we explore differences in communities between post authors and users who reposted them to identify which users receive information through implicit-links (Sect. 4.3).

Table 1. Overview of the Datasets

	Higgs	Nepal	Ordinary
Number of users	456,626	273,213	111,000
Number of follow links	14,855,842	17,818,902	3,130,963
Number of posts	41,426	49,098	10,000
Number of reposts	354,930	472,840	116,826

4 Analysis

4.1 Distance Between Source and Reposting Users

First, we analyze how non-trivial information diffusion occurs through reposts facilitated by implicit-links (**RQ1**). To examine which users engage in reposting via implicit-links, we investigate the distances on the follow network between the author of the original post and the users who reposted it. The distance on the follow network is presumed to reflect the closeness of relationships between users. For example, users with similar interests are likely to be closer to each other, whereas those with differing interests are expected to be more distant. Therefore, we hypothesize that users who are closer to the original post's author on the follow network are more inclined to repost. By evaluating repost frequency relative to user distance, we aim to characterize the attributes of users involved in creating implicit-links.

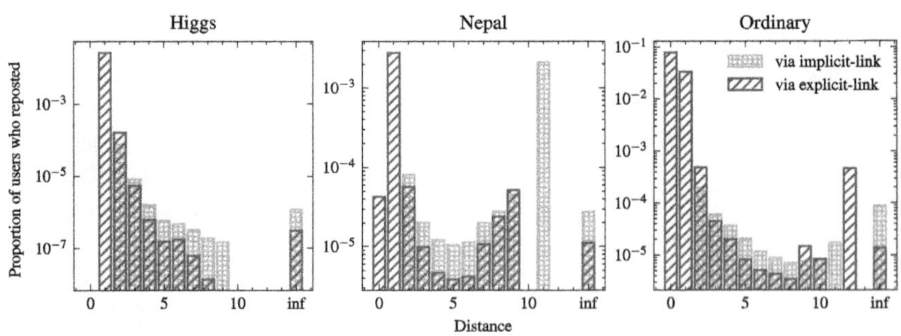

Fig. 3. The proportion of reposting users among users located at a certain distance from the source

Figure 3 shows the proportion of users who performed reposts via explicit-links and implicit-links among users located at a certain distance from the post's author. In contrast, Fig. 4 shows the proportion of reposts made via implicit-links among users at a given distances from the post's author. Note that, by definition, reposts via implicit-links cannot occur at a distance of 1. As anticipated, Fig. 3 demonstrates that users in closer proximity to the author are more likely to

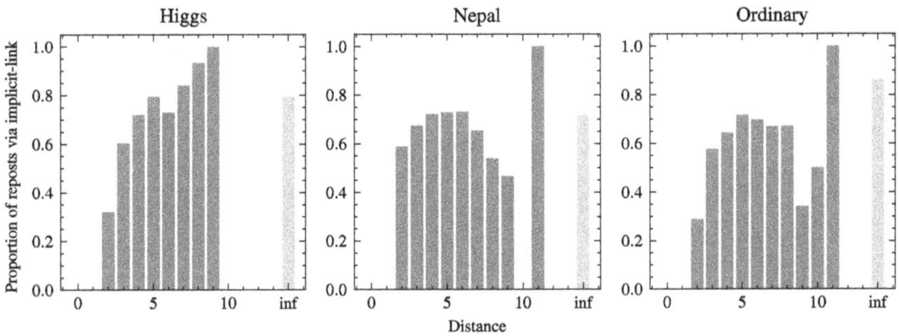

Fig. 4. The proportion of reposts via implicit-link by distance from the source

repost. In contrast, Fig. 4 indicates that users farther from the source are more likely to engage in information diffusion via implicit-links compared to users at closer distances.

The increasing proportion of implicit-links across larger distances between the source and reposting users implies that some implicit-links may originate outside the follow network. If we hypothesize that implicit-links exclusively occur within the follow network and are the result of user privacy settings or unrecorded interactions like likes and mentions, the proportion of implicit-inks would remain constant regardless of the distance between the source and reposting users. However, the proportion of implicit-links varies depending on the distance. Therefore, it is suggested that reposts via implicit-links occur due to information obtained through avenues other than social ties among users. For instance, users may acquire information through external websites, trending features, or searches. These external factors outside the follow network may potentially influence information diffusion via implicit-links.

4.2 Effects of Reposting on Future Cascade Sizes

To clarify the effects of reposts via implicit-links on the dynamics of information diffusion, we investigate how such reposts influence the final size of information diffusion cascade (**RQ2**). Previous studies on information diffusion dynamics and modeling have either implicitly or explicitly assumed that information spreads solely through the follow network. However, if information spreads extensively via implicit-links, it becomes imperative to incorporate them into models of information diffusion. Conversely, if information is seldom disseminated through implicit-links and diffusion is primarily driven by explicit-links, it suggests that there may be no necessity to include implicit-links in diffusion models.

We define the repost contribution index (RCI) and use it to measure the contribution of each repost on the final cascade size of a post. The RCI is defined as follows. Let $B(r(t, v))$ denote a set of reposts made prior to repost $r(t, v)$ by users followed by user v. Similarly, let $F(r(t, v))$ represent a set of reposts $r(t, w)$ that fulfill the condition $r(t, v) \in B(r(t, w))$. The RCI of repost $r(t, w)$ to the final cascade size of post t is determined by the following equation.

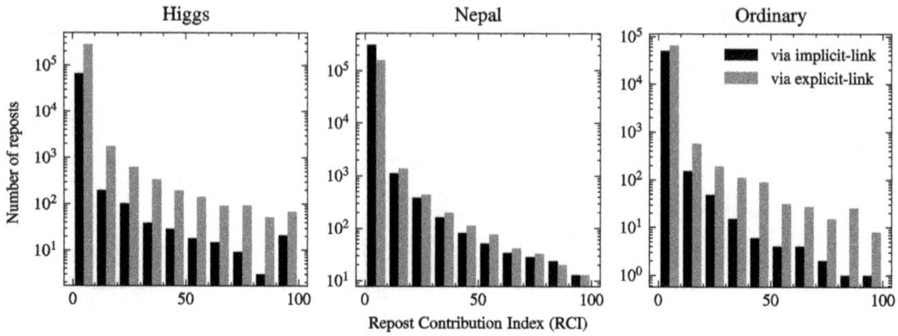

Fig. 5. Distribution of the repost contribution index (RCI)

$$\mathrm{RCI}(r(t,w)) = \sum_{r(t,v)\in F(r(t,w))} \frac{1+\mathrm{RCI}(r(t,v))}{|B(r(t,v))|} \tag{1}$$

Figure 5 compares the distribution of the RCI for reposts via implicit-links and reposts via explicit-links. The horizontal axis of the figure represents the RCI, while the vertical axis indicates the number of reposts associated with each RCI. From Fig. 5, we can find that majority of reposts, whether via explicit or implicit-links, have a low RCI. This finding aligns with the results of Goel et al. [8]. When comparing reposts via implicit-links and reposts via explicit-links, it is observed that reposts via explicit-links tend to have higher RCI. Consequently, this suggests that implicit-inks may be less effective in disseminating information compared to explicit-links.

4.3 Effects of Reposting on Inter-Community Diffusion

To examine the recipients of information through diffusion via implicit-links, we calculate the fraction of posts reposted by users belonging to communities different from the post's author. The findings from the previous section suggest that information diffusion via implicit-links has a weak effect on the diffusion scale. However, if implicit-links facilitate information dissemination across different communities, it can be deemed effective in reaching a diverse range of users.

For community detection, we utilized the Louvain algorithm [2], a method commonly employed in previous studies analyzing information diffusion on social media [20]. Community detection involves partitioning nodes within a network into multiple communities, characterized by a higher density of links among nodes within the same community [6]. It is expected that communities identified from social networks will reflect real-world groupings and node interests.

Figure 6 shows the proportion of reposts where the original post's author and the reposting user belong to the same community, alongside the proportion where they belong to different communities. In all results, reposts within the same community constitute the majority. However, the proportion of reposting

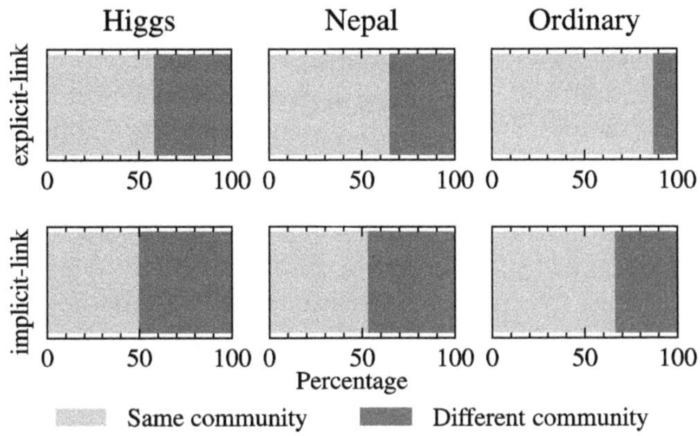

Fig. 6. The percentage of reposts where the author of the original post and reposting user belong to the same community and the percentage where they belong to different communities. The upper charts show the results for explicit-links whereas lower charts show the results for implicit-links.

users belonging to different communities from the post's author is higher for users reposting via implicit-links compared to those reposting via explicit-links. Consequently, it is suggested that while implicit-links may play a lesser role in large-scale diffusion, they have a tendency to disseminate information to users across diverse communities distinct from the source user.

5 Discussion

Our findings suggest that many information diffusion models, which assume only explicit links, may not adequately replicate the characteristics of real-world information spread. More than half of the reposts by users located three or more distances away from the original source are due to implicit links. Therefore, when using information diffusion models like the independent cascade model [11] or the linear threshold model [11] to simulate the spread of information, the probability of distant users receiving the information might diverge from real data. The insights from this study could be valuable in developing new information diffusion models that account for implicit links.

The results indicating that information can reach different communities through implicit links offer valuable insights for mitigating biased information spread, such as the echo chamber effect [3]. While it is known that most reposted information tends to stay within a single community [20], our findings suggest that implicit link diffusion helps spread information across different communities. If information recommendation systems can increase diffusion through implicit links, it might be possible to enhance information spread across communities. This could be beneficial for reducing the echo chamber effect and for implementing viral marketing strategies that reach a diverse audience.

Despite shedding light on these aspects of nontrivial diffusion via implicit-links, our study presents several limitations, suggesting the future works. While we have demonstrated that some implicit-links extend beyond the follow network, the precise factors driving nontrivial diffusion via implicit-links remain unclear. By incorporating interaction data beyond reposts and examining users' behavior outside of social media platforms, we anticipate clarifying the dynamics underlying the occurrence of implicit-links. Furthermore, our study solely analyzed the characteristics of nontrivial diffusion without deliberating on how to incorporate it into information diffusion models. Proposing novel information diffusion models that account for diffusion via implicit-links and exploring intervention strategies based on these models represent critical future works.

6 Conclusion

In this paper, we investigated the characteristics of nontrivial information diffusion via implicit-links among users of social media. Leveraging three Twitter datasets, we examined the characteristics of users who engage in reposts via implicit-links and investigated the subsequent impact of these reposts on diffusion. Our findings indicate that users located farther from the source are more inclined to involve in information diffusion via implicit-links, contrasting with those at closer proximities, suggesting that implicit-links arise from sources beyond direct social ties among users. Moreover, our results reveal that while implicit-links have a smaller effect on diffusion size compared to explicit-links, whereas implicit-links tend to disseminate information to users across diverse communities distinct from the source user. These observations suggest that future research should focus on developing information diffusion models that incorporate implicit links and clarifying the fundamental mechanisms underlying them.

Acknowledgments. This work was partly supported by JSPS KAKENHI JP23H03379 and JP22K11990.

Disclosure of Interests. The authors have no competing interests to declare that are relevant to the content of this article.

References

1. Bhowmick, A.K., Gueuning, M., Delvenne, J.C., Lambiotte, R., Mitra, B.: Temporal sequence of retweets help to detect influential nodes in social networks. IEEE Trans. Comput. Soc. Syst. **6**(3), 441–455 (2019). https://doi.org/10.1109/TCSS.2019.2907553
2. Blondel, V.D., Guillaume, J.L., Lambiotte, R., Lefebvre, E.: Fast unfolding of communities in large networks. J. Stat. Mech: Theory Exp. **2008**(10), P10008 (2008). https://doi.org/10.1088/1742-5468/2008/10/P10008

3. Cinelli, M., De Francisci Morales, G., Galeazzi, A., Quattrociocchi, W., Starnini, M.: The echo chamber effect on social media. Proc. Natl. Acad. Sci. **118**(9), e2023301118 (2021). https://doi.org/10.1073/pnas.2023301118
4. De Domenico, M., Lima, A., Mougel, P., Musolesi, M.: The anatomy of a scientific rumor. Sci. Rep. **3**(1), 2980 (2013). https://doi.org/10.1038/srep02980
5. Doerr, B., Fouz, M., Friedrich, T.: Why rumors spread so quickly in social networks. Commun. ACM **55**(6), 70–75 (2012). https://doi.org/10.1145/2184319.2184338
6. Fortunato, S.: Community detection in graphs. Phys. Rep. **486**(3), 75–174 (2010). https://doi.org/10.1016/j.physrep.2009.11.002
7. Furukawa, S., Tsugawa, S.: Empirical evaluation of link deletion methods for limiting information diffusion on social media. Soc. Netw. Anal. Min. **12**(1), 169 (2022). https://doi.org/10.1007/s13278-022-00994-6
8. Goel, S., Watts, D.J., Goldstein, D.G.: The structure of online diffusion networks. In: Proceedings of the 13th ACM Conference on Electronic Commerce, pp. 623–638. EC '12 (2012). https://doi.org/10.1145/2229012.2229058
9. Guille, A., Hacid, H., Favre, C., Zighed, D.A.: Information diffusion in online social networks: a survey. SIGMOD Rec. **42**(2), 17–28 (2013). https://doi.org/10.1145/2503792.2503797
10. Hughes, A., Palen, L.: Twitter adoption and use in mass convergence and emergency events. Int. J. Emergency Manage. **6**, 248–260 (2009). https://doi.org/10.1504/IJEM.2009.031564
11. Kempe, D., Kleinberg, J.M., Tardos, E.: Maximizing the spread of influence through a social network. In: Proceedings of the 9th ACM SIGKDD International Conference on Knowledge Discovery and Data Mining (KDD'03), pp. 137–146, August 2003. https://doi.org/10.1145/956750.956769
12. Marijke De Veirman, V.C., Hudders, L.: Marketing through instagram influencers: the impact of number of followers and product divergence on brand attitude. Int. J. Advertising **36**(5), 798–828 (2017). https://doi.org/10.1080/02650487.2017.1348035
13. Myers, S.A., Zhu, C., Leskovec, J.: Information diffusion and external influence in networks. In: Proceedings of the 18th ACM SIGKDD International Conference on Knowledge Discovery and Data Mining, KDD 2012, pp. 33–41 (2012). https://doi.org/10.1145/2339530.2339540
14. Narayanam, R., Narahari, Y.: A shapley value-based approach to discover influential nodes in social networks. IEEE Trans. Autom. Sci. Eng. **8**(1), 130–147 (2011). https://doi.org/10.1109/TASE.2010.2052042
15. Neumann, J., Fischer, P.M.: Inferring missing retweets in twitter information cascades. In: New Trends in Database and Information Systems, pp. 287–292 (2021). https://doi.org/10.1007/978-3-030-85082-1_25
16. Pourebrahim, N., Sultana, S., Edwards, J., Gochanour, A., Mohanty, S.: Understanding communication dynamics on twitter during natural disasters: a case study of hurricane sandy. Int. J. Disaster Risk Reduction **37**, 101176 (2019). https://doi.org/10.1016/j.ijdrr.2019.101176
17. Shioda, S., Nakajima, K.: Existence of twitter users with untraceable retweet paths and its implications. In: Proceedings of the 2020 Seventh International Conference on Social Networks Analysis, Management and Security (SNAMS), pp. 1–8 (2020). https://doi.org/10.1109/SNAMS52053.2020.9336544
18. Taxidou, I., Fischer, P.M., De Nies, T., Mannens, E., Van de Walle, R.: Information diffusion and provenance of interactions in twitter: is it only about retweets? In: Proceedings of the 25th International Conference Companion on World Wide Web,

WWW 2016 Companion, pp. 113–114 (2016). https://doi.org/10.1145/2872518.2889393
19. Tong, H., Prakash, B.A., Eliassi-Rad, T., Faloutsos, M., Faloutsos, C.: Gelling, and melting, large graphs by edge manipulation. In: Proceedings of the 21st ACM International Conference on Information and Knowledge Management, pp. 245–254. CIKM '12 (2012). https://doi.org/10.1145/2396761.2396795
20. Tsugawa, S.: Empirical analysis of the relation between community structure and cascading retweet diffusion. In: Proceedings of the International AAAI Conference on Web and Social Media **13**, pp. 493–504 (2019). https://doi.org/10.1609/icwsm.v13i01.3247
21. Tsugawa, S., Ohsaki, H.: Negative messages spread rapidly and widely on social media. In: Proceedings of the 2015 ACM on Conference on Online Social Networks, COSN 2015, pp. 151–160 (2015). https://doi.org/10.1145/2817946.2817962
22. Vosoughi, S., Roy, D., Aral, S.: The spread of true and false news online. Science **359**(6380), 1146–1151 (2018). https://doi.org/10.1126/science.aap9559
23. Yang, K.C., et al.: The COVID-19 infodemic: Twitter versus facebook. Big Data Soc. **8**(1), 20539517211013860 (2021). https://doi.org/10.1177/20539517211013861
24. Zhang, Z.K., Liu, C., Zhan, X.X., Lu, X., Zhang, C.X., Zhang, Y.C.: Dynamics of information diffusion and its applications on complex networks. Phys. Rep. **651**, 1–34 (2016). https://doi.org/10.1016/j.physrep.2016.07.002

Insights About Radicalism on the Digital Era: A Sociological Approach

Victoria Ferro[1], Athus Cavalini[2](✉), Fabio Goveia[1], and Fabio Malini[3]

[1] Postgraduate Program in Communication and Territorialities, Federal University of Espírito Santo, Vitória, Brazil
victorialferro99@gmail.com

[2] Postgraduate Program in Computer Science, Federal University of Espírito Santo, Vitória, Brazil
athus.cavalini@ifes.edu.br

[3] Department of Communication, Federal University of Espírito Santo, Vitória, Brazil

Abstract. The pervasiveness of new information flows and integration models stemming from the platformization of social life, which has been driven by the widespread adoption of novel communication technologies and digital social networks, has facilitated collective actions in terms of mobilization, reach, organization, and the emergence of new communication tools. Alongside the centrality of the internet in the dynamics of social life, extremist movements, which were already present in society, have gained new means to perpetuate and organize themselves. Telegram, as a digital platform with content moderation policies that are lenient compared to others, has become an open stage for the concentration of extremist users and groups. In view of this, the platform in question has also become a research object for studies on online radicalization, making it possible to monitor radical ideologies and the social consequences observed in the promotion of violence and hate speech. Seeking to identify the immaterial bases and social dynamics that compose extremist groups in Brazil, an analysis of data extracted from Telegram between May 1st, 2023, and August 30th, 2023, was conducted. For this purpose, a hybrid approach was adopted, combining big data techniques and sociological analyses. The results of this work shed light on how the opinions encouraged and expressed in radical digital communities contribute to the promotion of violence outside digital media platforms.

Keywords: Radicalism · Platformization · Social Networks

1 Introduction

Understanding people's opinions and feelings has been of interest to both public and private institutions [24]. This interest is evidenced by these organizations' need to comprehend whether the actions they take will be well-received by the target audience they seek to reach.

L. M. Aiello et al. (Eds.): ASONAM 2024, LNCS 15213, pp. 307–321, 2025.
https://doi.org/10.1007/978-3-031-78548-1_23

In today's context, starting from the year 2004, it is possible to assert that the world is experiencing the era of the internet known as "Web 2.0" [23], in which social media play a crucial role in expressing opinions and sentiments within communities [24], capable of directly influencing the opinions of other users on the social network. The fact that social networks have become a central tool in individuals' lives has made the digital public sphere a relevant space for debates on matters of public interest, capable of [...] convincing others, starting wars, invoking love and peace, overthrowing a government, destroying reputations, sharing emotions and opinions. Ideas, thoughts, ideologies, feelings, and failures can be expressed [28]. Thus, in contemporary times, the interest lies in determining whether people adopt the views expressed through social media as an indication of the general public's thinking, but also in identifying the behavioral consequences of these perceptions [7].

Communication mediated by digital platforms has gained significant importance in the analysis of the contemporary world, alongside its characteristics and the updates of cultural flows that intersect within the social sphere [10]. The expansion of the role of these digital platforms and their implications in certain social dynamics is precisely what characterizes the so-called "platformization" [25].

With a view to the insertion of messaging platforms and applications into everyday interpersonal communication, the study by [27] sought to analyze the dynamics of interaction between groups and communities, defining as a starting point the Covid-19 pandemic and the disinformation network observed during the period. The emergence of conspiracist and extremist communities during the global health crisis demonstrated the power of group organization, especially when observing that the anti-vaccine positions expressed by members of these groups unfolded into clear demonstrations of political alignment located at the far-right pole, at the same time in which the "constant promotion of aggressive behavior and violent public demonstrations" was noted [27].

The updating of current political agendas, especially considering the rise of identity discussions and social movements that fight for the achievement of universal rights, has created the emergence of an updated extreme right [1], bringing new approaches to the political debate. Respect for themes different from those called "classic" in political science literature. If, until then, politics had the economy and the welfare state as its central principle, what could be observed internationally was a shift in voters' concern with moral issues and values, such as abortion, adoption of children by gay couples, decriminalization of drugs, racial quotas, etc. [1].

Faced with the growth of violence and extremist behavior, fueled and encouraged through the use of digital communication platforms, [27] says that "[...] public pressure on social media companies is growing to decrease the influence and reach of extremist actors. Social media companies responded to these developments by deleting videos and accounts of alleged extremist protest groups".

Movements and actions combatting manifestations of violence on the networks, designed to stop the advancement of concrete manifestations of violence in social life, have unfolded into new policy parameters for use on most digital platforms. Since the United States election period, specifically in 2016, a series of studies have been carried out to observe the power of misinformation on social media. "[...] Apple's CEO Tim Cook call for a Public Service Announcement about dis-information, new labels to identify different types of content on social platforms, systematic programs for taking down bot accounts, the integration of critical media literacy programs in schools and best practices for making fact-checks and debunks shareable [...] Facebook and Google have announced methods for preventing fabricated sites from making money through their advertising platforms" [31].

In Brazil, Telegram has been increasingly used as an important communication tool. The mobile data analysis company, App Annie, conducted a survey on the proportion of Telegram usage in Brazil, finding that in January 2022, the platform amassed approximately 42 million active users in the country [13]. The conveniences of Telegram lie in offering various engagement options, including individual or group conversations, in private and public spaces controlled by administrators. Thus, the platform has become a subject of research in the quest to understand online radicalization behavior, through the identification of communities and content advocating ideologies and discourse of hatred and violence.

Thinking about the contemporary world, especially in the reality of Brazil, episodes of violence motivated by issues of extremist ideology can be seen, such as attacks on schools, which from 2002 to October 2023, totaled 36 attempts, according to a report published by the Federal Government's Ministry of Education in 2023 [3]. The data from the report highlight two key points: "extremism is the central element of the attacks" and "the attackers, in 100% of cases, were male and motivated by hate speech and/or online communities of extreme violence". Also, the report states that "the recruitment of these adolescents occurs primarily through virtual interactions, whose strategies include humor, aesthetics, and violent language, especially misogynistic, sexist, and racist, on Internet platforms used by extremist groups to organize communities of hate and mobilize attacks, resulting in impunity due to anonymity".

Considering discursive toxicity and its connection with dehumanization, the highlighted issues in the excerpt demonstrate the direction of these extremist communities toward attacking minority social groups. Not coincidentally, when filtering the dataset for narratives that exhibit toxicity, the same issues were found to be central to the messages circulated within Telegram extremist communities. Therefore, two analysis topics will be developed based on the data found, as a way to understand the communication methods used within these channels in the consolidation and diffusion of extremist ideologies.

2 Methods

From a methodological standpoint, this work was divided into three main stages, described in the following sections.

2.1 Data Collection

The first stage involved constructing an initial set of groups and channels on Telegram, whose thematic basis of communities demonstrated fidelity to the bias under analysis. This initial set comprised 8 channels and groups manually selected by the authors through platform searches.

Subsequently, the expansion of the channel network was carried out using the snowball technique [8], which entails leveraging references and mentions to other communities within the interactions of the groups and channels comprising the initial set. Thus, the technique relies on adding these references to the collection, thereby deepening the network expansion.

The dataset was extracted from Telegram using its official API with the assistance of the Telegram Observatory [8], a tool designed for structured collection and storage of Telegram communities from an initial set of channels and groups representing them. The tool allows to configure a time period for collection, that is, choose a start and end date. Given the objective and scope of this study, a four-month period was chosen, specifically from May 1st to August 30th, 2023.

After the collection process, the final dataset comprised 116 channels and groups, which shared a total of 1,630,870 messages during the specified period.

2.2 Data Pre-processing

The obtained data was submitted to the Perspective API [20], a well-known toxicity detection machine learning model, in order to obtain, for each message on the dataset, a score that indicates the toxicity level of its content.

Our target when performing the toxicity identification and filtering to delimit the degree of toxicity that deserved to be analyzed within this article was directly induced by the perception brought by [30] about the damage caused by these discourses in social life. The ability to affect individuals and entire communities in the face of the reinforcement of pre-determinisms historically postulated in Westernized everyday life is consolidated by episodes of violence and the development of extremist movements, which will be addressed and highlighted throughout the body of this article.

After the classification process, the dataset underwent two filtering stages to obtain filtered datasets without noise and more representative of the conversational patterns observed in the community. In the first stage, all URLs and emojis were removed, followed by the elimination of all empty messages. In the second stage, messages with significant toxicity (considering a score of 0.8) were separated from those with lower toxicity. The final result yielded two filtered datasets, labeled as toxic and non-toxic, respectively.

2.3 Analysis

The third stage involved producing visualizations of the analyzed data, focusing on demonstrating the key terms used by members of these communities during group interactions and indicating the messages that represent the main themes discussed in the group for qualitative analysis. This was achieved through the creation of a word co-occurrence network, based on the methodology proposed by [21] and generating a visualization using Gephi [2], a tool for manipulating complex networks. The visualization of terms aims to showcase the communicative specificities inherent to the analyzed groups, in addition to the main narrative focuses circulated within the community.

In parallel with the data analysis, the final stage included a literature review on extremism in Brazil, directing the discussion towards a reflection on the social ramifications of the phenomenon in question and the role of digital media in representing public opinion and its social implications outside the virtual environment.

3 Results and Discussion

This section presents the results obtained after the data processing and visualization generation stages, developing an extensive qualitative analysis of the content and making a historical parallel with the literature on radicalism, especially in the Brazilian context.

Within the non-toxic topics, shown in Fig. 1, it is possible to observe six main clusters, whose central subjects focus on four thematic fronts: the green cluster brings subjects of a Christian religious nature; the pink cluster deals with themes about Brazilian politics; the blue and yellow clusters represent a strong circulation of topics about armed conflicts and wars, mainly the Ukrainian War; the sea green and beige clusters contain conspiracy themes.

In relation to toxic topics, shown in Fig. 2, the graph highlights eleven clusters containing discursive toxicity. Most clusters have political agendas focused on social issues. Considering the proposed focus of analysis, four of these clusters will be worked on in a qualitative approach related to extremist phenomena evident in everyday life in Brazil. The clusters that will be worked on represent themes related to racism (dark orange), anti-Semitism (blue) and misogyny (green and light orange).

3.1 Speech, Toxicity, and Dehumanization

The new mode of informational and conversational flow "facilitated collective action in terms of mobilization, expansion, global organization, and generation of new channels" [7] and jointly created, alongside the democratization of information, new ways of exerting power.

The relationship between power and discourse is present in literature from political science and philosophy, which treat the two subjects as inseparable factors in a relationship of coexistence. In light of Foucault's ideas, discourse can be defined as a set of thoughts arising from power relations, thus capable of ideologically legitimizing political life. Discourse is considered a historical document that produces interpretive effects among the groups that make up the same society [17]. Mastery of language use and the reproduction of discourse cause the ability

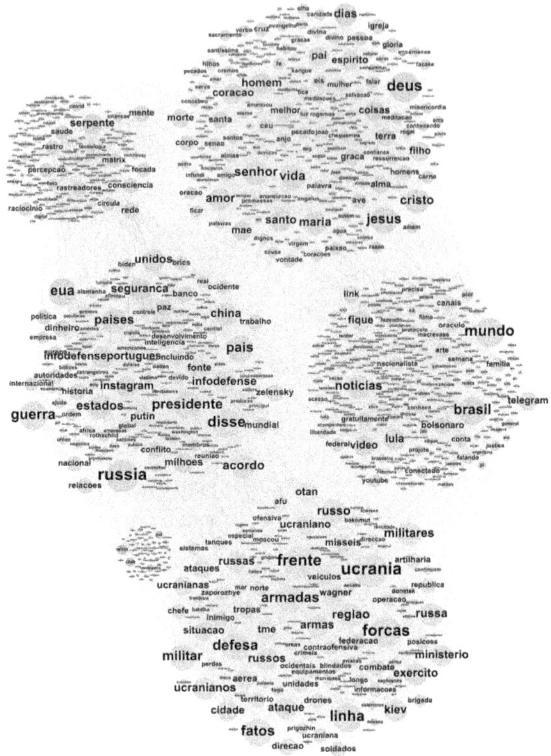

Fig. 1. Non-toxic topics.

to shape individuals through repetition, making the discursive wave mechanized and unconscious [30]. However, observing discourse as a political tool is not sufficient if we only focus on isolated expressions without understanding their real meaning, as they are shaped according to the historical and social contexts in which they arise.

The meaning attributed to language throughout history has been responsible for episodes both of conciliation and of the promotion of violence, capable of invoking both war and peace in their proper uses. Considering discursive language in a platformized society does not diverge from its power to impact the formation of public opinion and, consequently, the ways in which opinion transforms into social behavior [9], given that platformization has facilitated narrative flows beyond the barriers of orality [10].

This influence of speech is evident both in the individuality of the subject and collectively, as identification and belonging are the determining factors for the formation of collectivity – the sense of group [29]. [26] understand that "belonging has always been extremely important for the human race as a species, and this is due to instincts of social grouping, not being different with social media". The theory of social identity, initially formulated by [29], seeks to understand

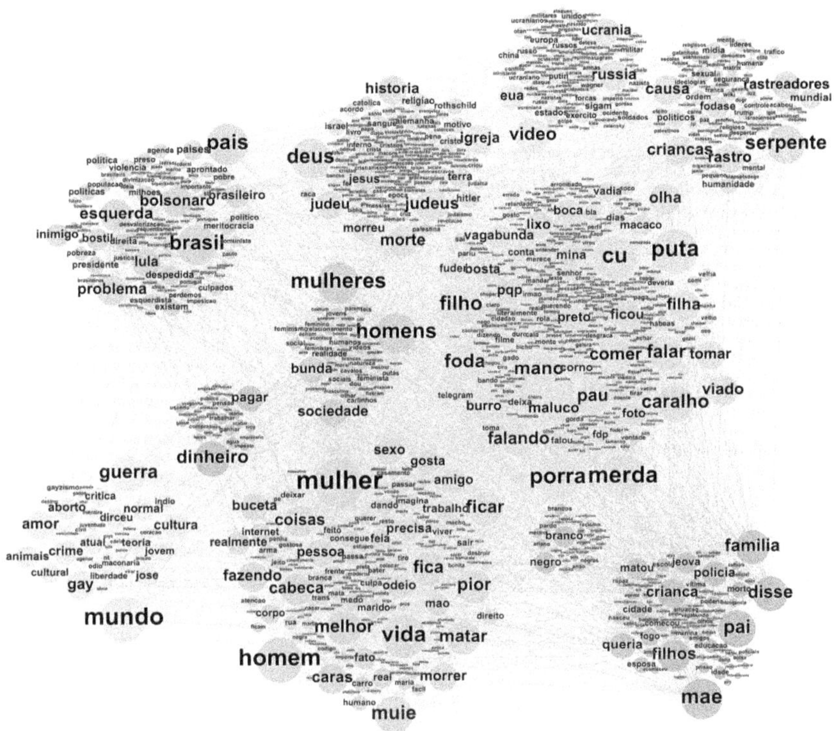

Fig. 2. Toxic topics.

the intergroup dynamics that cause members of a community to create bonds of identification with the group they are part of. This means that the feeling of belonging to a group is a source of pride and self-esteem for its members, leading to a process of social categorization, known as *in-group* (or us) and *out-group* (or them, the others) [22]. It is from this categorization that the tendency of group members to find negative aspects of an individual outside the group to which they belong appears.

It is based on group categorization and social segregation that toxicity is inserted into discursive power. Beyond an affective character in the polarization of political life, the creation of the collective imaginary about the Other is responsible for the unfolding of the process of demonizing those considered enemies [5]. Toxicity cannot (and should not) be reduced to the conception that it is merely a rude approach to something or some social group [4]. Toxicity often carries a biological determinism that disregards social and historical issues regarding the diversity of actors that make up society. The diffusion of negativity is precisely the striking characteristic of toxic discourse, and it is through it that the Other is deprived of their humanity, being seen as an individual who lacks the mental capacities of a human being [19].

At this moment, the concept of dehumanization cannot be overlooked when analyzing human history and its episodes of violence. Dehumanization has been, and continues to be, a widely used tool in favor of legitimizing violent processes, as a way to reduce moral concern for the individuals who make up the groups targeted by violence [19].

Examples of this usage are temporally close to us, especially if we consider the last century: the Holocaust in Nazi Germany, racial segregation in the United States and South Africa, the conflicts between the Tutsis and Hutus in Rwanda until the end of the 20th century, and other conflicts present in microcosmic realities belonging to the world we live in.

3.2 Racism and Racialism

The racist nature of the messages is consolidated with the presentation of pseudoscientific articles and materials by the actors. Such behavior resembles the racial anthropology of the 19th century, which justified white domination and persecution over other ethnicities based on a "science of races" and determined behaviors assigned to individuals as mere genetic factors inherent to their racial identities [12] (Fig. 3).

Fig. 3. Cluster of racism-related terms.

Within this context, the statements shown, as shown in Fig. 4 align with organic presuppositions, exposing as qualifications the supposed "science" that links black people to characteristics of violence and insanity in comparison to white people.

The postulates of racial anthropology, which sought to classify human beings according to differentiated hierarchies and analyze human races as superior or inferior [12], were carried out without the slightest scientific rigor, gradually leading to the abolition of this genre of study from anthropology, as well as from other areas of knowledge, due to its unsustainable theoretical nature.

Message 178461-1747293502

What precedes culture? Race. It's not culture that determines IQ. It's a damn mental gymnastics to justify the inferiority of Black people (both aesthetically and intellectually) that is downright shameful.

Fig. 4. Message based on racialism theory.

3.3 Antisemitism

Such "racial science," which emerged as a tool to support and justify the slave systems of Europe and America in the 19th century, was well received by Western academics and maintained its notoriety until the mid-20th century. Arthur de Gobineau, considered the "father of racism," was read and admired by various enthusiasts of the theory of racial inequality and very likely served as literary basis for the formulation of Nazi ideology during the Third Reich [12].

The alleged inequality between races is not the only postulate that links the extremist communities comprising the studied network to Nazi ideology. Within issues that are part of ethnic and cultural dynamics, it was also possible to identify anti-Semitic narratives in the exchanges of messages among the members of these communities, as shown in Fig. 5. Direct hate speech towards individuals of Jewish origin or who simply follow Judaism unfolds in debates about phenomena that are supposedly caused by Jews. All are linked to demonstrations of disgust by the actors in these networks.

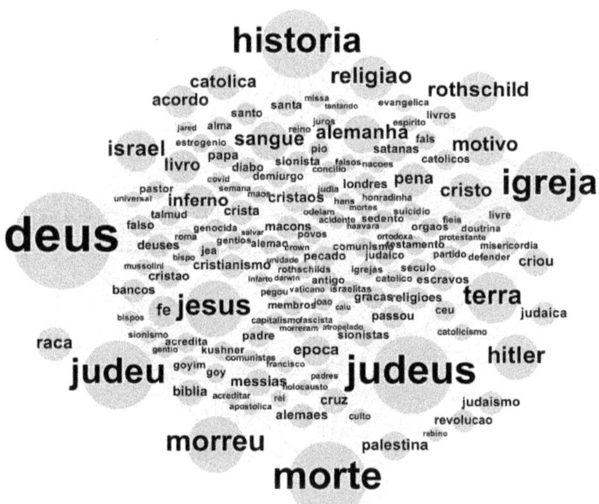

Fig. 5. Cluster of anti-Semitic related terms.

Anti-Semitism primarily appears in speculations about what is called the "New World Order" - a conspiracy theory that predicts that some governments, in partnership with certain political elites, work together to implement a totalitarian global government. For the actors of these discourses, people of Jewish origin are the main financiers of the alleged political scheme of world domination, and the collective imagination revolving around the idea that Jewish people hold endless financial resources is not a recent phenomenon. Figure 6 illustrates these theories.

Message 42737-1534437377

Jewish mafia has always been present in all major world events. This elite serves Jewish and satanic interests, pedophilia, diseases, deaths, wars, crises, vaccines, adrenochrome etc.. The true root of evil in this world are the Jewish-Zionist-Kabbalists. They have been expelled from over 100 countries over time, they are the owners of large companies. Master Jesus Christ warned about these vipers, he fought against them, the sword that Christ used was the Word.

Fig. 6. Message containing anti-Semitic conspiracy theory.

Regarding the representation of Jews as financiers and entrepreneurs by Brazilian intellectuals, it is worth mentioning the chronicle published by Humberto de Campos, a member of the Brazilian Academy of Letters, at the time of the rise of Nazism, titled "In Favor of Israel" [...]: the whole world rises up, at this moment, in favor of Israel, threatened in Germany by Hitler's anti-Semitism" [6]. Among his arguments in defense of Jews, however, we find the following highlight: "He does not offer his money to the Christian. The Christian is the one who asks for it on loan, submitting to the interest rates that Israel establishes, but does not impose. With these interest rates, Israel prospers, enriches, secretly dominates the world. The Jew never penetrated the palaces of Spain or Portugal asking the princes and nobles of the royal entourage to keep his sequins: princes and nobles were the ones who called for him and negotiated with him. And when they found themselves insolvent, they attributed to him all sorts of witchcraft and calamities" [6].

3.4 Misogyny

In addition to the perspectives of biological determinism based on ethnic attributes, a strong gender narrative was observed when analyzing the discourses circulated in the groups.

Misogyny, according to the Portuguese language dictionary, Michaelis, is the manifestation of "antipathy or morbid aversion to women". Taking this definition as a starting point, its presence in the communications of the groups studied becomes clear.

When observing the graph of words contained in Fig. 2, paying specific attention to the green and light orange clusters, it is possible to identify lexical clusters that, together, transform into violent manifestations of incitement to rape and the intellectual inferiority of women.

Observing the history of labor relations makes evident the categorization of the female role in society as reduced to social reproduction and servitude to men. The supposed lack of control over women's emotions makes it impossible for them to detach themselves from their luxuries, having to necessarily always be under male control [11]. The thought linked to primitive accumulation on the female body and its role of subservience to the male can be exemplified by the message shown in Fig. 7.

Message 185010-1092183763
THE ARIAN ALPHA MALE MANUAL Women are just extras, preventing crime is more important, but considering the possibilities we have, first we have the white woman. The white, virgin woman is the woman you will take seriously. This is the woman you will date, marry and have children with. The dirty brown woman is the woman you will just use, and then discard like an object, always being very careful not to get pregnant. And finally, we have the black woman. This is only for work and must be kept away from the white man.

Fig. 7. Message containing misogyny.

In addition to the reductionist thought and incitement to the sexual, marital and subjugated role of servitude attributed to women, the message brings with it a thought that encompasses an even deeper racial issue, which has been widely disseminated in the Brazilian scientific scene for many years, until the end of the 20th century. As in the text, Gilberto Freyre in his book Casa Grande & Senzala (1933), considered his *magnum opus*, when analyzing gender relations in Brazilian colonial society, makes the following observation:

"In relation to Brazil, let the saying go: 'White to marry, mulatta to f..., black to work'; a saying that is felt, alongside the social conventionalism of the superiority of white women and the inferiority of black women, the sexual preference for the mulatta. In fact, our loving lyricism reveals no other tendency than the glorification of the mulatto, the *cabocla*, the brunette celebrated for the beauty of her eyes, the whiteness of her teeth, her *dengues*, *quindins* and prettiness, much more than that, than the 'pale virgins' and the 'blonde maidens'" [14].

Despite the context of romanticization of Brazilian miscegenation that Freyre suggests within the functionalities of race and gender, it is possible to identify the contradiction in the appropriation of this thought by the members of these groups, who call themselves "Aryans".

[16] understands that it is essential to deal with race, class, and gender in an intersection that determines what is today the Brazilian social scenario. The reproduction of misogynistic discourse meets racist discourse in an inseparable

way, and its layers overlap in the consolidation of new prejudices promoted by hate speech.

3.5 Social Damages

Considering the large circulation of toxic narratives in the virtual environment of *Telegram*, the attempt to understand the developments of extremist content outside the virtual environment is necessary in view of the observation of news about the scenario of violence in Brazil in recent years.

When we think about racism, many questions arise in relation to the marks left by Brazil's slavery past, such as the glaring social inequality among black people (black and brown), who represent 71% of the population below the line of poverty, against the social situation of white people in the country [18]. However, not just covering structural conditions, racism manifests itself in violent ways in Brazil. According to a report carried out by [15], around 78% of the murders committed in Brazil in 2021 were against black people. In the same period, more than 20 thousand cases of racial insults and racism were recorded in Brazil.

Transposing the discussion to the intersection between race and gender, the report published by the Brazilian Public Security Forum in 2022 showed that more than 75 thousand rapes of girls and women were recorded in the country. Among these numbers, around 58.6% of the victims were black.

In addition to providing justifications for pre-existing violence through the use of digital media platforms, extremist groups also use communities as a means of organization.

As mentioned in the introduction, the number of attacks on schools has grown exponentially in the last five years, accumulating 24 of the 36 total records made by the Federal Government since 2002. Among the attacks, it was found in the report published by the Ministry of Education of the Federal Government of Brazil that "extremist groups recruit based on emotional resentments and reactionary values" [3]. The report also points out that "the lack of control over hate speech and practices, which spread even more easily through digital means" and "the exacerbation of extremism in Brazilian society" are considered causes of these episodes of violence.

4 Final Remarks

In light of the growth of extremist movements in Brazil and the pivotal role played by Telegram as a potential platform for orchestrating these groups, the objective of gathering recent data, alongside delineating the linguistic patterns adopted by community members and elucidating the pseudoscientific methodologies employed by actors to bolster toxic rhetoric and dehumanization, constitutes a vital initiative aimed at monitoring political discourse across digital platforms.

Social media platforms possess the capacity to provide real-time data, as well as other relevant information for potential mapping [32] and the continuity of studies on extremism. Updating these studies is constantly necessary, especially when considering the excessive dissemination of hate speech on digital platforms, which points to concrete episodes of violence and discrimination in Brazilian society.

This study offers a comprehensive sociological analysis, enriching our comprehension of the social and conversational dynamics of extremist communities in the digital environment, revealing how discourses continue to be based on hatred, discrimination, supremacy, and dehumanization of those who are considered, by them, to be different.

Drawing a historical parallel between contemporary discourse patterns in the digital age and those documented in existing literature, we demonstrate that, despite differences in medium, the same narratives prevalent in previous decades persist, underscoring the exigency for a more profound understanding of strategies aimed at curbing and mitigating the proliferation of these ideologies within digital spheres.

In this way, we hope that this study will catalyze further research in the realm of online radicalism, serving as a foundational framework for institutions and governments to formulate efficacious policies to counter radicalism, especially in the digital sphere, which continues to afford unparalleled dissemination avenues for the most different types of content.

Acknowledgments. This work was partially supported by the following funding agencies: *Coordenação de Aperfeiçoamento de Pessoal de Nível Superior – Brasil* (CAPES) – Finance Code 001; Fundação de Amparo à Pesquisa do Espírito Santo (FAPES) – 368/2022 - P: 2022-NGKM5, 129/2021 - P: 2021-GL60J; and Instituto Capixaba de Ensino Pesquisa e Inovação em Saúde (ICEPI).

References

1. Alkmim, A.C., Terron, S.L.: O brasil é realmente um país polarizado? análise das eleições presidenciais de 1989 a 2018. Estudos Avançados **36**(106), 7–32 (2022). https://doi.org/10.1590/s0103-4014.2022.36106.002
2. Bastian, M., Heymann, S., Jacomy, M.: Gephi: an open source software for exploring and manipulating networks (2009). http://www.aaai.org/ocs/index.php/ICWSM/09/paper/view/154
3. Brasil. Ministério da Educação: Ataques Às escolas no brasil: análise do fenômeno e recomendações para a ação governamental (2023)
4. Buell, L.: Toxic discourse. Critical Inquiry **24**(3), 639–665 (1998). http://www.jstor.org/stable/1344085
5. Caetano, R.A.P.: A desumanização do inimigo e seus reflexos em maus. Revista Temática **12**(10) (2016)
6. de Campos, H.: Os párias. No. 1, W M Jackson (1954)
7. Carrillo-Rodríguez, J., Toca-Torres, C.E.: La opinión pública en twitter análisis durante la contienda electoral en la capital colombiana. Opinião Pública **28**(3), 787–827 (2022). https://doi.org/10.1590/1807-01912022283787
8. Cavalini, A., Malini, F., Gouveia, F., Comarela, G.: Politics and disinformation: Analyzing the use of telegram's information disorder network in brazil for political mobilization. First Monday (2023). https://doi.org/10.5210/fm.v28i5.12901

9. Cervi, E.: Opinião Pública e Comportamento Político, January 2009
10. Faria Júnior, M.A.d., Silveira, S.A.d.: O whatsapp e a plataformização no brasil: uma descrição densa dos agentes articulados nas práticas de controle mediadas pela plataforma. Intercom: Revista Brasileira de Ciências da Comunicação **46** (2023). https://doi.org/10.1590/1809-58442023136pt
11. Federici, S.: Calibã e a Bruxa: mulheres, corpo e acumulação primitiva. No. **1**, 2017 (1954)
12. Fluehr-Lobban, C.: Anténor firmin: haitian pioneer of anthropology. Am. Anthropologist **102**(3), 449–466 (2000). http://www.jstor.org/stable/683404
13. Forbes: Em xeque no brasil, telegram lidera lista de apps mais baixados (2022). https://forbes.com.br/forbes-tech/2022/03/em-xeque-no-brasil-telegram-lidera-lista-de-apps-mais-baixados/. Accessed 10 May 2024
14. Freyre, G.: Casa-grande & senzala: formação da familia brasileira sob o regime de economia patriarcal. No. v. 1 in Casa-grande & senzala: formação da familia brasileira sob o regime de economia patriarcal, J. Olympio (1943). https://books.google.com.br/books?id=YsNFAAAAIAAJ
15. Fórum Brasileiro de Segurança Pública: Relatório Anual (2022)
16. Gonzales, L.: Racismo e sexismo na cultura brasileira. In: Revista Ciências Sociais Hoje, pp. 223–244. ANPOCS, Brasília (1984)
17. Gregolin, M.: Análise do discurso e mídia: a (re)produção de identidades. Comunicação Mídia e Consumo **4**(11), 11–25 (2008). https://doi.org/10.18568/cmc.v4i11.105
18. Instituto Brasileiro de Geografia e Estatística: Uma análise das condições de vida da população brasileira. IBGE (2022)
19. Landry, A.P., Orr, R.I., Mere, K.: Dehumanization and mass violence: a study of mental state language in nazi propaganda (1927–1945). PLOS ONE **17**(11), 1–14 (2022). https://doi.org/10.1371/journal.pone.0274957
20. Lees, A., et al.: A new generation of perspective api: efficient multilingual character-level transformers. In: Proceedings of the 28th ACM SIGKDD Conference on Knowledge Discovery and Data Mining, KDD 2022, pp. 3197–3207. Association for Computing Machinery, New York (2022). https://doi.org/10.1145/3534678.3539147
21. Malini, F.: Um método perspectivista de análise de redes sociais: cartografando topologias e temporalidades em rede. Encontro Anual da Compós. (2016). http://www.labic.net/wp-content/uploads/2016/06/compos_malini_2016.pdf
22. Mesquita, E., Lopes, E.L., Herrero, E.: "What if a person gets the vaccine and turns into an alligator?": a study of the effect of ideological polarization on purchase intention. BBR. Brazilian Bus. Rev. **19**(5), 565–583 (2022). https://doi.org/10.15728/bbr.2022.19.5.6.en
23. O'Reilly, T.: What is web 2.0: design patterns and business models for the next generation of software. University Library of Munich, Germany, MPRA Paper **65**, January 2007
24. Pang, B., Lee, L.: Opinion mining and sentiment analysis. Found. Trends Inf. Retr. **2**(1–2), 1–135 (2008). https://doi.org/10.1561/1500000011
25. Poell, T., Nieborg, D., Dijck, J.V.: Plataformização. Fronteiras - estudos midiáticos **22**(1), April 2020. https://doi.org/10.4013/fem.2020.221.01
26. Sandrini Bezerra, L., Gibertoni, D.: As mÍdias sociais durante a pandemia do covid-19: análise comportamental dos usuários durante este período e as possibilidades para o futuro. Revista Interface Tecnológica **18**(2), 144–156 (2021). https://doi.org/10.31510/infa.v18i2.1239

27. Schulze, H., Hohner, J., Greipl, S., Lechner, M., Gonzalo Nee Desta, I., Rieger, D.: Far-right conspiracy groups on fringe platforms: a longitudinal analysis of radicalization dynamics on telegram. Convergence: Int. J. Res. New Media Technol. **28**, 135485652211049 (2022). https://doi.org/10.1177/13548565221104977
28. Spenillo, G.: Perception and acceptance of social struggles – an increase of violence in brazilian public opinion on digital age. Filosofia Revista da Faculdade de Letras da Universidade do Porto **34**, 213–227 (2017). https://doi.org/10.21747/21836892/fil34a13
29. Tajfel, H., Turner, J.: An integrative theory of intergroup conflict. In: Organizational Identity: A Reader. Oxford University Press, March 2000. https://doi.org/10.1093/oso/9780199269464.003.0005
30. Tirrell, L.: Toxic speech: Toward an epidemiology of discursive harm. Philosophical Topics **45**(2), 139–162 (2017). https://www.jstor.org/stable/26529441
31. Wardle, C., Derakhshan, H.: Information Disorder: Toward an interdisciplinary framework for research and policy making. Council of Europe, September 2017
32. Xavier, F., Olenscki, J.R.W., Acosta, A.L., Sallum, M.A.M., Saraiva, A.M.: Análise de redes sociais como estratégia de apoio à vigilância em saúde durante a covid-19. Estudos Avançados **34**(99), 261–282 (2020). https://doi.org/10.1590/s0103-4014.2020.3499.016

Investigating Gender Euphoria and Dysphoria on TikTok: Characterization and Comparison

S. J. Dillon[1], Yueqing Liang[2], H. Russell Bernard[3], and Kai Shu[1(✉)]

[1] Emory University, Atlanta, GA 30322, USA
sarah.dillon@emory.edu, kai.shu@emory.edu
[2] Illinois Institute of Technology, Chicago, IL 60616, USA
yliang40@hawk.iit.edu
[3] Arizona State University, Tempe, AZ 85281, USA
asuruss@asu.edu

Abstract. With the emergence of short video-sharing platforms, engagement with social media sites devoted to opinion and knowledge dissemination has rapidly increased. Among these platforms, TikTok is one of the most popular globally and has become the platform of choice for transgender and nonbinary individuals, who have formed a large community to mobilize personal experience and exchange information. The knowledge produced in online spaces can influence the ways in which people understand and experience their own gender and transitions, as they hear about others and weigh experiential and medical knowledge against their own. This paper extends current research and past interview methods on gender euphoria and gender dysphoria to analyze what and how online communities on TikTok discuss these two types of gender experiences. Our findings indicate that gender euphoria and gender dysphoria are differently described in online TikTok spaces. These findings indicate similarities in the words used to describe gender dysphoria as well as gender euphoria in both the comments of videos and content creators' hashtags. Finally, our results show that gender euphoria is described in more similar terms between transfeminine and transmasculine experiences than gender dysphoria, which appears to be more differentiated by gendering experience and transition goals. We hope this paper can provide insights for future research on understanding transgender and nonbinary individuals in online communities.

Keywords: Transgender · Gender euphoria · Gender dysphoria · TikTok · Social media

1 Introduction

With the emergence of short video-sharing platforms, engagement with social media sites devoted to opinion and knowledge dissemination has rapidly

S. J. Dillon and Y. Liang—Equal contribution.

L. M. Aiello et al. (Eds.): ASONAM 2024, LNCS 15213, pp. 322–337, 2025.
https://doi.org/10.1007/978-3-031-78548-1_24

increased [28]. These videos are easy to access and understand, and can be quickly disseminated, driving community formation in online spaces. TikTok is one of the most popular of these platforms globally, with nearly 45.1 million daily active users[1], and has become the platform of choice for transgender and nonbinary individuals (TNB), who have formed a large community exchanging information and sharing knowledge [28]. Extensive online ethnography indicates that this knowledge comes from diverse sources, including medical experts, activists, and influencers, as well as from community members who share information about accessing medical services, personal experiences, what members of the community call "trans joy," traumatic encounters, and intimate changes in their bodies. The knowledge produced in online spaces can influence the ways in which people understand and experience their own gender and transitions, as they hear about others and weigh that personal knowledge against their own. Therefore, in this paper, we extend the current research on gender euphoria and gender dysphoria to reveal the ways in which online communities on TikTok discuss these two differing but related types of gender experience.

Dysphoria is in many ways considered foundational to the recognition of transgender people today, although this is a problematic definition [15]. Dysphoria is often defined as the distress that begets trans identity, more specifically as "clinically significant distress or impairment related to a strong desire to be of another gender, which may include the desire to change primary and/or secondary sex characteristics" [39]. Gender euphoria, on the other hand, is often considered gender dysphoria's inverse or corollary, and is defined as the heightened joy transgender people experience when their gender is recognized and affirmed [1,2,16,31]. Gender dysphoria is largely a medical term that grew from the efforts of trans activists to increase access to medical treatment in conversations with medical practitioners grappling with trans identity in the 1990s. In contrast, gender euphoria rose in use primarily in the past decade or so due to the efforts by community members to have the positive aspects of their gender transition and attendant experiences recognized. Gender dysphoria still largely serves as the label that can officially produce trans identity in medical spaces, because gender-affirming care (i.e., medical transition) is not available unless gender dysphoria has been diagnosed [33], although this does not capture the nuance of many trans and gender nonbinary people's experiences.

TikTok is a video-sharing platform with three main components: the video itself, the comments, and the hashtags that content creators apply to their videos. The video includes a combination of sounds (the auditory backdrop of the video) and the visual field in which the content creator appears. Videos are organized and searchable by the hashtags that content creators apply directly to their videos and are visible to viewers of the video within the TikTok application. The comments section is the space in which TikTok users can interact with the creators of the videos and with each other, to give feedback, have conversations, make jokes, and be in community. Figure 1 illustrates a typical TikTok discussing gender euphoria. This is an actual video that has been deidentified and displays

[1] https://www.demandsage.com/tiktok-user-statistics.

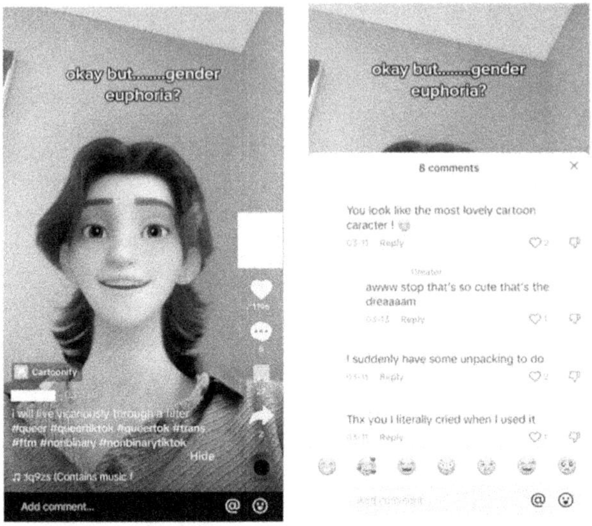

(a) TikTok video (b) Video comments

Fig. 1. An example of a TikTok video.

an individual using an anime filter, which obscures their face. Figure 1a illustrates what a video looks like when accessing TikTok, as well as the video description of a TikTok with its hashtags. Figure 1b illustrates the comments under this particular TikTok, and illustrates how users and content creators interact with another in conversation. We focus our analysis on the comments sections and the hashtags applied to TikToks such as this to investigate our central questions [28]. This research will analyze how transgender communities on TikTok define, use, and experience gender euphoria and gender dysphoria to support data-driven and nuanced consideration of these terms across a wide bandwidth of people. This research ultimately works to respect that transgender people are the experts of their own diverse experiences [37].

Here we investigate: (1) how communities of transgender people organize their experiences of gender dysphoria and gender euphoria using those algorithmic structures, and what kinds of emotional weight those terms carry for both content creators and commentors; and (2) how the organization of transmasculine and transfeminine[2] communities on TikTok intersect or do not around

[2] Transfeminine and transmasculine are terms used to describe transgender experiences. Transmasculine refers to people who are transitioning to a more masculine gender and/or presentation, including trans men, FTM (female-to-male) transsexuals, transmasculine nonbinary people, and so on. Transfeminine refers to people who are transitioning to a more feminine gender and/or presentation, including trans women, MTF (male-to-female) transsexuals, transfeminine nonbinary people, and the like. Our data do not currently get at the experiences of transgender nonbinary people who do not identify as transmasculine or transfeminine.

the topics of gender dysphoria and gender euphoria. According to the above, we could ask the following research questions (RQs):

- **RQ1**: What types of words are used and how are those words correlated for TikTok comment sections relating to the gender dysphoria and gender euphoria hashtags?
- **RQ2**: What types of words are used and how are those words correlated for TikTok descriptions relating to the gender dysphoria and gender euphoria hashtags?
- **RQ3**: What sentiments are associated with gender euphoria and gender dysphoria hashtags?
- **RQ4**: What types of topics are discussed in the gender dysphoria and gender euphoria TikTok online space?

2 Dataset

To explore how people express their gender euphoria and dysphoria in TikTok videos, we scraped data from TikTok[3] using the TikTok API tool[4]. We extracted videos marked with the hashtag #gendereuphoria or #genderdysphoria, and then collected the video descriptions and comments for each video. Our dataset comprises 116 and 110 videos for each of the two main hashtags and includes 1 description and 100 comments for each video. The statistics of the dataset are in Table 1.

Table 1. Statistics of the collected data.

Hashtag	#Videos	#Video description	#Cmts.
#Gendereuphoria	116	116	12,647
#Genderdysphoria	110	110	12,033

3 Analysis

We group the data into: (1) video descriptions for #gendereuphoria; (2) comments for #gendereuphoria; (3) video descriptions for #genderdysphoria; and (4) comments for #genderdysphoria and analyze the data with WORD CLOUDS, SEMANTIC NETWORK ANALYSIS, SPEARMAN'S RANK-ORDER CORRELATION, SENTIMENT ANALYSIS, and LATENT DIRICHLET ALLOCATION (LDA) ANALYSIS.

[3] https://www.tiktok.com/.
[4] https://davidteather.github.io/TikTok-Api/.

(a) #Gendereuphoria (b) #Genderdysphoria

Fig. 2. Word cloud of comments for different hashtags.

RQ1: What types of words are used and how are those words correlated with each other for TikTok comments about gender dysphoria and gender euphoria hashtags?

To answer **RQ1**, we first use WORD CLOUDS to visualize which words people usually use to describe these two hashtags. A word cloud is a visualization of word frequency in a set of texts. Figure 2 shows the words TikTokers typically use in describing gender euphoria or gender dysphoria.

There are wide similarities in the words used to describe gender dysphoria and gender euphoria in the comments on videos. The comments often focus on visual aspects of the videos, with common terms such as “love”, “look”, and “hair” appearing in both word clouds. In the gender euphoria word cloud, positive terms like “good”, “amazing”, “happy”, “pretty”, “handsome”, and “beautiful” are prominent. In contrast, the gender dysphoria word cloud features more ambiguous terms like “one”, “want”, “girl”, and “boy”, with “look” and “feel” being the largest, highlighting the importance of visual and affective responses in the comments.

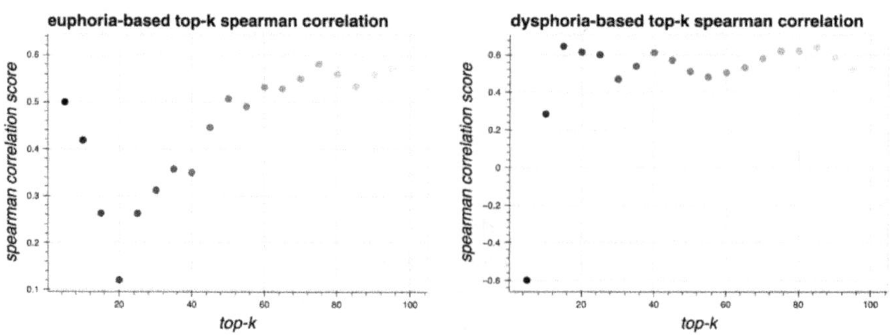

(a) #Gendereuphoria (b) #Genderdysphoria

Fig. 3. Spearman’s correlation between the top words of comments for different hashtags, by changing the k.

While word clouds provide an overall view of word frequency differences, they lack quantitative detail. Therefore, we use SPEARMAN'S RANK-ORDER COEFFICIENT to numerically distinguish word frequencies. We calculated Spearman's coefficient by varying k from 5 to 100 to observe changes in correlation between the top-k words in comments for each hashtag. Figure 3(a) and Fig. 3(b) show that the top-k words for gender euphoria and gender dysphoria comments generally have similar ranks. However, Fig. 3(a) demonstrates that rank similarity varies with k, decreasing until $k = 20$, then stabilizing. This suggests a latent discrepancy in how TikTok commenters view gender euphoria and dysphoria, particularly in the top 20 words. In contrast, the trend in Fig. 3(b) is more stable. This difference likely arises because commenters exhibit more excitement about gender euphoria but avoid disparaging experiences of gender dysphoria. Sentiment analysis (Table 2) supports this interpretation, showing positive attitudes towards gender euphoria and neutral sentiments towards gender dysphoria.

We also conducted SEMANTIC NETWORK ANALYSIS to examine word co-occurrence patterns. Analyzing the top-15 words for both hashtags, the heat maps in Fig. 4 reveal relationalities among words. In #gendereuphoria com-

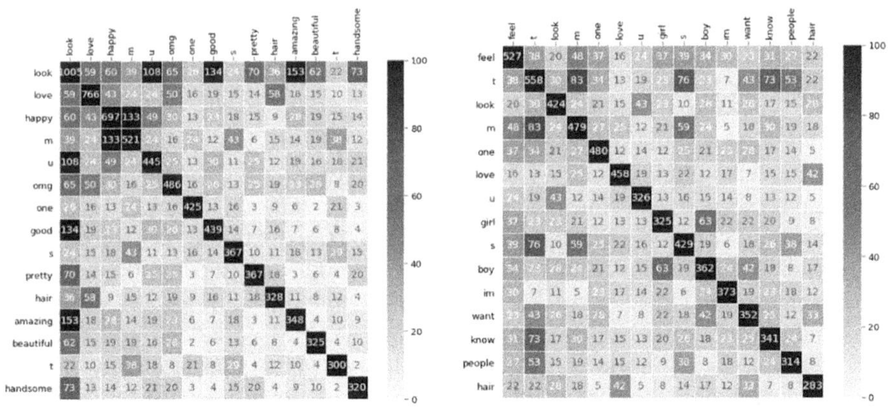

(a) #Gendereuphoria (b) #Genderdysphoria

Fig. 4. Semantic Network Analysis for word-pairs in comments of different hashtags.

ments, words like "look" and "love" co-occur in 59 comments, while "look" and "good" co-occur in 134 comments, indicating a strong correlation between "look" and "good". People often use phrases like "look good" or "look amazing" to describe gender euphoria. "Hair" is also a significant term in these comments. For #genderdysphoria comments, the focus is more on feelings, with words like "feel" being prominent. The term "t" (testosterone) suggests a dominance of transmasculine conversations. Words like "m" (masculine) and "know" also show high correlation frequencies.

RQ2: What types of words are used, and how are those words correlated with each other for TikTok descriptions under the gender dysphoria and gender euphoria hashtags?

To explore the words that video creators use to title their videos (**RQ2**), we apply the same methods as in RQ1 to TikTok video descriptions under the gender dysphoria and gender euphoria hashtags.

(a) #Gendereuphoria (b) #Genderdysphoria

Fig. 5. Word cloud of video descriptions for the gender euphoria and gender dysphoria hashtags.

Figure 5(a) and Fig. 5(b) show that *words used under both hashtags show broad similarity.* Common terms include "trans", "LGBTQ", "nonbinary", "for you page", "fem", "genderenvy", "they/them", "enby", and "transmac". Transmasculine terms, such as "masc", "trans boy", "transmasc", "transman", "transguy", and "top surgery", are more common than transfeminine terms like "fem". This could be due to our sample or a higher number of TikTok videos focusing on transmasculine experiences. As expected, "gender dysphoria" and "dysphoria" are more common in the dysphoria sample, and "gender euphoria" and "euphoria" in the euphoria sample. Interestingly, "gender euphoria" appears in the dysphoria cloud, but "gender dysphoria" does not appear in the euphoria cloud. This suggests a recognition of euphoria during dysphoria but not vice versa.

Spearman's coefficient for video descriptions shows a positive correlation with k (see Fig. 6). When k is small, the top-k words in one dataset have very different ranks in the other dataset. As k increases, Spearman's correlation score also increases, indicating that on a larger scale, the difference diminishes. This shows broad similarities in how creators title TikToks about gender dysphoria and gender euphoria.

Figure 6(a) and 6(b) demonstrate that while the top-k words for gender euphoria and gender dysphoria are different at smaller k values, the correlation increases with larger k values. This indicates that although specific terms differ initially, broader similarities emerge as more terms are considered, reflecting common themes in how creators title their TikToks about gender experiences.

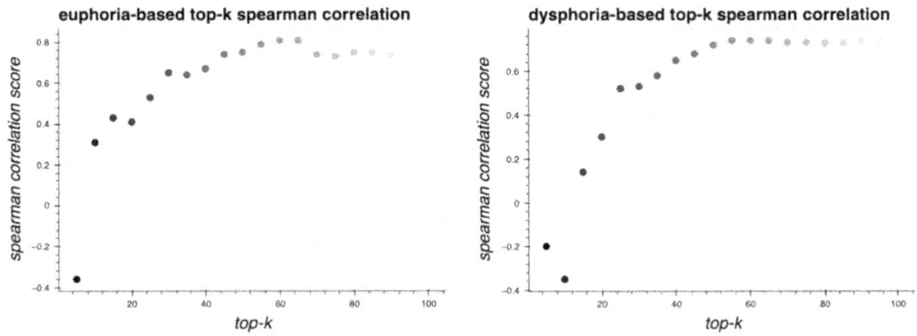

(a) #Gendereuphoria (b) #Genderdysphoria

Fig. 6. Spearman's correlation between the top words of video descriptions for different hashtags, by changing the k.

The data suggest that the frequency range in video descriptions for both hashtags tends to converge, highlighting the interconnected nature of gender-related content on TikTok.

RQ3: What sentiments are associated with gender euphoria and gender dysphoria hashtags?

To investigate **RQ3**, we conducted SENTIMENT ANALYSIS with the Vader[5] sentiment analysis tool, which is specifically designed for online social networks, to estimate whether the opinions expressed in a given text are positive, negative, or neutral. We ran a sentiment analysis for each comment/video description and report the mean and standard deviation of sentiment scores for each dataset. Scores < -0.05 imply negative sentiment; scores between -0.05 and 0.05 imply a neutral sentiment; and scores > 0.05 imply a positive sentiment. The results are shown in Table 2. From the mean of each dataset, we observe the overall sentiment of users about gender euphoria and gender dysphoria, as expressed in their comments and video descriptions. In contrast to the diversity in the comments, the descriptions for the two hashtags were quite similar. Creators of videos experiencing euphoria and dysphoria themselves expressed little difference when describing the two experiences, at least in the descriptions for the videos. This contrasts with our findings in Spearman's k correlation analysis of comments for those same videos. We conclude that creators and viewers use different expressions of language when focusing on gender euphoria and gender dysphoria. However, this difference could be because video descriptions serve as branding or searchable signposts by which users can find content. In contrast, comment sections can allow for more conversation about a topic.

In our sentiment analysis, the positive associations with gender euphoria and the more neutral emotional weight of gender dysphoria, more negative than that of gender euphoria, are clear in both the comment sections and content

[5] https://github.com/cjhutto/vaderSentiment.

creator descriptions for TikToks tagged with #genderdysphoria and #gendereuphoria. The sentiment means of the comment section for both gender dysphoria and gender euphoria are much more positive than those for the hashtags of the content creators. This indicates that the emotional weight of TikToks relating to gender dysphoria was much more negative and the positive implications of gender euphoria were significantly less positive for content creators than for commenters. Though comment sections for TikToks related to gender euphoria were statistically much more positive than for those related to gender dysphoria, comment sections related to gender dysphoria remained positive in sentiment and were significantly more positive than were the sentiment analysis for the content creators' descriptions for those videos. This may indicate commenters' attempt to allay the severely negative sentiment of content creators through uplifting comments and positive feedback on their videos relating to their gender dysphoria. Likewise, commenters displayed statistically more positive sentiment toward gender euphoria than did content creators, perhaps again seeking to support and affirm the gender euphoria of content creators.

Table 2. Sentiment analysis of comments and descriptions for the gender euphoria and gender dysphoria hashtags. Mean < 0.05: neutral, mean > 0.05: positive.

	Hashtag	Mean	Std
Comments	*#Genderdysphoria*	0.171	0.421
	#Gendereuphoria	0.333	0.422
Descriptions	*#Genderdysphoria*	0.040	0.387
	#Gendereuphoria	0.266	0.413

RQ4: What types of topics are mainly being discussed in the gender dysphoria and gender euphoria TikTok online space?

To answer **RQ4**, we used Latent Dirichlet Allocation (LDA), an unsupervised topic modeling method to obtain the most representative words in the comments we collected [3]. We focus on 5 topics for each hashtag, visualizing the inter-topic distance in two dimensions, and illustrating the top-5 most relevant terms for each topic within the LDA graphs. Based on these rankings, we can compare how people express their views of gender euphoria and gender dysphoria and the terms they used in comment sections to describe those experiences.

Figure 7 illustrates our LDA analysis clusters of the comments with the hashtag of gender dysphoria. Clusters 1, 3, and 5 contained the most salient information. Cluster 3 is best represented by the term "girl' referencing the more transfeminine experience of gender dysphoria (i.e. contains most references to and is, therefore, most different in its focus on that topic). This is also evidenced by the presence of the feminine gendered term "pretty", alongside "girl" among

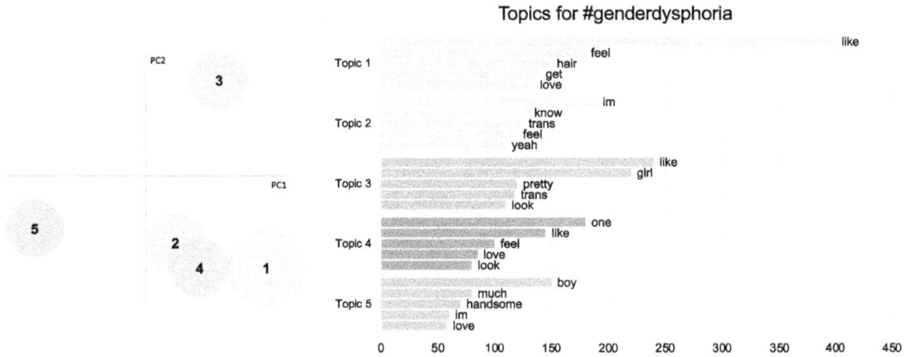

Fig. 7. LDA analysis for #Genderdysphoria comments. For each topic, we show the top 5 words.

the top 5 terms used. Just as in our word clouds relating to both gender dysphoria and gender euphoria (Figs. 2, 5), terms referencing the visual field were quite prominent. Further, the words "look" and "want", relating to desire, are common in the comment section of Cluster 3.

Cluster 5, in contrast, illustrates comments largely represented by the term "boy" along with the masculine gendered term "handsome", in the top five terms. Cluster 5, then, likely depicts the more transmasculine experience on TikTok. The term "want" is the sixth most commonly used term in these comments, but unlike Cluster 3, "look" is not among the most common words. Clusters 3 and 5 appear to be maximally distant in Fig. 5, indicating that the topics centered in the comment sections of TikToks relating to gender dysphoria are very different for the transmasculine and transfeminine experiences.

Cluster 1 is equidistant from both Cluster 3 and Cluster 5 and focuses predominantly on the visual field in these transgender dysphoric experiences. Cluster 1 is most represented by the term "hair" in the comments, reinforcing the data in the word clouds shown in Fig. 2(b) and 5(b) about the importance of hair appearance for those experiencing gender dysphoria. Terms such as "like", "feel", "get", and "love" all hint at a kind of personal emotive experience and desire in videos relating to gender dysphoria. Overall, these results support our findings about gender dysphoria (see Table 2) from the sentiment analysis of comments: The terms used do not connote strong negativity, though the topic of TikTok is not a positive one.

Figure 8 illustrates the clusters found by the LDA analysis of TikTok comments tagged gender euphoria. The 5 clusters focus on different positive terms, bolstering the results of our sentiment analysis, indicating an overall positive weight for comments pertaining to gender euphoria.

Cluster 1 is most illustrated by the terms "like", "good", and "look", as well as "happy", indicating the importance of visuality and the joy visible to commenters or experienced by commenters watching these TikToks. Cluster 2's most significant term is "beautiful", and overlaps significantly with Cluster 1,

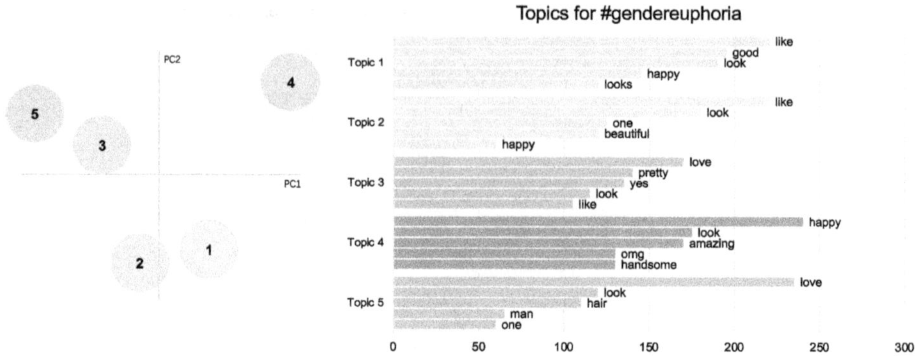

Fig. 8. LDA analysis for #Gendereuphoria comments. For each topic, we show the top 5 words.

which also centers words such as "like", "look", and "happy". Both these two clusters demonstrate a strongly positive sentiment overall.

Clusters 3 and 5 sit in a different quadrant from Cluster 1 and 2, and likewise overlap significantly. Both Clusters 3 and 5 indicate that terms such as "love", "look", and "like" appear as the top five most used words in the comment sections analyzed. Cluster 3 also includes the term "pretty", whereas Cluster 5 includes terms such as "man" and "hair", indicating that Cluster 3 likely represents comments more focused on transfeminine experiences of gender euphoria, in contrast to Cluster 5, which likely represents a more transmasculine experience. In addition to Cluster 5, Cluster 4 also seems to represent a transmasculine experience, with the term "handsome" being most used in its clustering of commented words. This could indicate, just as in our word clouds (Figs. 2, 5), a greater transmasculine presence in TikTok spaces relating to gender euphoria. Further, if Cluster 5 largely relates to comments focusing on transmasculine gender euphoria, it raises the question of whether the word "hair" is a less gender neutral term in practice within TikTok comments, and whether it tends to be associated, as our results indicate, with conversations about transmasculinity.

Cluster 4 dovetails with other clusters in terms of its overall positive word associations, and includes words such as "look", "happy", and "amazing". There were far fewer overarching differences in the clusters associated with gender euphoria than with those associated with gender dysphoria. Gender dysphoria LDA analysis delineated clear transmasculine and transfeminine clusters of comments, whereas the differences in comment sections relating to gender euphoria were much less salient. Though gender dysphoria significantly differentiates between transmasculine and transfeminine experiences, the terms used to describe gender euphoria appear to have wide resonance across transgender lives.

4 Discussion

In answering **RQ1–RQ4**, the synthesis of our results across methodologies was quite telling. As expected, the two gender experiences, gender dysphoria and gender euphoria, are characterized by different terms in the description section of TikTok videos. Our Spearman's k Correlation (Fig. 6) analyzed the description section of TikToks under the hashtags for #genderdysphoria and #gendereuphoria and indicated that when people describe their own experiences of gender dysphoria and gender euphoria, they use different terms to do so. This analysis found that these two experiences are differentiated by TikTok users and was expected based on clinical research into gender dysphoria and gender euphoria [8,15,16,18,22,28,30].

Our data support gender diverse content creators' claims on TikTok that these spaces are very uplifting, and indicates the positive impacts of community building for gender diverse individuals online [36]. Because commenters are primarily able to comment on the visual field of a video, which is a limited foray into the personal world of another person, the comments tended to focus on optical materialities, evidenced by the results of our word clouds (Fig. 2) and LDA analysis (Fig. 7 and 8) evaluating comment sections which showed consistent centering of terms such as "look". Our semantic network analysis indicated that "look" was a term used most commonly in comment sections in regards to gender euphoria, whereas in comment sections in regards to gender dysphoria, terms such as "feel" tended to be used more often. Words such as "love" and "like" also appeared consistently in results focused on comment sections, including word clouds (Fig. 2 and 5), semantic network analysis (Fig. 4), and LDA analysis (Fig. 7 and 8) of both gender dysphoria and gender euphoria. Our sentiment analysis of hashtags and comments may point to why: commenters were consistently more positive than content creators, and the presence of positive terms such as "love" and "like" in comments may indicate commenters' attempts to uplift or otherwise positively influence transgender content creators talking about their gender dysphoria or gender euphoria. Also, our Spearman's k correlation analysis of TikTok comments (Fig. 3) found a wide difference between comment sections for videos about gender euphoria and gender dysphoria, such that commenters were significantly more positive about gender euphoria, but not significantly negative about gender dysphoria. This finding is supported by our sentiment analysis on comments regarding gender dysphoria, in which commenters were also significantly more positive than content creators.

Further, our findings reflect the experience of some transfeminine voices, which argue that transmasculine perspectives on TikTok tend to dominate online conversations. Our word clouds (Fig. 2 and 5), semantic network analysis (Fig. 4), and LDA analysis (Fig. 7 and 8) all indicate that transmasculine voices may dominate TikTok spaces devoted to gender euphoria and gender dysphoria based on the increased presence of terms and the heightened influence of clusters relating to the transmasculine gender experience. However, these findings are not conclusive. Further research using these tools are needed to determine whether this is a fact of our sampling or a fact of the TikTok online space.

Finally, our research indicates that while gender dysphoria may be more delineated by transmasculine versus transfeminine gender trajectories and experiences, gender euphoria may be a far more broad-based experience across transgender lives. Our LDA analysis further indicates that in comment sections the terms transmasculine and transfeminine people use to describe their experiences of gender dysphoria are more divergent than the words used to describe experiences of gender euphoria. Clusters from the gender euphoria LDA analysis (Fig. 8) showed more consistent overlap in the graphs as well as more similar word implications than that of the LDA analysis focused on gender dysphoria (Fig. 7). Also, the LDA analyses focused on gender dysphoria clearly delineated transmasculine and transfeminine clusters and terms, which was a pattern absent in the LDA analysis focused on gender euphoria. This is a fascinating finding that deserves greater research.

5 Related Work

There has been significant research into how TikTok operates and how knowledge dissemination tends to operate on social media sites [6,9,13,21]. [38] have done model research on how user preferences function on TikTok, for example. [40] have measured the influence of user-generated content and its helpfulness for viewers, and [26] have researched how rumor functions across these online community spaces. The importance of knowledge dissemination is critical, as [23] have analyzed in the context of community health interventions and improving the lives of patients. Social media sites can be a powerful force in support of knowledge sharing [4,5,14,20], including for the sharing of scientific information and medical publications [17]. [28], in an important study, has analyzed how content about transgender communities can move across the app, but was limited to the analytics of MacKinnon's TikTok account itself, and not a wider net of analysis across TikTok using different approaches. Our research complements this approach by analyzing content about trans people broadly across TikTok.

Social media, particularly TikTok and YouTube, have become important sites of community building and interaction among LGBTQ individuals [7,27,29,32]. Social media sites are driven by algorithms that filter, manage, and direct content from creators in positive and negative ways, which users can resist or at other times cannot [10–12,19,34,35] (Gillespie 2015). This is especially the case on TikTok, where every user has access to a highly personalized For You Page on which in-app algorithms determine what content will appear. Thus, analyzing these connections using different methodologies is a critical next step in research about LGBT+ communities online, particularly transgender communities. There is a dearth of such approaches, although [25] study on the sentiment analysis of transgender people's social media posts about mental health conditions is extremely relevant to such a research agenda. Studies like [25] and this research demonstrates the potentially surprising associations of trans communities in virtual spaces and are critical to supporting this growing minority community.

Previous studies on gender dysphoria and gender euphoria specifically have tended to utilize social media as a method for recruitment [2,8,15,16,18,22,28,

30], but have relied on interview methods, which are limited in terms of the population of participants that can be included. Topical and sentiment methodological analysis allows for a wider engagement with different users about personal experiences, such as gender dysphoria, and allows the data on that topic to arise not from specific questions posed, which may affect the type of data collected, but rather from the discourse already occurring between trans people. Our research intervenes into the study of gender dysphoria using topic analysis and illustrate how trans communities discuss gender dysphoria, gender euphoria, and the expected and potentially surprising ways in which these conversations align.

6 Conclusion and Future Work

This study highlights the effectiveness of topical and sentiment analysis in understanding interactions among transgender communities on TikTok. Unlike previous research focusing on user-based analytics [28], our analysis of video content itself provides new insights into community formation and engagement.

Our methods reveal interaction patterns, content types, and terminology that are critical for researchers studying transgender and other marginalized communities on TikTok. These findings offer a framework for understanding how these communities discuss and interact with various topics.

Future research could investigate the extent to which gender-diverse communities on TikTok support and affirm creators. Key areas of interest include the support received by different creators, the expression of gender euphoria versus dysphoria, and the potential dominance of transmasculine voices. Longitudinal sentiment analysis, comparing comments with content hashtags and descriptions, could provide deeper insights into these dynamics and self-perception in virtual environments [24].

Acknowledgments. This material is based upon work supported by NSF (SaTC-2241068 and IIS-2339198), and a Cisco Research Award.

References

1. Austin, A., Papciak, R., Lovins, L.: Gender euphoria: a grounded theory exploration of experiencing gender affirmation. Psychol. Sex. **13**(5), 1406–1426 (2022)
2. Beischel, W.J., Gauvin, S.E., van Anders, S.M.: "A little shiny gender breakthrough": community understandings of gender euphoria. Int. J. Transgend. Health **23**(3), 274–294 (2022)
3. Blei, D., Ng, A.Y., Jordan, M.I.: Latent Dirichlet allocation. J. Mach. Learn. **3**, 993–1022 (2003)
4. Castillo, L.I., Hadjistavropoulos, T., Brachaniec, M.: The effectiveness of social media in the dissemination of knowledge about pain in dementia. Pain Med. **22**(11), 2584–2596 (2021)
5. Chan, A.K., Nickson, C.P., Rudolph, J.W., Lee, A., Joynt, G.M.: Social media for rapid knowledge dissemination: early experience from the COVID-19 pandemic. Anaesthesia **75**(12), 1579 (2020)

6. Fiallos, A., Fiallos, C., Figueroa, S.: TikTok and education: discovering knowledge through learning videos. In: 2021 Eighth International Conference on EDemocracy & EGovernment (ICEDEG), pp. 172–176. IEEE (2021)
7. Fredenburg, J.: YouTube as an ally of convenience: the platform's building and breaking with the LGBTQ+ Community. Georgetown University (2020)
8. Galupo, M.P., Pulice-Farrow, L., Lindley, L.: "Every time i get gendered male, i feel a pain in my chest": understanding the social context for gender dysphoria. Stigma Health **5**(2), 199 (2020)
9. Garcia, M.B., Juanatas, I.C., Juanatas, R.A.: TikTok as a knowledge source for programming learners: a new form of nanolearning? In: 2022 10th International Conference on Information and Education Technology (ICIET), pp. 219–223. IEEE (2022)
10. Gillespie, T.: Platforms intervene. Soc. Media Soc. **1**(1), 1–2 (2015)
11. Grandinetti, J.: Examining embedded apparatuses of AI in Facebook and TikTok. AI Soc. **38**, 1–14 (2021)
12. Grandinetti, J., Bruinsma, J.: The affective algorithms of conspiracy TikTok. J. Broadcast. Electron. Media **67**, 1–20 (2022)
13. Hayes, C., Stott, K., Lamb, K.J., Hurst, G.A.: "Making every second count": utilizing TikTok and systems thinking to facilitate scientific public engagement and contextualization of chemistry at home (2020)
14. Huang, L., Clarke, A., Heldsinger, N., Tian, W.: The communication role of social media in social marketing: a study of the community sustainability knowledge dissemination on linkedin and twitter. J. Mark. Anal. **7**, 64–75 (2019)
15. Jacobsen, K., Devor, A., Hodge, E.: Who counts as trans? A critical discourse analysis of trans tumblr posts. J. Commun. Inq. **46**(1), 60–81 (2022)
16. Jacobsen, K., Devor, A., et al.: Moving from gender dysphoria to gender euphoria: Trans experiences of positive gender-related emotions. Bull. Appl. Transg. Stud. **1**(1–2), 119–143 (2022)
17. Johannsson, H., Selak, T.: Dissemination of medical publications on social media–is it the new standard? (2020)
18. Johnson, A.H.: Rejecting, reframing, and reintroducing: Trans people's strategic engagement with the medicalisation of gender dysphoria. Sociol. Health Illness **41**(3), 517–532 (2019)
19. Karizat, N., Delmonaco, D., Eslami, M., Andalibi, N.: Algorithmic folk theories and identity: how TikTok users co-produce knowledge of identity and engage in algorithmic resistance. Proc. ACM Hum.-Comput. Interact. **5**(CSCW2), 1–44 (2021)
20. Khan, M.N., Ashraf, M.A., Seinen, D., Khan, K.U., Laar, R.A.: Social media for knowledge acquisition and dissemination: the impact of the COVID-19 pandemic on collaborative learning driven social media adoption. Front. Psychol. **12**, 648253 (2021)
21. Kong, W., Song, S., Zhao, Y.C., Zhu, Q., Sha, L.: TikTok as a health information source: assessment of the quality of information in diabetes-related videos. J. Med. Internet Res. **23**(9), e30409 (2021)
22. Konnelly, L.: Transmedicalism and 'trans enough': linguistic strategies in talk about gender dysphoria. Gender Lang. **16**(1) (2022)
23. Langley, J., Wolstenholme, D., Cooke, J.: 'Collective making' as knowledge mobilisation: the contribution of participatory design in the co-creation of knowledge in healthcare. BMC Health Serv. Res. **18**, 1–10 (2018)
24. Lee, A.Y., Mieczkowski, H., Ellison, N.B., Hancock, J.T.: The algorithmic crystal: conceptualizing the self through algorithmic personalization on TikTok. Proc. ACM Hum. Comput. Interact. **6**(CSCW2), 1–22 (2022)

25. Li, M., Wang, Y., Zhao, Y., Li, Z.: Transgender community sentiment analysis from social media data: A natural language processing approach. CoRR **abs/2010.13062** (2020)
26. Li, Z., Du, X., Zhao, Y., Tu, Y., Lev, B., Gan, L.: Lifecycle research of social media rumor refutation effectiveness based on machine learning and visualization technology. Inf. Process. Manag. **59**(6), 103077 (2022)
27. Lovelock, M.: 'Is every youtuber going to make a coming out video eventually?': Youtube celebrity video bloggers and lesbian and gay identity. Celebr. Stud. **8**(1), 87–103 (2017)
28. MacKinnon, K.R., Kia, H., Lacombe-Duncan, A.: Examining TikTok's potential for community-engaged digital knowledge mobilization with equity-seeking groups. J. Med. Internet Res. **23**(12), e30315 (2021)
29. Miller, J.F.: Youtube as a site of counternarratives to transnormativity. J. Homosex. **66**(6), 815–837 (2019)
30. Pulice-Farrow, L., Cusack, C.E., Galupo, M.P.: "Certain parts of my body don't belong to me": trans individuals' descriptions of body-specific gender dysphoria. Sex. Res. Soc. Policy **17**, 654–667 (2020)
31. Rachlin, K.: Medical transition without social transition: expanding options for privately gendered bodies. Transgender Stud. Q. **5**(2), 228–244 (2018)
32. Rodriguez, J.A.: LGBTQ incorporated: Youtube and the management of diversity. J. Homosex. **29**, 1–22 (2022)
33. Shuster, S.M.: Trans Medicine: The Emergence and Practice of Treating Gender. New York University Press, New York (2021)
34. Siles, I., Valerio-Alfaro, L., Meléndez-Moran, A.: Learning to like TikTok... and not: algorithm awareness as process. New Media Soc., 14614448221138973 (2022)
35. Simpson, E., Hamann, A., Semaan, B.: How to tame "your" algorithm: LGBTQ+ users' domestication of TikTok. Proc. ACM Hum.-Comput. Interact. **6**(GROUP), 1–27 (2022)
36. Singh, A.A.: Transgender youth of color and resilience: negotiating oppression and finding support. Sex Roles **68**, 690–702 (2013)
37. Solutions Not Punishments Collaborative: Deeper than visibility (2022)
38. Tao, Z., Wei, Y., Wang, X., He, X., Huang, X., Chua, T.S.: MGAT: multimodal graph attention network for recommendation. Inf. Process. Manag. **57**(5), 102277 (2020)
39. Turban, J.: What is gender dysphoria? (2020). https://www.psychiatry.org/patients-families/gender-dysphoria/what-is-gender-dysphoria. American Psychiatric Association
40. Zhuang, W., Zeng, Q., Zhang, Y., Liu, C., Fan, W.: What makes user-generated content more helpful on social media platforms? Insights from creator interactivity perspective. Inf. Process. Manag. **60**(2), 103201 (2023)

PRAGyan - Connecting the Dots in Tweets

Rahul Ravi(✉), Gouri Ginde, and Jon Rokne

University of Calgary, Calgary, AB T2N 1N4, Canada
{rahul.ravi,gouri.deshpande,rokne}@ucalgary.ca

Abstract. Understanding the underlying reasons behind events and statements is crucial for businesses, policymakers, and researchers. In this research the data available from social media platform is analyzed, aiming to extract these reasons. An integration of Knowledge Graphs (KGs) with Large Language Models (LLMs) is used to perform causal analysis of tweets dataset. The LLM aided analysis techniques often lack depth in uncovering the causes driving observed effects. By leveraging KGs and LLMs, which encode rich semantic relationships and temporal information, this study aims to uncover the complex interplay of factors influencing causal dynamics and compare the results obtained using GPT-3.5 Turbo. We employ a Retrieval-Augmented Generation (RAG) model, utilizing a KG stored in a Neo4j (a.k.a PRAGyan) data format, to retrieve relevant context for causal reasoning. Our approach demonstrates that the KG-enhanced LLM RAG can provide improved results when compared to the baseline LLM (GPT-3.5 Turbo) model as the source corpus increases in size. Our qualitative analysis highlights the advantages of combining KGs with LLMs for improved interpretability and actionable insights, facilitating informed decision-making across various domains. Whereas, quantitative analysis using metrics such as BLEU and cosine similarity show that our approach outperforms the baseline by 10%.

Keywords: Knowledge Graphs (KG) · Natural Language Processing (NLP) · Causal Analysis · Large Language Models (LLM) · Relation Extraction · Retrieval-Augmented Generation (RAG) · Embeddings

1 Introduction

Understanding the underlying reasons behind events and statements is of great interest for businesses, policymakers, and researchers alike. These reasons may frequently be found from available textual data. An important source is inline social media which has become widely used and as such it has developed into a rich source for such data [3]. Large Language Models (LLMs) such as GPT-3.5 Turbo and BERT provide deep, context-aware language understanding, capturing contextual and temporal information critical for deciphering the nuances of social media discourse [5]. However, these techniques often offer

L. M. Aiello et al. (Eds.): ASONAM 2024, LNCS 15213, pp. 338–354, 2025.
https://doi.org/10.1007/978-3-031-78548-1_25

surface-level insights and lack the depth to uncover root causes [17]. In this paper, Knowledge Graphs (KG), based on graph theory fundamentals, storing data as nodes and links, offering structured representations of domain-specific knowledge and capturing rich semantic relationships are considered for data representation [6,19,29]. Using this representation temporal information, which is also crucial for understanding causal relationships in social media data, is used since the sequence and timing of events provide essential insights into causation [15]. **Our aim is therefore** to extract causal information from social media content, leveraging both KGs and LLMs with temporal knowledge for enhanced causal analysis.

Thus, in this research, we explore the effectiveness of integrating knowledge graphs (KGs) and large language models (LLMs) in performing causal analysis of text-based social media data such as tweets from the X (formerly Twitter) platform [27]. We hypothesize that combining these two approaches (LLM and KG) can enable an effective detection of causes and their underlying effects, providing a richer understanding of the dynamics in social media interactions [6]. Thus, **in this study** we explore and compare a hybrid method, the combination of KG and LLM, to enhance the interpretability and actionable insights derived from causal analysis by delving deeper into the underlying drivers of events and statements in time series tweets. As such, we compare the GPT model with our hybrid approach PRAGyan for causal analysis on the COVID-19 tweets dataset. **Hence, we propose PRAGyan**[1], a hybrid KG-RAG-LLM model with Retrieval-Augmented Generation (RAG) as a key component, binding KG and LLM together. RAG combines retrieval mechanisms with generative models to enhance the generation process by providing relevant contextual information. Taking inspiration from [7], we use RAG as a retriever model to fetch pertinent documents or knowledge snippets along with a generator model to produce responses conditioned on both the query and the retrieved information. This ensures contextually accurate and informative outputs, leveraging the vast knowledge available from the KGs together with the sophisticated language understanding acquired from using LLMs [9]. This dynamic integration makes RAG particularly well-suited for tasks requiring up-to-date information extraction and complex reasoning, such as causal analysis acquired from social media data [21].

The main contributions in this study are:

- We propose a novel methodology to derive actionable insights from a time series analysis of a COVID-19 tweets dataset [24] to derive deeper insights to events for causal analysis.
- The proposed approach can trace back the list of tweets which lead to such causations thus bringing in interoperability and transparency which is currently missing from LLM only solutions (such as GPT models) [18,20].
- Our method can be used on a continuously growing dataset and not just on a snapshot of a dataset.

[1] PRAGyan means wisdom and gyan means knowledge in Sanskrit language.

– We evaluate our method using both qualitative and quantitative techniques.

The paper is structured as follows: Sect. 2 sets the context with a motivating example. Section 3 details various elements of this study. Section 4 explains the details of the usage of the methodology and dataset. Section 5 showcases the main results and Sect. 6 captures related work. Section 7 lists the limitations of the study and Sect. 8 highlights future works and conclusion.

2 Motivating Example

During the COVID-19 pandemic, we noticed a recurring theme in tweets where people frequently mentioned an increased appreciation for simple things in life. This observation sparked our curiosity, and we attempted to manually understand the reasons behind this sentiment. However, this process proved to be extremely challenging due to the scattered and unstructured nature of the relevant tweets.

To answer a question like, "What caused people to appreciate simple things around them during the COVID-19 pandemic?" we had to sift through countless tweets. These tweets were not posted sequentially nor were they directly related to each other. The information was dispersed across different parts of the dataset, making it difficult to piece together a comprehensive answer manually.

After spending significant manual effort we found that people's appreciation for simple things was influenced by various interconnected factors. The lockdown measures led to isolation, which in turn heightened people's awareness of their immediate surroundings. The restriction of activities forced individuals to spend more time at home, allowing them to notice and value the simple things that were always around them. This combination of factors provided a sense of comfort and appreciation during challenging times.

This manual process highlighted the need for a more efficient solution. The difficulty in manually analyzing the vast and scattered data from social media underscored the necessity of developing a method that could automatically integrate and analyze this information to uncover deeper insights. Thus, we aimed to understand and interpret the complex dynamics of social media interactions represented on a temporal graph and to automate this difficult task further.

3 Key Concepts

In this section, we discuss the essential preliminaries of our research.

Knowledge Graphs (KGs): KGs provide semantic frameworks for organizing domain-specific information, facilitating causal analysis of social media data [6, 19,29,41]. By enriching textual data with structured semantic information, KGs enhance context and relationship understanding, aiding in identifying underlying drivers of events and causal relationships. They encode entity relationships and attributes, offering deeper insights, in particular, into social media discussions and integrating diverse information sources, thereby improving the robustness of causal analysis [4]. Neo4j, which is a leading graph database platform, excels in managing complex relationships within large text corpora, including temporal information on edges, which allows accurate representation of event chronology and causal links [34]. Its flexible schema and query language make it ideal for dynamic, heterogeneous data like social media content, enabling timely analysis and sophisticated queries for deeper insights.

Large Language Models: LLMs such as GPT and BERT have revolutionized natural language understanding tasks [25]. Trained on vast corpora of textual data, LLMs capture intricate semantic nuances within text, enabling reasoning and inference tasks [5]. GPT-3.5 Turbo, an advanced version of GPT models by OpenAI, is used as the baseline in this study. It excels in comprehending queries and context, producing accurate and contextually relevant responses [43]. Its ability to handle longer and complex contexts is crucial for our study, and it is accessible through OpenAI's API for high-quality causal reasoning.

Embedding and Encoding Techniques: Dimension reduction transforms high-level representations of text or graph into a lower dimensional 2D space as vectors [35]. When performed on a graph with a list of nodes and edges, it is referred to as "embedding" and when performed on a sequential text, it's known as "encoding". This research employs both KG and LLM for optimized context extraction:

- *Node2Vec:* Node2Vec generates node embeddings by simulating random walks on the graph and applying the Skip-Gram model, capturing semantic similarities based on structural context [11,39]. These embeddings capture semantic and temporal relationships between nodes, essential for understanding connections in social media data. They facilitate the identification of underlying factors driving events and statements, enabling efficient transfer learning and informed decision-making.
- *Sentence-BERT:* BERT (Bidirectional Encoder Representations from Transformers) is used for text comprehension [31,38]. Input data is converted lowercase and tokenized it into subword units before generating encodings by summing token, segment, and position encodings. Final encodings for each token are then obtained based on the NLP task.

LLM-Enhanced KG: The process begins with a fine-tuned relation extraction LLM (LLAMA3 [26]), which identifies entities and relationships from text to generate triples which are used to construct a KG. This integration enhances the ability to identify and articulate the intricate web of correlations between events and statements, allowing for a more structured exploration of complex interactions within the data.

Retrieval-Augmented Generation (RAG): The KG, in conjunction with Sentence-BERT LLM is used to retrieve relevant context through Retrieval-Augmented Generation (RAG). After the graph has been embedded, the LLM is used to identify contextually similar subsets of corresponding nodes and edges with query encoding, which are then expanded based on the graph structure for obtaining both contextually and semantically relevant context.

KG-Enhanced LLM: Finally, the retrieved context is used by GPT-3.5 Turbo to reason about the causes behind user-defined queries. Traditional language models like GPT excel in generating context-aware responses but often lack the structured knowledge to fully grasp complex relationships in social media data [1]. Hence, the context enhances the model's responses with structured correlation knowledge from the KG. This integration improves the model's ability to identify and analyze causal relationships, improving the accuracy and depth of causal analysis.

The combined approach outperforms using standalone LLMs (the baseline) or KGs, leveraging the strengths of both techniques. The upcoming sections discusses an implementation strategy for this approach, in PRAGyan, and the results highlight the efficiency and effectiveness of this integrated method in uncovering causal factors driving events and statements on social media platforms through comparative analysis.

4 Methodology

4.1 Research Question (RQ)

We have formulated a key research question focused on evaluating the effectiveness of combining large language models (LLMs) with knowledge graph (KG) embeddings for causal analysis:

Does the integration of KG and LLM through RAG enhance causal reasoning capabilities on social media data as opposed to using LLMs without context?

Explanation: This question investigates whether integrating knowledge graph embeddings with LLMs enhances the context and accuracy of causal reasoning in social media data. By combining the structured knowledge in KGs with the deep contextual understanding of LLMs, we hypothesize that this approach will yield superior results for identifying and reasoning about causation. The expected inputs include raw social media text data, knowledge graph embeddings, and LLMs for encoding and prompt engineering. The goal is to demonstrate improved performance in causal reasoning tasks using the combined approach compared to using LLMs alone. The study design that follows encompasses key steps involved in creation of the proposed hybrid model (PRAGyan) and to answer our RQ.

4.2 Dataset

X (formerly Twitter) provides rich source of first hand experience details and information reporting as it is frequently used by lay people and hence it is one of the primary platforms for individuals where they share real-time updates, personal experiences, and opinions on various topics. We obtained the COVID19

Tweets dataset from Kaggle [30]. The dataset had thirteen features/fields. However, we only used the "text" that contains all the tweets along with "date" that contains timestamps for each tweet. Future work might include "user_location" (geography based analysis of causes) and "date" (temporal parameter that directly affects causation). This is a popular dataset with more than 22k downloads, mainly for data cleaning and learning purposes as stated on the website [30].

The tweets were gathered between 02/28/2020 and 07/23/2020 with a coverage across the world using Twitter API with the hashtag, "#covid19". It's size is 68.71 MB and contains over 35k rows. On average each tweet has around 17 words after pre-processing.

4.3 Study Design

Our study design comprises two main sections: the baseline approach and the proposed model. Each section tackles the problem of causal reasoning in social media data, specifically focusing on tweets, but they employ different methodologies.

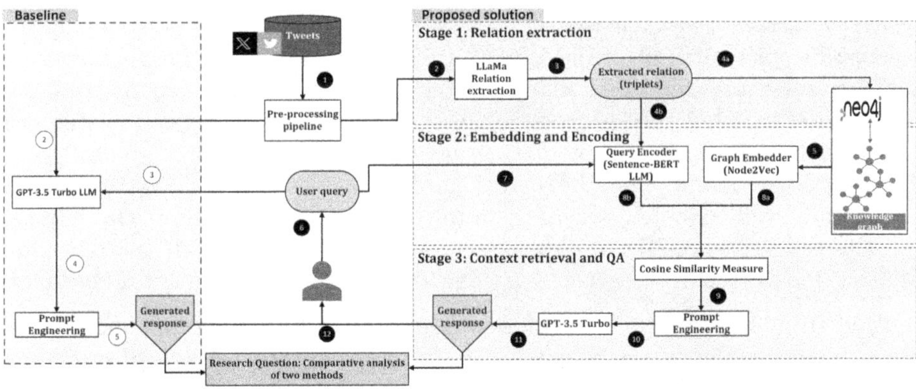

Fig. 1. Our study design: On the left, the baseline approach and corresponding workflow is depicted. On the right, our proposed model: PRAGyan and its corresponding workflow is depicted. Various elements in the middle section are common to both the methods - Steps ❶, ❻ & 12

Pre-processing: Figure 1 shows the overall architecture of our study. The first step (❶) is the data pre-processing which is common to both the baseline method and our proposed method. In addition to the tweets' text, we also extract timestamps from tweets and ensure their accuracy by handling different time zones, correcting any inconsistencies, and standardizing the format. The tweets strings are highly unstructured, ambiguous and full of noise (special characters, acronyms and URL's), Thus we first utilize a dictionary to expand contractions of words often used in social media slang replacing them with their appropriate representation. For example, "aren't": "are not/am not", "can't've": "cannot have", etc. We also remove special characters, emojis, URLs and Hashtags.

The Baseline: The baseline model is using the GPT-3.5 Turbo LLM directly for querying the entire preprocessed text corpus (Step ②) through prompt engineering for reasoning causation (Step ④) when a user query is passed to it (Step ③) and result is generated in Step ⑤. Unlike the proposed model, this method does not utilize a knowledge graph for context retrieval. Instead, it relies solely on the LLM's ability to process and generate responses based on the raw text data.

Proposed Model (PRAGyan): The proposed PRAGyan model follows a three-stage process: Relation extraction, Embedding/Encoding, and Context Retrieval and QA, as illustrated in Fig. 1. Each stage leverages specific technologies chosen for their capabilities to enhance the overall process.

Relation Extraction: In the proposed model, relation extraction is performed using the LLaMa3 model (Step ❷ in the diagram), specifically the fine-tuned variant from Hugging Face (solanaO/llama3-8b-sft-qlora-re [26]). This model identifies entities and their relations (together known "triplets") from the preprocessed tweets (Step ❸). LLaMa3 [37] enhances traditional relation extraction methods through supervised fine-tuning by leveraging advanced natural language understanding. The process begins by loading the pre-trained model with a PEFT adapter for efficient memory usage. The tokenizer creates prompts from the preprocessed text, and the model generates outputs, forming the triples. They are stored as a KG within a graph database, Neo4j (Step ❹a). Subsequently, the Sentence-BERT encodings are also obtained for all the relational triples (Step ❹b) which are stored as edge property information in the database.

Embedding and Encoding: In this stage, first the KG stored in Neo4j is embedded using Node2Vec (Step ❺). This method captures both local and global structural information of the graph, creating high-quality vector representations stored in Neo4j for efficient retrieval of contextually relevant data. The objective is to maximize the probability of observing a node's neighbors given its embedding through the Node2Vec algorithm:

$$\max \sum_{u \in V} \sum_{v \in N(u)} \log \Pr(v|u)$$

where $\Pr(v|u)$ is the probability of node v being a neighbor of node u, computed using the softmax function:

$$\Pr(v|u) = \frac{\exp(\mathbf{z}_v^\top \mathbf{z}_u)}{\sum_{w \in V} \exp(\mathbf{z}_w^\top \mathbf{z}_u)}$$

Here, $\mathbf{z}_u$ and $\mathbf{z}_v$ are the embeddings of nodes u and v, respectively. The learned embeddings are stored in the graph database for efficient retrieval and analysis.

In a dynamic graph situation where new events continuously occur, it is running periodically while incorporating the latest temporal data ensures the embeddings stay up-to-date with the evolving graph.

When a user inputs a query (Step ❻) to reason causation, the query is encoded using an LLM (Sentence-BERT) to create a vector representation (Step ❼). This representation is mapped onto the corresponding Node2Vec embeddings to identify the most relevant context in the next stage.

Context Retrieval and QA: This step involves retrieving the most relevant contexts from the knowledge graph to engineer prompts, a process known as prompt engineering (Step ❾). These prompts are fed into a GPT-3.5 Turbo model (Step ❿) to generate a response. The response, enriched with relevant context from the knowledge graph, is provided to the user, ensuring a detailed and comprehensive answer. This is known as RAG. It involves:

- *Identifying Contextually Relative Subgraph*: The Sentence-BERT encoding of the query is compared with the encodings calculated from the triples to choose a highly similar set through Cosine similarity (Step ❽a). This set is then looked up on the Node2Vec embedding space with their corresponding embeddings. The resulting set of node embeddings are contextually similar to the query (Step ❽b) which are passed on to the next step for further refinement to extract contextual information.
- *Retrieving Final Context - Contextually and Semantically Relevant*: With this set, using Cosine similarity, we identify it's top semantically similar subsets of neighbouring nodes, edges in the Node2Vec embedding space (Step ❾). Both similarity measures (this and previous step) are validated for a threshold of 0.35, which was determined through trials and inspections of contexts with various thresholds. This ensures the right amount of context that could potentially provide the true cause. The subgraphs are mapped onto the preprocessed sentences, retrieving the top-ranked sentences (contextually and semantically most similar to the input statement) that serve as the source context for prompting with LLM (Step ❿). Including temporal constraints ensures the retrieved context respects the chronological order of events.

The retrieved context is then used by the GPT-3.5 Turbo LLM to generate responses with an effective prompt instructing the model to infer the cause for the input query (Step **11**). In the last step (Step **12**) a comparative analysis of the two methods is then performed.

4.4 Evaluation and Comparative Analysis

The performance of the integrated approach (PRAGyan) is evaluated and compared with the baseline approach (GPT3.5 Turbo). The evaluation process is conducted as follows:

Qualitative Evaluation: According to [9], qualitative metrics like faithfulness, answer relevance, and context relevance provide a nuanced understanding of how well the models capture and convey meaningful information. The paper [8] further emphasizes the importance of automated qualitative evaluation, highlighting challenges such as ensuring consistency and aligning automated metrics with human judgment. These qualitative assessments are crucial for determining the effectiveness of models for generating accurate and contextually relevant responses, which is vital for the success of our research.

The baseline processes the entire corpus without specific context from the knowledge graph, generating responses based solely on the raw text data. The evaluation process involves manually designed queries from the tweets and correct causes created for validation. The goal is to determine which approach yields the closest truth or most relevant cause statements to the right answers based on specific metrics. The baseline is not fine-tuned for this specific task; instead, it utilizes the pre-trained model to generate responses based on the provided context.

Quantitative Evaluation: Quantitative evaluation is performed utilizing two metrics as follows:

- **BiLingual Evaluation Understudy (BLEU) Score**: BLEU measures the precision of n-grams in generated text against reference texts, evaluating word or phrase matches with reference causes [28]. It's simple, easy to understand, and reliable for large-scale evaluations. Although it doesn't capture word importance or synonyms, BLEU provides a quick, reproducible measure of text accuracy, which is crucial for assessing if the model generates causes close to the ground truth [32].
- **Cosine Similarity on Sentence-BERT Encodings**: The BERT Similarity metric, based on pre-trained BERT embeddings, captures contextual meanings and semantic similarity between texts, making it ideal for assessing the conveyed information rather than exact structure [31]. SentenceBERT, optimized for sentence embeddings, shows a high correlation with human judgments of similarity, surpassing BLEU and better reflecting the quality and relevance of generated summaries [2]. The similarity score ranges from 0 to 1, with 1 being the highest.

Together, these metrics ensure a comprehensive evaluation of both structural and semantic quality in generated text, as highlighted in [14]. Hence, they are well suited for evaluating performance for this use case.

5 Results

To create the proposed model, we used Google Colab Pro equipped with A100 GPU and a high RAM setting of about 80 GB and Python 3.11 as the programming environment. We also utilized OpenAI APIs, and libraries from Huggingface, NLTK, SkLearn packages [40] for generating the results.

Table 1. Evaluation Metrics for the Baseline and PRAGyan Models

Metric	The Baseline	PRAGyan
BLEU	0.45	**0.49**
Cosine Sim with LLM Encoder	0.86	**0.88**

5.1 Quantitative Evaluation

The two average metric scores for both models observed over multiple queries are presented in Table 1. PRAGyan outperforms the baseline model across both the metrics, indicating its superior performance in generating responses that are both accurate and comprehensive. Based on the box plots in Fig. 2, we can infer the following about the performance of the two models:

1. **BLEU score**
   - *PRAGyan*: It shows a wider range of BLEU scores, with some instances achieving very high scores, suggesting the model's capability to produce highly relevant and accurate text in some cases. The higher variability, indicated by the wide interquartile range and the position of the average score above the median, suggests that while the model can perform very well, it is not consistently high-performing across all instances.
   - *The Baseline*: The BLEU scores for the baseline are more tightly clustered, indicating more consistent performance. However, the lower average and median scores suggest that, on average, this model does not achieve the same high level of relevance and accuracy as PRAGyan. The consistency is beneficial for tasks requiring stable performance but may lack the peak performance seen in PRAGyan.
2. **Cosine similarity**
   - *PRAGyan*: The BERT Encoding Cosine Similarity scores for PRAGyan are consistently high, with a median score of approximately 0.89. This indicates that at least half of the similarity scores are at or above this value, highlighting the model's effectiveness in capturing semantic relationships. The small interquartile range (IQR) suggests that the scores are tightly clustered around the median, demonstrating reliable performance in terms of semantic similarity with minimal variation.
   - *The Baseline*: The BERT Encoding Cosine Similarity scores for the baseline model are generally lower compared to PRAGyan, with a median score of approximately 0.86. The wider range and greater variation in scores suggest that the baseline model struggles more with certain instances, leading to less consistent and reliable performance in capturing semantic relationships.

Our proposed model, leverages the strengths of both KG and RAG, resulting in better overall performance. It showed significant advantages for the COVID-19 tweets dataset, generating contextually and semantically relevant and accurate information, thus, improving the quality of insights derived from this social media content. However, higher score variations suggest room for improvement in the knowledge graph's construction, as it currently only conveys correlation information. In contrast, while the baseline provides more consistent performance, its generally lower scores indicate less accuracy. As the data environment grows, this increases the overhead for the baseline, potentially affecting results and processing times. Hence, maintaining a graph database helps with dynamic real-time social media updates and faster querying.

> PRAGyan demonstrates superior average performance and the capability to produce highly relevant and accurate text, as indicated by higher metric scores. This makes it particularly beneficial for analyzing social media data, where context and relevance are crucial. The variability in its performance highlights its potential to excel in many instances, suggesting that integrating LLM with KG through RAG enhances the model's ability to generate high-quality text.

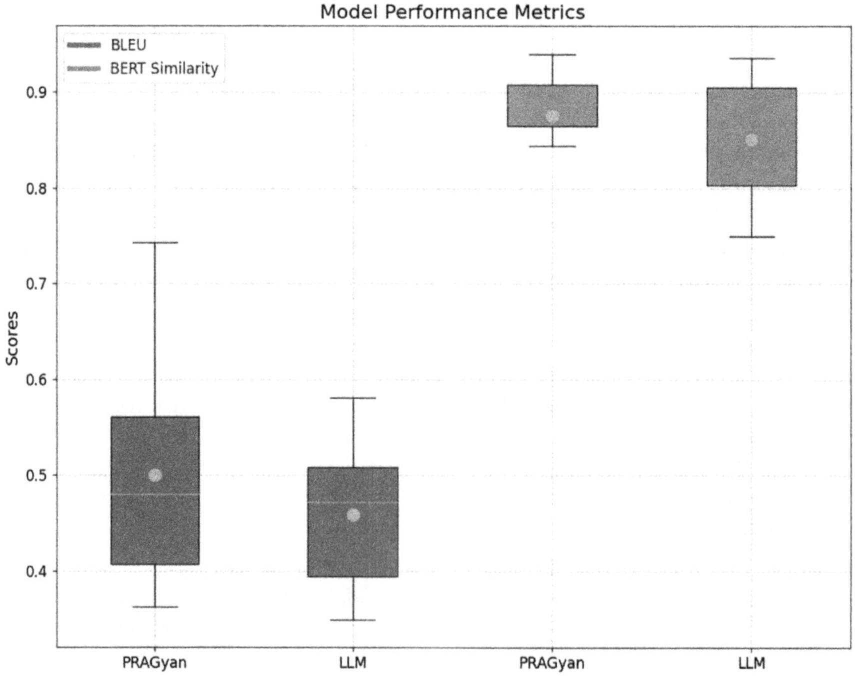

Fig. 2. Box plot showing metrics between the Baseline and PRAGyan

5.2 Qualitative Evaluation

Table 2 illustrates a comparative analysis of the responses generated by PRAgyan and the baseline in addressing questions about the COVID-19 pandemic.

For the first question regarding measures to support COVID-19 patients, PRAGyan identifies specific actions such as the setup of a plasma bank, prayer support, and updates on ICU patients, highlighting concrete steps taken to enhance treatment outcomes. In contrast, the baseline provides a general summary, lacking specific details.

Similarly, for the second question about the impact of the pandemic on grocery shopping behaviors and store operations, PRAGyan details instances of panic buying, dedicated shopping hours for the elderly, and increased demand for online shopping. It also mentions specific challenges faced by stores, like shortages of essential items and the pressure on employees to keep shelves stocked. On the other hand, the baseline gives a more generalized response, focusing broadly on fear of shortages and stockpiling without specific contextual examples.

The baseline model generates responses based solely on the raw text data, leading to variations in responses for the same set of queries when run multiple times. This is a characteristic of generative AI models like LLMs, which can produce different outputs depending on the input context and randomness inherent in the generation process, unless fine tuned comprehensively for datasets or most efficient prompt engineering. In contrast, PRAGyan enhances the generation process by providing specific context derived from the knowledge graph, resulting in more consistent and accurate responses.

This comparison underscores the enhanced ability of PRAGyan to provide richer, context-aware responses by utilizing structured knowledge from the KG, thereby offering deeper insights and more actionable information.

6 Related Work

The growing interest in computational methods for analyzing dynamic text data, such as social media, has led to significant advancements in data representation and analysis techniques, with KGs and LLMs at the forefront.

Advanced Graph Techniques and NER in Social Media: [6] explore combining contextual word encodings with KG embeddings, enhancing sentiment analysis through integrated representations. Effective Named Entity Recognition (NER) in social media texts, detailed by [22,23], informs our preprocessing steps. Additionally, a fine-tuned LLAMA3 model for relation extraction (triples) demonstrates efficiency in training on data [26].

Causal Knowledge Graphs and Explainability: [10,16,33] delve into constructing causal KGs from varied data sources and enhancing explainability through causal reasoning. These studies provide a framework for our work in creating KGs that not only predicts relations and similarities but also causation, offering actionable insights into factors driving information on social media.

Table 2. Responses from PRAGyan (Context from Embeddings through Selective Tweets) and the Baseline (Entire Corpus as the Text Source)

Question	Summary from Baseline	Summary with Context (PRAGyan)	Tweets
What measures are being taken to support patients recovering from COVID-19?	Measures are being taken to support patients recovering from COVID-19 due to the impacts of the pandemic and the need to address the health and well-being of individuals affected by the virus.	Efforts to enhance treatment, such as the setup of a plasma bank, prayer support, and sharing of updates on ICU patients indicate a focus on improving recovery outcomes for COVID-19 patients.	[2020-07-25] [12:21 PM] Rajasthan government today started plasma bank at sawai man singh hospital in jaipur for treatment of covid 19. [2020-07-25] [12:25 PM] Mom is in icu due to covid just want prayer from you and everyone who is listening you covid19. [2020-07-25] [12:27 PM] Praying for good health and recovery from covid19.
How did the COVID-19 pandemic affect grocery shopping behaviors and store operations?	Fear of shortages and uncertainty due to the COVID-19 pandemic led to panic buying and stockpiling of essential items, such as food, toilet paper, and sanitizing products, in anticipation of potential disruptions in supply chains and daily routines.	The COVID-19 pandemic led to panic buying, resulting in empty shelves and increased demand for online shopping. Grocery stores implemented measures like dedicating shopping hours for the elderly and encouraging online orders. The pandemic caused shortages of essential items like toilet paper and hand sanitizer, and store employees faced increased pressure to keep shelves stocked.	[2020-03-15] [09:00 AM] Coronavirus Australia: Woolworths to give elderly, disabled dedicated shopping hours amid COVID-19 outbreak [2020-03-16] [09:25 AM] Due to the Covid-19 situation, we have increased demand for all food products... The wait time may be longer for all online orders, particularly beef share and freezer packs. [2020-03-16] [10:00 AM] As news of the region's first confirmed COVID-19 case came out of Sullivan County last week, people flocked to area stores to purchase cleaning supplies, hand sanitizer, food, toilet paper and other goods. [2020-03-16] [11:00 AM] All month there hasn't been crowding in the supermarkets or restaurants, however, reducing all the hours and closing the malls means everyone is now using the same entrance and dependent on a single supermarket.

Causality with LLMs: [18,20] explore the capabilities of LLMs in causal reasoning. [24] introduce TimeLMs, which are diachronic language models that adapt to the evolving linguistic landscape of social media, enhancing our ability to capture and analyze the nuances of cause-effect relationship over time.

RAG Process: The RAG process plays a pivotal role in our methodology by using LLMs and KGs to rank similarities between input text and a list of sentences. This approach provides the top 5 ranked sentences containing causal information, thus reducing the context for further LLM-based cause extraction. [27] propose frameworks for unifying LLMs and KGs, which we build upon to enhance predictive power and semantic richness.

Combining LLMs and KGs: The integration of KGs with LLMs signifies a transformative shift in computational methods. [27] outline a comprehensive roadmap for merging LLMs and KGs, which we build upon. This unification harnesses the predictive power of LLMs and the rich semantic structures of KGs, overcoming limitations found in each model when used in isolation. Our research leverages these frameworks to construct an ideal pipeline that better suits our goal.

Through the integration of these pioneering studies, our research leverages cutting-edge methodologies to advance the field of causal reasoning.

7 Limitations

Models like Node2Vec, Sentence-BERT, and GPT-3.5 Turbo can present challenges when working with other domains, differently formatted data and working environment. They were chosen for their accessibility, and compatibility with the existing resources and environment, making them practical for this research.

Evaluating RAG and LLM models is complex due to their hybrid nature. RAG systems need assessment of both retrieval and generation components, focusing on dynamic knowledge bases and subjective measures of relevance and coherence as explained in [42]. Similarly, [12] highlights that evaluating LLMs demands a comprehensive approach to address data leakage, inappropriate content, and safety standards.

Metrics like BLEU and Cosine Similarity provide quantitative assessments but may not fully capture qualitative aspects of causal reasoning, necessitating human evaluation. Tailoring the framework with appropriate models and metrics ensures robust and contextually relevant evaluation across different datasets and use cases.

8 Conclusion and Future Work

In this research we demonstrated the effectiveness of integrating Knowledge Graphs (KGs) with Large Language Models (LLMs) for causal analysis of social media data. The combined approach significantly enhances the depth and accuracy of understanding the factors driving social media interactions.

Our proposed model has shown interesting results for the COVID19 dataset, and we believe that this approach can be further used to pinpoint a wide variety of information in the tweets dataset specific to a problem. For example: 1) Such an approach might be useful to identify misinformation and the tweets that could be giving rise to it in a region or country. 2) It could be useful for identifying the most sought after development woes by the citizens based on the tweets mined for particular hashtags 3) Our approach relies on a database which forms the crux of the solution, thus, it allows us to identify original tweets which could be a problematic one in election or sensitive times such as riots and wars.

The RAG method used in our framework improves the quality of causal inferences, as validated by higher similarity scores compared to the baseline model. This integration leverages the strengths of both KGs and LLMs, providing detailed and contextually enriched insights that are crucial for informed decision-making. The added layer of temporal information further ensures that only temporally plausible relationships are considered, and it enriches the context for queries by capturing the dynamic evolution of events over time. Overall, the integration of KGs with LLMs offers a powerful methodology for analyzing the complex dynamics of social media data, yielding more precise and actionable insights.

In the future, we aim to refine the construction of the KG further by including causation information along with correlation to ensure better retrieval of context for the LLM query. This enhancement will provide a more robust understanding of causal relationships in the data. Additionally, we will explore alternative embedding techniques such as GraphSAGE or GAT (Graph Attention Networks) to capture richer semantic information [13,36].

References

1. Chan, C., Jiayang, C., et al.: Exploring the potential of ChatGPT on sentence level relations: a focus on temporal, causal, and discourse relations. In: Findings of the Association for Computational Linguistics: EACL 2024, pp. 684–721 (2024)
2. Choi, H., Kim, J., Joe, S., Gwon, Y.: Evaluation of BERT and albert sentence embedding performance on downstream NLP tasks. In: 2020 25th International Conference on Pattern Recognition (ICPR), pp. 5482–5487. IEEE (2021)
3. Deshpande, G., et al.: User feedback from tweets vs app store reviews: an exploratory study of frequency, timing and content. In: 2018 5th International Workshop on Artificial Intelligence for Requirements Engineering (AIRE), pp. 15–21. IEEE (2018)
4. Dessì, D., et al.: Generating knowledge graphs by employing natural language processing and machine learning techniques within the scholarly domain. Futur. Gener. Comput. Syst. **116**, 253–264 (2021)
5. Devlin, J., Chang, M.-W., Lee, K., Toutanova, K.: BERT: pre-training of deep bidirectional transformers for language understanding. arXiv preprint arXiv:1810.04805 (2018)
6. Dieudonat, L., Han, K., et al.: Exploring the combination of contextual word embeddings and knowledge graph embeddings. arXiv preprint arXiv:2004.08371 (2020)
7. Ding, Y., Fan, W., et al.: A survey on rag meets LLMs: towards retrieval-augmented large language models. arXiv preprint arXiv:2405.06211 (2024)
8. Es, S., James, J., Espinosa-Anke, L., Schockaert, S.: Ragas: automated evaluation of retrieval augmented generation. arXiv preprint arXiv:2309.15217 (2023)
9. Gao, Y., Xiong, Y., et al.: Retrieval-augmented generation for large language models: a survey. arXiv preprint arXiv:2312.10997 (2023)
10. Gopalakrishnan, S., et al.: Text to causal knowledge graph: a framework to synthesize knowledge from unstructured business texts into causal graphs. Information **14**(7), 367 (2023)
11. Grover, A., Leskovec, J.: node2vec: Scalable feature learning for networks. In: Proceedings of the 22nd ACM SIGKDD International Conference on Knowledge Discovery and Data Mining, pp. 855–864 (2016)

12. Guo, Z., et al.: Evaluating large language models: a comprehensive survey. arXiv preprint arXiv:2310.19736 (2023)
13. Hamilton, W., Ying, Z., Leskovec, J.: Inductive representation learning on large graphs. In: Advances in Neural Information Processing Systems, vol. 30 (2017)
14. Haque, S., Eberhart, Z., Bansal, A., McMillan, C.: Semantic similarity metrics for evaluating source code summarization. In: Proceedings of the 30th IEEE/ACM International Conference on Program Comprehension, pp. 36–47 (2022)
15. Hong, S., Wu, M., Li, H., Wu, Z.: Event2vec: learning representations of events on temporal sequences. In: Chen, L., Jensen, C.S., Shahabi, C., Yang, X., Lian, X. (eds.) APWeb-WAIM 2017, Part II. LNCS, vol. 10367, pp. 33–47. Springer, Cham (2017). https://doi.org/10.1007/978-3-319-63564-4_3
16. Jaimini, U., Sheth, A.: Causalkg: causal knowledge graph explainability using interventional and counterfactual reasoning. IEEE Internet Comput. **26**(1), 43–50 (2022)
17. Jin, Z., Chen, Y., et al.: Cladder: assessing causal reasoning in language models. In: Thirty-Seventh Conference on Neural Information Processing Systems (2023)
18. Jin, Z., Liu, J., et al.: Can large language models infer causation from correlation? arXiv preprint arXiv:2306.05836 (2023)
19. Kejriwal, M., et al.: Knowledge Graphs: Fundamentals, Techniques, and Applications. MIT Press, Cambridge (2021)
20. Kıcıman, E., Ness, R., et al.: Causal reasoning and large language models: opening a new frontier for causality. arXiv preprint arXiv:2305.00050 (2023)
21. Lewis, P., Perez, E., et al.: Retrieval-augmented generation for knowledge-intensive NLP tasks. In: Advances in Neural Information Processing Systems, vol. 33, pp. 9459–9474 (2020)
22. Liu, X., et al.: Recognizing named entities in tweets. In: Proceedings of the 49th Annual Meeting of the Association for Computational Linguistics: Human Language Technologies, pp. 359–367 (2011)
23. Liu, X., et al.: Named entity recognition for tweets. ACM Tran. Intell. Syst. Technol. (TIST) **4**(1), 1–15 (2013)
24. Loureiro, D., et al.: TimeLMs: diachronic language models from Twitter. arXiv preprint arXiv:2202.03829 (2022)
25. Minaee, S., Mikolov, T., et al.: Large language models: a survey. arXiv preprint arXiv:2402.06196 (2024)
26. Onofrei, S.: Relation extraction with llama3 models. https://towardsdatascience.com/relation-extraction-with-llama3-models-f8bc41858b9e. Accessed 26 Apr 2024
27. Pan, S., Luo, L., et al.: Unifying large language models and knowledge graphs: a roadmap. IEEE Trans. Knowl. Data Eng. (2024)
28. Papineni, K., et al.: BleU: a method for automatic evaluation of machine translation. In: Proceedings of the 40th Annual Meeting of the Association for Computational Linguistics, pp. 311–318 (2002)
29. Peng, C., et al.: Knowledge graphs: opportunities and challenges. Artif. Intell. Rev. **56**(11), 13071–13102 (2023)
30. Preda, G.: Covid19 tweets (2020). https://www.kaggle.com/gpreda/covid19-tweets. Accessed 21 May 2021
31. Reimers, N., Gurevych, I.: Sentence-BERT: sentence embeddings using Siamese BERT-networks. arXiv preprint arXiv:1908.10084 (2019)
32. Reiter, E.: A structured review of the validity of bleu. Comput. Linguist. **44**(3), 393–401 (2018)
33. Tan, F.A., Paul, D., et al.: Constructing and interpreting causal knowledge graphs from news. In: Proceedings of the AAAI Symposium Series, vol. 1, pp. 52–59 (2023)

34. Tjortjis, C.: Graph Databases: Applications on Social Media Analytics and Smart Cities. CRC Press, Boca Raton (2023)
35. Van Der Maaten, L., Postma, E.O., van den Herik, H.J., et al.: Dimensionality reduction: a comparative review. J. Mach. Learn. Res. **10**(66–71), 13 (2009)
36. Velickovic, P., et al.: Graph attention networks. stat **1050**(20), 10–48550 (2017)
37. Wadhwa, S., et al.: Revisiting relation extraction in the era of large language models. In: Proceedings of the Conference on Association for Computational Linguistics. Meeting, vol. 2023, pp. 1:55–1:66. NIH Public Access (2023)
38. Wang, J., Huang, J.X., et al.: Utilizing BERT for information retrieval: survey, applications, resources, and challenges. ACM Comput. Surv. **56**(7), 1–33 (2024)
39. Wang, Y.Q., Dong, L.Y., et al.: Kg2vec: a node2vec-based vectorization model for knowledge graph. PLoS ONE **16**(3), e0248552 (2021)
40. Wolf, T., et al.: Transformers: state-of-the-art natural language processing. In: Proceedings of the 2020 Conference on Empirical Methods in Natural Language Processing: System Demonstrations, pp. 38–45 (2020)
41. Zhaoguang, X., Dang, Y.: Data-driven causal knowledge graph construction for root cause analysis in quality problem solving. Int. J. Prod. Res. **61**(10), 3227–3245 (2023)
42. Yu, H., Gan, A., et al.: Evaluation of retrieval-augmented generation: a survey. arXiv preprint arXiv:2405.07437 (2024)
43. Zeidi, S.S.: Synergizing UML class modeling and natural language to code conversion: a GPT-3.5-powered approach for seamless software design and implementation. Leiden Institute of Advanced Computer Science (LIACS), universiteit leiden, theses.liacs.nl (2023)

Identifying Cyberbullying Roles in Social Media

Manuel Sandoval[1(✉)], Mohammed Abuhamad[1], Patrick Furman[1], Mujtaba Nazari[1], Deborah L. Hall[2], and Yasin N. Silva[1]

[1] Loyola Chicago University, Chicago, IL 60660, USA
{msandovalmadrigal,mabuhamad,pfurman,mnazari,ysilva1}@luc.edu
[2] Arizona State University, Glendale, AZ 85306, USA
d.hall@asu.edu

Abstract. Social media has revolutionized communication, allowing people worldwide to connect and interact instantly. However, it has also led to increases in cyberbullying, which poses a significant threat to children and adolescents globally, affecting their mental health and well-being. It is critical to accurately detect the roles of individuals involved in cyberbullying incidents to effectively address the issue on a large scale. This study explores the use of machine learning models to detect the roles involved in cyberbullying interactions. After examining the AMiCA dataset and addressing class imbalance issues, we evaluate the performance of various models built with four underlying LLMs (*i.e.* BERT, RoBERTa, T5, and GPT-2) for role detection. Our analysis shows that oversampling techniques help improve model performance. The best model, a fine-tuned RoBERTa using oversampled data, achieved an overall F1 score of 83.5%, increasing to 89.3% after applying a prediction threshold. The top-2 F1 score without thresholding was 95.7%. Our method outperforms previously proposed models. After investigating the per-class model performance and confidence scores, we show that the models perform well in classes with more samples and less contextual confusion (*e.g.* Bystander Other), but struggle with classes with fewer samples (*e.g.* Bystander Assistant) and more contextual ambiguity (*e.g.* Harasser and Victim). This work highlights current strengths and limitations in the development of accurate models with limited data and complex scenarios.

Keywords: cyberbullying · role detection · social media · LLM

1 Introduction

With the rise of social media, cyberbullying has become a widespread issue that affects young people worldwide. Bullying, both conventional and online, has been the focus of numerous studies in the social sciences. Cyberbullying takes a variety of forms, such as spreading rumors, negative statements about race, gender, physical appearance, disability, or religion, humiliation, and threats of violence in public posts and comments. Cyberbullying is a serious issue with known harmful consequences, including psychological and social problems [21]. Whereas many studies have focused on detecting cyberbullying incidents (*e.g.* content), relatively fewer have been conducted to identify the roles of individuals/users involved in cyberbullying interactions [11,16]. Previous research has considered the following main roles: victim, bully, bystander assistant, bystander defender, and outsider. There are several critical benefits of accurate

L. M. Aiello et al. (Eds.): ASONAM 2024, LNCS 15213, pp. 355–370, 2025.
https://doi.org/10.1007/978-3-031-78548-1_26

cyberbullying role detection: (1) it enables psychology researchers to better understand the dynamics of cyberbullying, *e.g.* studying the underlying motivations and behaviors of bullies and temporal patterns of bystander-defender activity, (2) it can provide key information to help social media platforms implement targeted measures, *e.g.* providing support and counseling resources for victims, (3) it can help social media platforms implement educational initiatives that raise awareness among bystanders about their role in enabling/resolving cyberbullying incidents and the importance of reporting, and (4) a better understanding of role dynamics enables the development of effective detection models for cyberbullying and anti-bullying.

To enable accurate role identification on a large scale, machine learning techniques could help identify patterns and indicators of cyberbullying behavior. The challenges of implementing such techniques include the need for large amounts of labeled data and addressing many scenarios where the roles of cyberbullying can overlap or the context is limited.*This study addresses the task of identifying cyberbullying roles in social media interactions and sheds light on the merits and limitations of existing methods and datasets, paving the way for future research in this area.* To this end, we examine and process the AMiCA dataset [22] and employ oversampling methods to address the challenge of the imbalanced nature of the dataset. We then develop and evaluate the performance of various machine learning models that are based on four large language models (LLMs): BERT, RoBERTa, T5, and GPT-2. Moreover, we compare the models we implemented with previously proposed role-detection approaches.

Our results indicate that providing context and employing oversampling significantly enhance the performance of models. Among other models, the fine-tuned RoBERTa model trained on oversampled data achieves an F1 score of 83.5% and a top-2 F1 score of 95.7%. The top-2 result indicates the probability of having the correct class in the top two predictions. Achieving this high top-2 F1 score prompted the investigation of the models' performance on a per-class granularity and the analysis of common cases of wrong prediction. For example, using a 25-th percentile confidence score of the victim samples as a threshold for valid predictions increased the F1-score of fine-tuned RoBERTa to 89.3% (*i.e.* $89.3 - 83.5 = 5.8\%$ improvement), with a rejection rate of 16.4% for comments.

Our analysis also emphasizes the importance of training data and the embedded context within samples when building models to detect cyberbullying roles. The implemented models tend to exhibit strong performance when there are a large number of samples for a particular class, but encounter difficulties when there are fewer samples available.

Contributions. The contributions of this study are twofold:

1. Implementing and evaluating different strategies to build machine learning models for identifying the roles of individuals involved in cyberbullying incidents. These strategies involve processing the data, handling the class imbalance, training various models, and evaluating and analyzing their performance.
2. Providing insights into the challenging nature of the cyberbullying role identification task and the limitations of the AMiCa dataset in addressing role overlap and extended conversational context.

2 Related Work

Cyberbullying, in its many forms, has received considerable empirical attention within the social sciences, with much of the focus on the detrimental impact of cyberbullying on psychological and social outcomes of those involved [21]. To protect internet users, particularly youth and adolescent users, from significant negative mental and psychosocial consequences, researchers have begun to develop frameworks for understanding and identifying the roles of different users who engage in or witness cyberbullying.

Cyberbullying Roles in the Social Sciences. Research on both traditional bullying and cyberbullying has identified several distinct roles, including *victim*, *bully*, *bully assistant*, *defender of the victim*, and *bystander* [18], each of which carries out a specific behavior that can influence the cyberbullying interaction. For instance, bystanders can reinforce the bully's actions, given that inaction can convey explicit or implicit cues that bullying is acceptable, funny, or even entertaining [17]. Crucially, in an online setting, the cyberbullying-bystander feedback loop can manifest in actions specific to the platform, such as by providing reinforcing comments or by utilizing platform-specific features (*e.g.* likes (Facebook, Instagram), upvotes (Reddit), or re-blogging (X, formerly known as Twitter). Garnering more followers can also function as a behavior-affirming signal for the cyberbully. Indeed, in previous research, cyber-bystanders who encouraged cyberbullying by reinforcing or assisting the aggressors ranked higher in the justification of violence than cyber-bystanders who defended or supported the victim [13]. Additionally, users who helped reinforce a cyberbully had the highest scores in a measure of cyberbullying perpetration, indicating that those who support the aggressors are also likely to be the perpetrators in other cyberbullying interactions [13]. Researchers have proposed that the choice to reinforce the cyberbully or support the victim is determined by a mix of personal and societal norms [7]. In terms of prevention, cyber-bystanders, *i.e.* users who witness cyberbullying interactions, can play an active and key role in potential intervention. That is, bystanders have the capacity to intervene to support the victim and help alleviate the negative effects of bullying [1].

Cyberbullying Detection via Machine Learning. Many studies have been proposed that apply off-the-shelf solutions, *e.g.* SVM, Naïve Bayes, and Logistic Regression, to binary classification (bullying versus non-bullying) [5,24]. Dadvar and Eckert studied four deep learning architectures, CNN, LSTM, BiLSTM, and BiLSTM with attention on a cyberbullying-labeled YouTube dataset that included 54k posts and 4k users [6]. Cheng *et al.* included network-related content such as user profile information, likes, and follows to identify cyberbullying [4]. In both studies by Cheng *et al.* [2] and [3], they modeled temporal dynamics using a hierarchical representation of social media sessions, where a session is composed of a sequence of comments, and a comment is a sequence of words. Other researchers have integrated network-related content, video, images, and time-related components into all-in-one deep learning architectures [20, 23]. Ziems *et al.* [25] collected a new dataset for cyberbullying detection based on X (formerly Twitter) that attempts to apply the definition of offline bullying, *i.e.* the bullying interactions should contain aggressive language, be repetitive, contain harmful intent, be visible to peers, and present a power imbalance between the attacker and target.

Cyberbullying Role Detection via Machine Learning. The area of cyberbullying role detection is relatively unexplored. To the best of our knowledge, the only models that address this problem are the ones by Jacobs *et al.* [11] and Rathnayake *et al.* [16], each of which is included in the present performance evaluation. Rathnayak *et al.* [16] used the AMiCA dataset [22] to develop a DistilBERT-based ensemble model [19] to classify cyberbullying roles. While the authors report that their algorithm (*OffensEval*) achieves an F1 score of 83%, their evaluation only considered 4 of the 5 roles in the AMiCA dataset. We found that OffensEval's performance decreases significantly when considering all of the roles. Jacobs *et al.* [11] used the AMiCA data to investigate multiple algorithmic configurations including single-algorithm classifiers (reporting 55% as the best F1 result with English data and Logistic Regression and 54% as the best score with Dutch data and Logistic Regression, SVM was the second best performing classifier), ensemble classifiers (reporting 55% as the best F1 result with English data using the Cascading approach), and transformer-based pretrained language models (reporting 55% as the best F1 score using RobBERT with Dutch data and 60% as the best score using RoBERTa with English data).

The AMiCA dataset (Question-Answer pairs from AskFM) is the only labeled social media dataset that includes cyberbullying role labels. This dataset considers 5 cyberbullying roles: Harasser, Victim, Bystander Defender, Bystander Assistant, and Bystander Other. More recently, Hamlett *et al.* [9] proposed an Instagram dataset that includes a wide array of labels (including cyberbullying roles). This dataset, however, only contains 100 social media sessions making it difficult to integrate in robust ML model development.

3 Methods

Figure 1 provides an outline of our framework. This section describes the methods and design considerations.

Problem Definition. Let $C \in \{p_1, p_2, p_3, \dots, p_n\}$ be a corpus of n samples, where a given i-th sample $p_i = \{c_i, t_i\}$ is a pair of context c_i and target t_i text comments (corresponding to a Q&A pair from the dataset). Let $c_i = \{w_1^{c_i}, w_2^{c_i}, w_3^{c_i}, \dots, w_l^{c_i}\}$ and $t_i = \{w_1^{t_i}, w_2^{t_i}, w_3^{t_i}, \dots, w_l^{t_i}\}$ be the token representations of c_i and t_i, respectively, for a length of l tokens. Let $Y \in \{y_1, y_2, y_3, \dots, y_n\}$ be the associated labels for samples in C, where a given i-th sample $p_i \rightarrow y_i$ such that $y_i = \{y_i^c, y_i^t\}$ (*i.e.* y_i^c and y_i^c correspond to the labels associated with c_i and t_i, respectively). For any given label $y_i^x \in \{0, 1, 2, 3, 4\}$ with values of 0, 1, 2, 3, and 4 representing the labels *harasser*, *victim*, *bystander-defender*, *bystander-assistant*, and *bystander-other*, respectively.

Let $f(p) \longrightarrow y$ be a classifier function for cyberbullying roles, where an input p is assigned to y. Since p is a pair of c and t that are individually assigned to y^c and y^t, we present these pairs to a model with a single label that corresponds to the first text comment in the pair p (*e.g.* $f(\{c, t\}) \longrightarrow y^c$ and $f(\{t, c\}) \longrightarrow y^t$). This reversal of c and t doubles the size of the corpus. This is done to classify either c or t while presenting the other comment as a context.

Machine Learning Model. We use LLM model $\mathcal{E}$ to generate embeddings of any sample x, *i.e.* $\mathcal{E}(\{w_1^x, w_2^x, w_3^x, \dots, w_l^x\}) \rightarrow \{e_1^x, e_2^x, e_3^x, \dots, e_l^x\}$. Using the embeddings

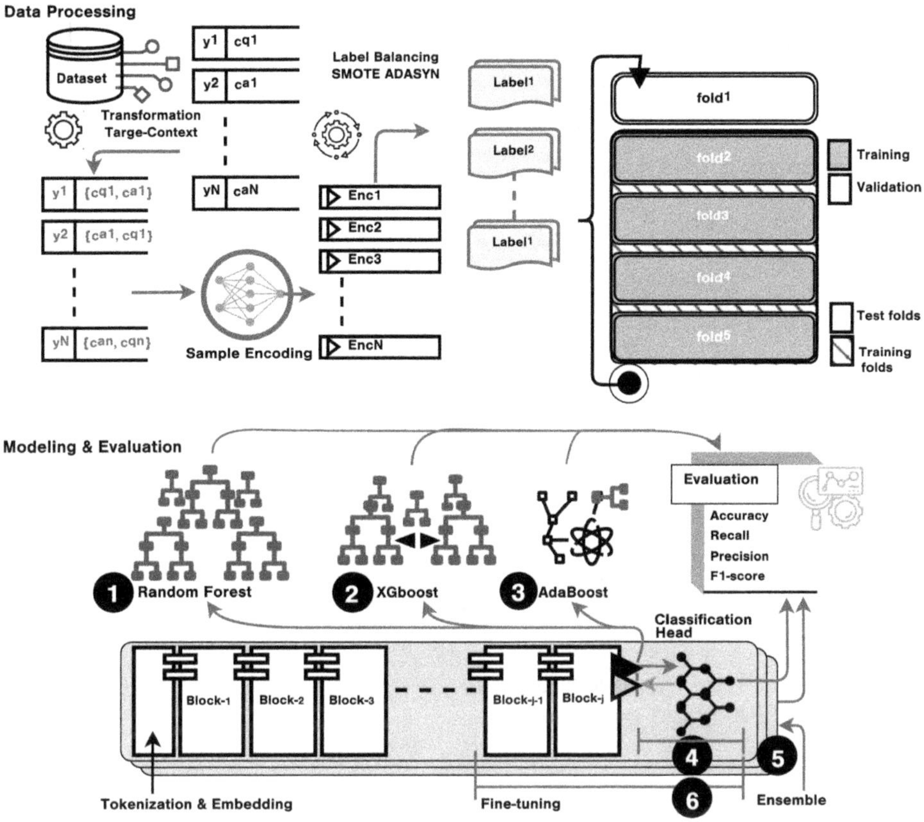

Fig. 1. Modeling Pipeline: Dataset is processed, and samples are transformed to target/context form and then processed to handle class imbalance. Using 10-fold cross-validation, LLMs are employed/evaluated using various methods for role detection.

of the first token $\mathsf{inp} = e_1^x$ (*i.e.* corresponding to the token $\langle \mathsf{bos} \rangle$) or the average of all embeddings ($\mathsf{inp} = \frac{1}{l}\sum_i(e_i^x)$) as input to an ML model f, the model learns the cyberbullying roles of the input x. Since we use t and c pairs, we obtain inp as $\mathsf{inp} = \mathsf{inp}^t + \lambda\ \mathsf{inp}^c$, where λ is a scaling constant for the effect of c on the estimation of $y_x^t = f(\mathsf{inp})$. In our experiments, we use $\lambda = 0.5$ to train Random Forest, AdaBoost, and XGBoost models.

LLMs. LLMs are also used to build role classification models. For an input $\{t_i, c_i\}$, we present tokens $\{w_1^{t_i}, w_2^{t_i}, \ldots, w_l^{t_i}, \langle \mathsf{sep} \rangle, w_1^{c_i}, w_2^{c_i}, \ldots, w_l^{c_i}\}$ to learn y_i^t. We modify the classification head of the LLM to scale the effect of the context c_i on the output. Considering $f(.)$ as consecutive m blocks, *i.e.* $f = b_m(b_{m-1}, \ldots (b_0(.)))$, where b_m is the classification head block, we modify the input to b_m to receive $\{b_{m-1}(b_{m-2} \ldots (b_0(\{w_1^{t_i}\})) + \lambda\ b_{m-1}(b_{m-2} \ldots (b_0(\{w_1^{c_i}\}))\}$, where $\lambda = 0.5$. We apply the same approach when fine-tuning the models.

Table 1. Samples of Q&A pairs and their labels.

Comment Text	Role	Harm
Q: [Well your the true b [redacted]]$^{\text{Gen. Insult}}$ no it not a joke when someone cuts but why would you cut in the first place [b[redacted]]$^{\text{Gen. Insult}}$	Harasser	1
A: [Are you really doing this right now?]$^{\text{Assertive Self. Def.}}$ [Please leave me alone.]$^{\text{Assertive Self. Def.}}$	Victim	1
Q: [YOURE NOT EVEN UGLY OMFGGGGGGGGGGGGGGGGG]$^{\text{Good Characteristics}}$ [I SHALL STAB THAT PERSON]$^{\text{Threat/Blackmail}}$ OXOXOXOXOXOOXO	Bystander Defender	2
A: ILY XOOXXOXOXO	Bystander Other	0

Table 2. AMiCA dataset statistics and comments lengths. Classes 0, 1, 2, 3, and 4 representing *harasser*, *victim*, *bystander-defender*, *-assistant*, and *-other*, respectively.

Class	# Comments (%)	Median Length (Tokens)			
		RoBERTa	GPT-2	T5	BERT
0	3574 (**2.918%**)	11	9	13	12
1	1354 (**1.105%**)	15	13	17	16
2	424 (**0.346%**)	32	30	38	33
3	24 (**0.020%**)	17.5	15.5	20	16.5
4	117126 (**95.612%**)	9	7	10	10
Total	122502 (**100%**)	**9**	**7**	**10**	**10**

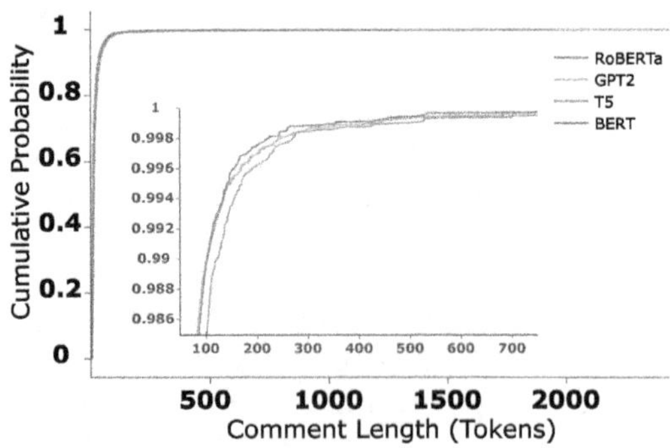

Fig. 2. ECDF of comment lengths using various tokenizers with zoomed-in chart in the center. The max-length is set as the 99-th percentile, *i.e.* 103, 101, 121, and 101 tokens for RoBERTa,GPT2, T5, and BERT tokenizers, respectively.

Data Collection and Processing. The AMiCA dataset [22] is, to our knowledge, the only available dataset of adequate size that is explicitly labeled with cyberbullying roles. This dataset was gathered from ASKfm, a social networking site where users can anonymously ask and answer questions. The Q&A pairs were collected by crawling through a set of seed profiles between April and October 2013. The corpora contains 113, 698

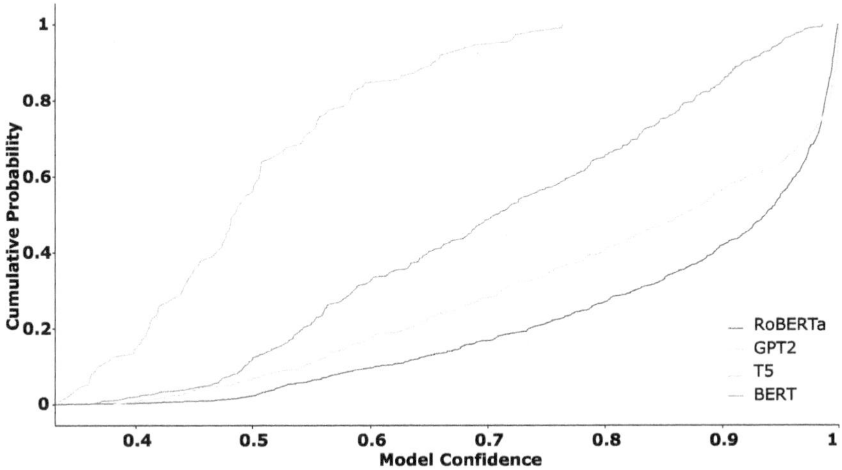

Fig. 3. ECDF of model confidence for correctly identified victim comments for fine-tuned LLMs on oversampled data. 25-th percentile thresholds are: 0.7863, 0.6737, 0.4198, and 0.5611 for RoBERTa, GPT2, T5, and BERT, respectively.

English posts and 78, 387 Dutch posts, of which we excluded the Dutch posts. The dataset includes role labels for each question and answer, and sub-sentence behavior labels. An example of these annotations is provided in Table 1. Table 2 provides an overview of the distribution of classes in the dataset. Although the dataset does contain detailed annotations including the cyberbullying severity and labeled text-spans within each comment, we chose not to utilize this additional information in training our models and instead rely solely on the raw text of the comments, given that this information would not be readily available in real-world scenarios. The anonymity and nature of AMiCA make it impossible to utilize a large amount of context when identifying roles, and the small dataset size makes it difficult to train large models effectively. We investigate several strategies to overcome this challenge.

Transforming Q&A to Context-Target. The AMiCA dataset is structured as Q&A pairs with labels for each comment. To maintain the connection between the comments, we generate two samples for each Q&A pair in the original dataset, utilizing both labels. In the first sample, the question is used as the context and the answer as the target for role identification. In the second sample, the answer is used as the context and the question as the target. Therefore, the original 61, 251 English Q&A pairs are transformed into 122, 502 samples. Because 95% of comments contain *bystander-other* responses, we randomly sample 5, 000 pairs with *bystander-other* label.

Data Imbalance in Latent Space. We use ADASYN [10] with its default parameters and n-neighbors $=$ 15 to oversample the minority classes, *Harasser*, *Victim*, *Bystander Defender*, and *Bystander Assistant*, in the latent space. Providing raw samples to LLMs to obtain representations for ADASYN, we use the following models, BERT [8], RoBERTa [12] base model (*i.e.* trained and fine-tuned on 124 million tweets for sentiment analysis), T5 [15], and GPT-2 [14]. Using the respective tokenizer, LLMs tokenize each comment separately to a specific length that is determined by the 99-th percentile of the length of the observed comment in the data. Comment length varies widely, but the vast majority of comments are short with the median comment length being only 10 tokens when using RoBERTa's tokenizer (see Table 2). *Bystander-defender* comments tend to be the longest, with a median length of 32 tokens, while the *bystander-other* comments, which are the least related to cyberbullying instances, are the shortest, with a median length of only 9 tokens. The empirical cumulative distribution function (ECDF) of the length of the comment is shown in Fig. 2.

Comment pairs are padded or truncated to the same length, and then forwarded together to the model to obtain the final embeddings. For example, an input pair $\{t_i, c_i\}$ with the 99-th percentile of length l is tokenized and passed as $\{w_1^{t_i}, w_2^{t_i}, \ldots, w_l^{t_i}, \langle \mathsf{sep} \rangle, w_1^{c_i}, w_2^{c_i}, \ldots, w_l^{c_i}\}$, where $w_1^{t_i}$ and $w_1^{c_i}$ are the $\langle \mathsf{bos} \rangle$ tokens for t_i and c_i, respectively. The final representation is the weighted sum vector of the t_i and c_i embeddings, which is calculated as $e_1^{t_i} + \lambda\, e_1^{c_i}$, where $e_1^{t_i}$ and $e_1^{c_i}$ correspond to $w_1^{t_i}$ and $w_1^{c_i}$, respectively, and λ is a factor to balance their weights (λ=0.5 in our experiments).

Model Selection. For role detection in cyberbullying interactions, we consider six models to train on two different types of data obtained using four different types of embedding strategies. Using four LLMs, *i.e.* BERT [8], RoBERTa [12], T5 [15], and GPT-2 [14], we train various ML models (*i.e.* Random Forest, AdaBoost, and XGBoost), and fine-tune and ensemble LLMs to perform the role detection task. Next, we outline the settings and the fine-tuning process.

Machine Learning Models: Random Forest is constructed with 300 decision trees grown to the maximum extent and without bootstrapping (*i.e.* all samples used for each tree). The final output is done by a majority vote. The AdaBoost classifier starts by training a decision tree (with a maximum depth of 1) using the dataset and then trains additional trees after adjusting the weights of incorrectly classified samples, so more attention is given to difficult cases. The XGBoost classifier uses 100 boosted decision trees grown to a depth of 6 and trained using a uniform sampling method on half the dataset for each boosting iteration with a learning rate of 0.3. The tree construction algorithm is the faster histogram optimized approximate algorithm with 256 histogram bins.

Dense Neural Network Classification Head: A classification head is added on top of the LLM, which consists of a two-layer dense feedforward neural network. The first and second layers of this network have 2,048 and 1,024 units, respectively, and use ReLU activation functions. The network is then connected to a final softmax output layer, which is responsible for estimating the probability of each role. To address the issue of overfitting, we use L2 regularization with a strength coefficient of 0.09 and a dropout rate of 0.3 for each layer. The training process terminates when the validation loss ceases to decrease for 40 consecutive epochs.

Ensemble of LLMs with Dense Classification Head: We build the ensemble classifier using the weights of dense classification heads trained with different epochs. The minimum number of training epochs to consider model weights in the ensemble is set to 275. Then, we consider the weights after five-training-epoch intervals. The stopping

Table 3. Model performance of baselines and proposed models. The proposed models use oversampled data via ADASYN. **A**ccuracy, **R**ecall, **P**recision, **F1** score, and Rejection Rate (RR) are provided with/without confidence threshold using 10-fold stratified cross-validation. Best results per LLM are in bold.

LLM	Model	Overall Metrics					Metrics with Threshold				
		A	R	P	F1	Top-2 F1	A	R	P	F1	RR
Baseline	Base-RoBERTa [11]	0.676	0.676	0.509	0.580	0.669	—	—	—	—	—
	OffensEval Filtered [16]	0.540	0.540	0.641	0.561	0.623	—	—	—	—	—
	OffensEval Unfiltered [16]	0.264	0.264	0.258	0.217	0.282	—	—	—	—	—
BERT	Random Forest	0.544	0.544	0.522	0.527	0.807	0.573	0.573	0.554	0.556	0.101
	XGBoost	0.614	0.614	0.588	0.593	0.833	0.640	0.640	0.622	0.621	0.085
	AdaBoost	0.340	0.340	0.430	0.368	0.651	0.429	0.429	0.533	0.463	0.284
	Classification Head	0.486	0.486	0.491	0.413	0.790	0.547	0.547	0.590	0.509	0.189
	Ensemble	0.444	0.444	0.498	0.406	0.748	0.529	0.529	0.621	0.529	0.259
	Fine-tuned LLM	**0.736**	**0.736**	**0.727**	**0.721**	**0.903**	**0.761**	**0.761**	**0.757**	**0.748**	**0.069**
RoBERTa	Random Forest	0.507	0.507	0.502	0.496	0.785	0.534	0.534	0.528	0.523	**0.112**
	XGBoost	0.558	0.558	0.545	0.549	0.809	0.597	0.597	0.585	0.588	0.129
	AdaBoost	0.388	0.388	0.429	0.404	0.706	0.436	0.436	0.478	0.452	0.175
	Classification Head	0.662	0.662	0.700	0.664	0.910	0.768	0.768	0.821	0.775	0.303
	Ensemble	0.647	0.647	0.705	0.649	0.900	0.750	0.750	0.819	0.758	0.300
	Fine-tuned LLM	**0.835**	**0.835**	**0.834**	**0.835**	**0.957**	**0.894**	**0.894**	**0.892**	**0.893**	0.164
T5	Random Forest	0.521	0.521	0.516	0.512	0.784	0.550	0.550	0.545	0.540	0.109
	XGBoost	0.581	0.581	0.570	0.574	0.811	0.621	0.621	0.610	0.614	0.133
	AdaBoost	0.365	0.365	0.428	0.386	0.695	0.396	0.396	0.458	0.417	0.106
	Classification Head	0.541	0.541	0.559	0.527	0.834	0.628	0.628	0.707	0.626	0.278
	Ensemble	0.538	0.538	0.580	0.537	0.831	0.644	0.644	**0.739**	0.654	0.322
	Fine-tuned LLM	**0.709**	**0.709**	**0.681**	**0.669**	**0.888**	**0.718**	**0.718**	0.700	**0.681**	**0.030**
GPT2	Random Forest	0.512	0.512	0.495	0.496	0.794	0.546	0.546	0.532	0.530	**0.114**
	XGBoost	0.568	0.568	0.549	0.556	0.817	0.605	0.605	0.588	0.593	0.116
	AdaBoost	0.340	0.340	0.430	0.368	0.651	0.429	0.429	0.533	0.463	0.284
	Classification Head	0.427	0.427	0.430	0.379	0.701	0.497	0.497	0.536	0.475	0.333
	Ensemble	0.352	0.352	0.461	0.324	0.628	0.407	0.407	0.541	0.380	0.247
	Fine-tuned LLM	**0.737**	**0.737**	**0.739**	**0.733**	**0.896**	**0.783**	**0.783**	**0.785**	**0.780**	0.130

condition is met once the validation loss stops decreasing for 40 epochs. The number of models are 26, 563, 25, and 776 (using augmented data), and 9, 22, 124, and 50 (using oversampled data) for BERT, RoBERTa, T5, and GPT-2, respectively.

Fine-Tuned LLM: We fine-tune the last two blocks and the classification head of LLMs using our dataset to perform the role detection task. The classification head consists of a single dense layer with 2048 units, a ReLU activation function, and L2 regularization with a strength coefficient of 0.09 and dropout with a rate of 0.3. This layer is connected to the output layer with five units and a softmax activation function to generate the probabilities of each role. The classification head is modified to receive the weighted sum of the BOS tokens of the target and context pair.

Compute/GPU Settings: ML models are trained on a local workstation with an Intel Xeon Gold 6230R 2.1GHz CPU and 256 GB of RAM. The training and fine-tuning LLMs are conducted using GPU-enabled VMs on Colab and Lambda Cloud. The resources are selected for performance and efficiency and should not affect the results of the experiment.

Performance Evaluation. We use weighted average F1-score and F1-scores per class to determine the performance of the model. We also record the overall accuracy and the weighted average of the recall and precision for each experiment. The confusion matrices of the models are analyzed to evaluate their performance and gain a better understanding of the challenges differentiating among distinct classes. A common instance of confusion is mistaking harassers for victims and vice versa, but there are several other instances of overlap, such as harassers and bystander assistants, which have very similar behavior in cyberbullying interactions. This can be qualitatively explained as victims aggressively defending themselves, which can make it appear, especially with limited context, that they are the harassers. To provide a meaningful quantitative metric that accounts for these types of confusion, we also calculate thresholds for predicting a class and top-2 metrics.

Prediction Threshold. Prediction threshold is adopted after the experiments are complete and the threshold used is standardized as the 25-th percentile of correctly classified victim comments. This cut-off is based on the ECDF of model confidence for correctly identified comments, shown in Fig. 3, and the fact that predicted victim comments typically have the lowest associated model confidence. This ensures thresholding is only applied when the model is relatively uncertain by its standards to predict the correct class. For thresholded metrics, an adjusted model prediction is used. If the probability of the model's original prediction exceeds the threshold or is correct, then the original prediction is used. If the model is not confident in its prediction and the probability is below the threshold, then the model's second choice class with the next highest probability is used instead. This provides a view of how often the correct class is in the top two choices of the model and shows how frequently two classes are confused with one another.

4 Experiments and Results

In this section, we compare the results of the proposed models and the models implemented to serve as baselines. All models were evaluated on the same AMiCa [22] dataset.

Baseline 1: OffensEval. OffensEval [16] is an LLM based model composed of three DistilBERT models functioning as an ensemble for cyberbullying role classification. In the original authors' configuration of the dataset, each question and answer is a standalone post/training sample. The function of the 'outer' model is to determine whether or not a post is cyberbullying, *i.e.* binary classification. If the post is classified as bullying, the 'bully' model determines whether the role is Harasser or Bystander Assistant.

Table 4. Per-class results for best models of each type. The base LLMs used are BERT for Random Forest and XGBoost, GPT-2 for AdaBoost, and RoBERTa for Classification Head, Ensemble, and Fine-tuned LLM. Augmented data generally yields the best results, although, in the case of RoBERTa Fine-tuned LLM, the best results are achieved using oversampled data.

Model	Class	Overall Metrics				Metrics with Threshold				
		R	P	F1	Top-2 F1	R	P	F1	RR	Support
Random Forest	Harasser	0.479	0.507	0.493	0.850	0.521	0.559	0.539	0.136	3574
	Victim	0.195	0.322	0.243	0.462	0.242	0.377	0.295	0.175	1354
	Bystander Defender	0.380	0.222	0.280	0.515	0.436	0.272	0.335	0.222	424
	Bystander Assistant	0	0	0	0	0	0	0	0.375	24
	Bystander Other	0.706	0.649	0.676	0.899	0.726	0.661	0.692	0.090	5000
XGBoost	Harasser	0.599	0.575	0.587	0.886	0.655	0.631	0.643	0.169	3574
	Victim	0.312	0.399	0.350	0.611	0.390	0.495	0.436	0.243	1354
	Bystander Defender	0.410	0.290	0.340	0.557	0.467	0.357	0.404	0.231	424
	Bystander Assistant	0	0	0	0	0	0	0	0.208	24
	Bystander Other	0.735	0.736	0.736	0.912	0.768	0.762	0.765	0.105	5000
AdaBoost	Harasser	0.279	0.416	0.334	0.728	0.343	0.500	0.407	0.167	3574
	Victim	0.182	0.162	0.171	0.401	0.219	0.211	0.215	0.168	1354
	Bystander Defender	0.474	0.127	0.200	0.494	0.479	0.131	0.206	0.127	424
	Bystander Assistant	0	0	0	0	0	0	0	0.125	24
	Bystander Other	0.567	0.582	0.574	0.836	0.619	0.624	0.621	0.172	5000
Classification Head	Harasser	0.755	0.654	0.701	0.965	0.828	0.739	0.781	0.204	3574
	Victim	0.376	0.568	0.452	0.805	0.495	0.690	0.576	0.278	1354
	Bystander Defender	0.509	0.579	0.542	0.744	0.620	0.709	0.661	0.323	424
	Bystander Assistant	0	0	0	0.074	0.042	0.100	0.059	0.292	24
	Bystander Other	0.878	0.883	0.881	0.955	0.904	0.902	0.903	0.084	5000
Ensemble	Harasser	0.728	0.657	0.691	0.968	0.797	0.742	0.769	0.198	3574
	Victim	0.393	0.562	0.462	0.803	0.488	0.656	0.560	0.261	1354
	Bystander Defender	0.588	0.560	0.574	0.743	0.654	0.642	0.648	0.294	424
	Bystander Assistant	0.045	0.048	0.047	0.077	0.045	0.063	0.053	0.364	24
	Bystander Other	0.882	0.881	0.881	0.955	0.909	0.894	0.901	0.074	5000
Fine-tuned LLM	Harasser	0.824	0.802	0.813	0.967	0.894	0.875	0.885	0.204	3574
	Victim	0.654	0.665	0.660	0.907	0.775	0.805	0.790	0.310	1354
	Bystander Defender	0.639	0.727	0.680	0.831	0.759	0.826	0.791	0.344	424
	Bystander Assistant	0	0	0	0.167	0.042	0.091	0.057	0.458	24
	Bystander Other	0.913	0.916	0.914	0.978	0.941	0.937	0.939	0.079	5000

Table 5. Examples of harasser and victim comment confusion.

Question	Answer	Q Pred. Role	Q Actual Role	A Pred. Role	A ActualClass
come visit me [redacted]	busy [redacted]	4	0	0	1
sit on my face	[redacted]	1	0	0	0
wants some butter u [redacted] CRACKER	Who the [redacted] even puts butter on crackers you [redacted]	0	0	0	1
Pretty sure you told me you weren't a virgin awks	Pretty sure you [redacted] ur dad awks	1	0	0	1

Likewise, if the outer model determines a post is not bullying, the 'defender' model determines if the role is Victim or Bystander Defender. To overcome class imbalance, this approach uses 10 stratified fold cross validation paired with weighted random sampling when training the outer model. To the best of our understanding, OffensEval was not trained to classify Bystander Other. To more comprehensively evaluate OffensEval with our approaches, which consider the full set of 5 roles in the dataset, we considered two approaches, OffensEval Filtered and OffensEval Unfiltered. For OffensEval Filtered we trained and tested the model on a reduced dataset with only the samples of roles Harasser, Victim, Bystander Defender, and Bystander Assistant. This approach directly replicates the model implemented by the original authors as they did not consider the Bystander Other role. For OffensEval Unfiltered, we expanded the testing set of each fold to include Bystander Other. This second approach enables a more direct comparison with the approaches we propose. Observe that considering real-world scenarios, OffensEval Filtered assumes that existence of a previous model that will filter the Bystander Other interactions from the dataset.

Baseline 2: Base-RoBERTa. To serve as a baseline for comparison, we implemented the RoBERTa-based model proposed in [11]. The authors of this paper stated that this model achieved an F1 score of 0.6. To the best of our understanding, the original authors interpreted the 'Bystander Other' role as their 'Not Bullying' class. Moreover, due to the significant class imbalance presented by the few occurrences of 'Bystander Assistant', they merged 'Bystander Assistant' into 'Harasser'.

Model Performance. Table 3 shows the performance of different models trained using oversampled, as well as the performance for the baseline models. Regarding the performance of baseline methods, we can observe that SAC-LR achieved a F1 score of 67%, SAC-SVM obtained a F1 score of 65%, OffensEval filtered obtained an F1 score of 53%, and finally, OffensEval unfiltered, *i.e.* OffensEval predicting all 5 classes, achieved an F1 score of 21%. Of our proposed methods, fine-tuning the last two blocks of RoBERTa proved to be uniformly the best model for every metric when using oversampling. It achieves an accuracy of 83.5%, which increases roughly 6% after thresholding with a rejection rate of 16.4%. This model also outperforms all the baseline methods. Generally, there is a notable increase in the top-2 F1 score compared to the F1 score. This indicates the challenge that the model encounters in accurately determining the top prediction and resolving confusion. Applying a confidence threshold helps the model reject uncertain predictions due to either limited context or class ambiguity.

Per-class Model Performance. Detailed per-class metrics are also reported in Table 4 for the model that performs the best for each type, measured by the weighted average F1 score without thresholding. Although there is some variation among models, generally, the *bystander-other* and *harasser* comments are the easiest to distinguish, with similar levels of recall and precision and high F1 scores around 0.9 and 0.8 for the best models, respectively. *Victim* and *bystander defender* comments lie in the middle with F1 scores slightly below 0.7. Although the sample size of *victim* comments is larger, they prove to be slightly more difficult to recognize compared to the comments of *bystander defender*. This could be due to the observed overlap between victims and the more frequent harasser class when victims defend themselves using aggressive language. Most models tend to ignore or incorrectly recognize the *bystander-assistant* class. This is likely due to a combination of only having 24 total occurrences of the role in the entire dataset and the similarity of this role to harassers, particularly without more available context. The top-2 F1 scores per-class show that the models frequently choose between two similar classes. Although the top-1 F1 scores for victims and bystander defenders are 0.66 and 0.68, respectively, they increase to 0.91 and 0.83 when considering the top-2 predicted classes. The increase in F1 scores is the largest by far for victims, most likely due to their confusion with harassers. The top-2 metrics provide insight into the potential performance of the proposed models when using different datasets that have more diverse/distinct classes or clearer guidelines for annotating samples, *e.g.* when aggressive victims use the same language as harassers.

5 Discussion

Class Confusion. The fine-tuned RoBERTa model performs well in most situations, but struggles in some surprising situations such as distinguishing between harassers and victims. For seemingly opposite roles, having 8.22% of harasser comments mistakenly labeled as victims and 23.12% of victims labeled as harassers is unexpected. Table 5 shows a sample of Q&A pairs with harassers and victims mistaken for one another. In several cases, victims mimic the language used by the harasser in their defensive response or otherwise aggressively defend themselves with language typically associated with harassment. This makes it difficult, even for a human familiar with the subject reading these comments, to identify with a high degree of certainty who the victim is in these interactions. Although the exact reasoning behind the model's struggles remains unclear, by analyzing the model's mistakes, it is clear that the task is not trivial and that before substantial improvement in role prediction performance can be achieved, we likely need to either clarify role definitions or expand the number of possible roles for each comment.

Model Application. While the proposed models consider the Ask.fm Question-Answer pair format, the models could be easily applied to data formats used in other platforms. For instance, a majority of platforms utilize a thread based discussion focused around a single post, *e.g.* Reddit, Youtube, Facebook, and Instagram. To adapt the thread-based format, we could build a Q-A pair using the initial post text as the question and each individual comment as an answer. Another approach would be to consider each instance of a user mention, often seen as one user tagging another via the '@' symbol. Each instance of a comment with a user mention and the response by the mentioned user can constitute a Q-A pair.

6 Conclusion

Our work investigates cyberbullying role detection in social media interactions using an imbalanced dataset with five classes (AMiCA dataset), *i.e.* Harasser, Victim, Bystander Assistant, Bystander Defender, and Bystander Other. The issue of cyberbullying has frequently posed a challenge, with over half of adolescents reporting instances of bullying while using social networks or engaging in online chats. Having a model for identifying the roles of cyberbullying instances would be beneficial for adolescents and parents, as this model could enable the implementation of more effective anti-bullying tools. Previous studies have focused on determining whether a post exhibits bullying behavior or not. More recently, some research has begun exploring the detection of different roles involved in cyberbullying, including victim, bully, and bystander. Contributing to the detection of cyberbullying roles in social media comments, this study explores the performance of various models with different training strategies and sheds light on the strengths and shortcomings of the employed methods. We plan to publish the code upon paper acceptance.

An important task for future work is the development of a more comprehensive labeled dataset that enables better detection models by more accurately capturing the roles linked to a given comment. This could be achieved, for instance, by enabling the assignment of multiple roles to a single comment (*e.g.* a single comment could be labeled as a Bystander Defender and a Harasser). Moreover, the labeling approach could be extended by including a degree of each role (*e.g.* mild vs. severe harasser). Both mechanisms would help capture more accurately complex instances of role overlap.

Acknowledgments. This work was supported by NSF Awards #2227488 and #1719722 and a Google Award for Inclusion Research.

References

1. Bastiaensens, S., Vandebosch, H., Poels, K., Van Cleemput, K., DeSmet, A., De Bourdeaudhuij, I.: 'Can i afford to help?' how affordances of communication modalities guide bystanders' helping intentions towards harassment on social network sites. Behav. Inf. Technol. **34**(4), 425–435 (2015)
2. Cheng, L., Guo, R., Silva, Y., Hall, D., Liu, H.: Hierarchical attention networks for cyberbullying detection on the instagram social network. In: Proceedings of the 2019 SIAM International Conference on Data Mining (SDM), pp. 235–243. Society for Industrial and Applied Mathematics (2019). https://doi.org/10.1137/1.9781611975673.27
3. Cheng, L., Guo, R., Silva, Y.N., Hall, D., Liu, H.: Modeling temporal patterns of cyberbullying detection with hierarchical attention networks. ACM/IMS Trans. Data Sci. **2**(2) (2021)

4. Cheng, L., Li, J., Silva, Y., Hall, D., Liu, H.: PI-bully: personalized cyberbullying detection with peer influence. In: Electronic Proceedings of IJCAI 2019, pp. 5829–5835 (2019)
5. Dadvar, M., de Jong, F., Ordelman, R., Trieschnigg, R.: Improved cyberbullying detection using gender information. In: Proceedings of the Twelfth Dutch-Belgian Information Retrieval Workshop (DIR 2012), pp. 23–25. Ghent University, Belgium (2012)
6. Dadvar, M., Eckert, K.: Cyberbullying detection in social networks using deep learning based models; a reproducibility study. arXiv preprint arXiv:1812.08046 (2018)
7. Dang, J., Liu, L.: Me and others around: the roles of personal and social norms in Chinese adolescent bystanders' responses toward cyberbullying. J. Interpersonal Violence **37**(9-10), NP6329–NP6354 (2022). https://doi.org/10.1177/0886260520967128. pMID: 33073678
8. Devlin, J., Chang, M.W., Lee, K., Toutanova, K.: Bert: pre-training of deep bidirectional transformers for language understanding (2019)
9. Hamlett, M., Powell, G., Silva, Y.N., Hall, D.: A labeled dataset for investigating cyberbullying content patterns in instagram. In: Proceedings of the International AAAI Conference on Web and Social Media, vol. 16, no. 1, pp. 1251–1258 (2022). https://doi.org/10.1609/icwsm.v16i1.19376
10. He, H., Bai, Y., Garcia, E.A., Li, S.: Adasyn: adaptive synthetic sampling approach for imbalanced learning. In: 2008 IEEE International Joint Conference on Neural Networks (IEEE World Congress on Computational Intelligence), pp. 1322–1328 (2008). https://doi.org/10.1109/IJCNN.2008.4633969
11. Jacobs, G., Van Hee, C., Hoste, V.: Automatic classification of participant roles in cyberbullying: can we detect victims, bullies, and bystanders in social media text? Nat. Lang. Eng. **28**(2), 141–166 (2022)
12. Loureiro, D., Barbieri, F., Neves, L., Anke, L.E., Camacho-Collados, J.: Timelms: diachronic language models from twitter (2022)
13. Orue, I., Fernández-González, L., Machimbarrena, J.M., González-Cabrera, J., Calvete, E.: Bidirectional relationships between cyberbystanders' roles, cyberbullying perpetration, and justification of violence. Youth Soc. **55**(4), 611–629 (2023). https://doi.org/10.1177/0044118X211053356
14. Radford, A., Wu, J., Child, R., Luan, D., Amodei, D., Sutskever, I.: Language models are unsupervised multitask learners (2019)
15. Raffel, C., et al.: Exploring the limits of transfer learning with a unified text-to-text transformer (2023)
16. Rathnayake, G., Atapattu, T., Herath, M., Zhang, G., Falkner, K.: Enhancing the identification of cyberbullying through participant roles. In: Proceedings of the Fourth Workshop on Online Abuse and Harms, pp. 89–94. Association for Computational Linguistics (2020)
17. Salmivalli, C.: Participant roles in bullying: how can peer bystanders be utilized in interventions? Theory Into Pract. **53**(4), 286–292 (2014)
18. Salmivalli, C., Lagerspetz, K., Björkqvist, K., Österman, K., Kaukiainen, A.: Bullying as a group process: participant roles and their relations to social status within the group. Aggressive Behav. **22**(1), 1–15 (1996)
19. Sanh, V., Debut, L., Chaumond, J., Wolf, T.: Distilbert, a distilled version of bert: smaller, faster, cheaper and lighter (2020)
20. Singh, V.K., Ghosh, S., Jose, C.: Toward multimodal cyberbullying detection. In: Proceedings of the 2017 CHI Conference Extended Abstracts on Human Factors in Computing Systems, pp. 2090–2099. Association for Computing Machinery (2017)
21. Teng, T.H., Varathan, K.D.: Cyberbullying detection in social networks: a comparison between machine learning and transfer learning approaches. IEEE Access (2023)
22. Van Hee, C., et al.: Automatic detection of cyberbullying in social media text. PLoS ONE **13**(10), e0203794 (2018)

23. Wang, K., Xiong, Q., Wu, C., Gao, M., Yu, Y.: Multi-modal cyberbullying detection on social networks. In: 2020 International Joint Conference on Neural Networks (IJCNN), pp. 1–8 (2020)
24. Xu, J.M., Jun, K.S., Zhu, X., Bellmore, A.: Learning from bullying traces in social media. In: Proceedings of the 2012 Conference of the North American Chapter of the Association for Computational Linguistics: Human Language Technologies, pp. 656–666. NAACL HLT 2012. Association for Computational Linguistics, USA (2012)
25. Ziems, C., Vigfusson, Y., Morstatter, F.: Aggressive, repetitive, intentional, visible, and imbalanced: Refining representations for cyberbullying classification. CoRR abs/2004.01820 (2020). https://arxiv.org/abs/2004.01820

Exploration of Hugging Face Models by Heterogeneous Information Network and Linking Across Scholarly Repositories

Muhammad Asif Suryani(✉), Saurav Karmakar, and Brigitte Mathiak

Knowledge Technologies for the Social Sciences, GESIS-Leibniz-Institut für Sozialwissenschaften, Köln, Germany
{asif.suryani,saurav.karmakar, brigitte.mathiak}@gesis.org

Abstract. With the pervasive integration of Machine Learning (ML) focusing complex tasks across various domains, generally respective models and datasets are made available in numerous scientific repositories such as Hugging Face, GitHub for recognition and understanding within the research communities hence supporting open science initiative. However, the adaptability of these repositories is increasing among users hence raising a concern about the usability of these models and datasets effectively. Therefore, it is necessary to explore these repositories and compile comprehensive information that could facilitate users as well as repositories itself. Hugging Face, a leading repository, aims to furnish a platform that organizes and presents detailed information on models and datasets employed in research. As its adoption escalates within the research communities, the necessity to delve into such repositories becomes crucial to offer researchers valuable insights and promote efficient knowledge dissemination. This study focuses on exploring Hugging Face, particularly its machine learning models by exploiting various relevant features and the insights are presented in a Heterogeneous Information Network (HIN). Our research not only demonstrates the effectiveness of the available models on Hugging Face but also highlights potential links to relevant scholarly repositories by highlighting the significance of exploration which could contribute to the future integration of repositories, facilitating a more unified and accessible framework for scientific research.

Keywords: Data Exploration · Hugging Face · Machine Learning Models · Heterogeneous Information Network · Repository Exploration

1 Introduction

The exponential growth in Artificial Intelligence (AI) and Machine Learning (ML) research has significantly impacted the scientific domains, leading to an

L. M. Aiello et al. (Eds.): ASONAM 2024, LNCS 15213, pp. 371–386, 2025.
https://doi.org/10.1007/978-3-031-78548-1_27

unprecedented increase in computing-related scientific publications and nourish the interdisciplinary research paradigms. These publications not only contribute in acquiring knowledge but also introduce innovative models and datasets which could be crucial for future research directions in respective domains. As metadata within these publications serves as the critical feature for cataloging, searching, and referencing these contributions, enabling researchers to look for their relevant information efficiently [9,10].

However, these resources are effectively organized and made available by various repositories to facilitate the research communities. There have been numerous repositories each serving distinct roles within the research ecosystem. Kaggle[1], renowned for hosting competitions and datasets, provides a platform for practical data science and ML challenges. GitLab[2], on the other hand, offers a more comprehensive code repository service that facilitates version control and collaborative project management, including the sharing of AI models and datasets. Hugging Face[3] stands out by specifically targeting the AI research community, offering a specialized repository for AI models and datasets. As these repositories are playing a significant role in providing a state of the art platform to disseminate the models and datasets conveniently with the research communities which generally seems beneficial for the models itself and also for the communities by supporting the open science initiative [3].

Hugging Face has rapidly gained prominence within the AI research community, attributed to its user-friendly interface and comprehensive collection of models and datasets. Unlike traditional repositories, Hugging Face focuses on democratizing access to state-of-the-art AI technologies, encouraging open-source contributions, and facilitating easy implementation and bench-marking of models. Its significance extends beyond a mere repository and it acts as a hub for collaboration, knowledge exchange, and innovation in respective domains. Hugging Face aims to effectively comprehend their models by prompting curators for certain features which could be beneficial in categorizing the models, besides maintaining certain informative features. So collectively these diverse features are made available against each of the model card [2].

These features could provide essential information such as number of downloads, likes, library names, pipeline, citation and tags. Exploration of these potential features for model cards could be interesting and may bring insights by highlighting potential research trends to bridge the repositories and support repositories harmonization. However, to explore these models cards and their potential relationships, a diverse data model and heterogeneous representation is essential. Heterogeneous Information Network (HIN) has been used in various studies modelling diverse features and their relationships. In this problem setting, complex interconnections between features of model cards and respective links to publications can be effectively represented and explored through network representations of repositories such as Hugging Face and arXiv. The

[1] https://www.kaggle.com/.
[2] https://github.com/.
[3] https://huggingface.co/.

network enable a novel approach to navigate the large repository landscape, hence uncovering hidden patterns, and facilitating targeted searches. As sample studies utilizing HIN for repository exploration have demonstrated its potential in enhancing and understanding the data relationships, contributing to more efficient and insightful research methodologies. [7,12].

However, exploring relevant features of Hugging Face models could bring more descriptive insights at the repository level, which defiantly will be beneficial for the communities and enhance the overall user experience. So to address this challenging problem, it is necessary to consider wide range of diverse available features from these scientific models to further broaden the information extraction spectrum. As it will bring desired information alongside respective contextual information, which could be modelled into an information network. In this paper, we conducted a study to explore machine learning research model available in Hugging Face, which aims to provide insights to users by modelling diverse features into a Heterogeneous Information Network. Primarily, here we address certain research questions, including but not limited to:

- How can descriptive statistics of Hugging Face models (downloads, likes) reveal trends in the adoption and usefulness of machine learning models?
- What are the most influential models and libraries in Hugging Face, and whether it contributes toward cross repository linkages.
- How does representing the model information of Hugging Face repository as a Heterogeneous Information Network (HIN) facilitate the exploration of relationships between models, and research articles?
- What insights can be acquired by analyzing the network structure, and how does this structure reflects the contributions of researchers to open science initiatives?
- Does any collaborations between different research entities reflect and flourish the interdisciplinary research by model citations?
- What role does metadata (e.g., model information, tasks, tags) plays in enhancing the usability and discoverability of models within the Hugging Face?

This study is organized as: following section will present the related work in this problem setting. The third section presents the exploratory study by providing the data model for heterogeneous information network. Later sections provide discussion on the collected insights followed by conclusion and future directions.

2 Related Work

Over the course of interdisciplinary research, the involvement of machine learning models has showcased substantial adaptation. Such initiatives are predominantly facilitated by open-source repositories. These repositories provide a platform for researchers to understand and utilize machine learning models in their research. Numerous research studies aim to highlight the usefulness of these repositories.

In addition, effective relevant repositories exploration may contribute towards enhanced user experience.

Open-Source Software (OSS) projects are widely available on various social coding platforms such as GitHub and GitLab. However, with the emergence of machine learning models for interdisciplinary research tasks, specialized platforms like Hugging Face Hub (HFH) have become essential by focusing on sharing datasets, pre-trained models, and applications. HFH is rapidly growing, currently maintaining more than 400K repositories accessible via an API. Besides the provided API, there is a need for solutions to facilitate data collection and the exploration of HFH's different facets. To address these tasks, researchers introduced HFCommunity, an extraction process for HFH data and a relational database designed to facilitate the empirical analysis of the growing number of ML projects [1].

Hugging Face (HF) has emerged as a central repository for sharing and developing machine learning and artificial intelligence models. This study provides insights into various aspects of HF focusing on the carbon emissions and ML model maintenance. It aims to guide researchers in mining software repositories within the HF ecosystem, offering a framework that promotes the responsible and sustainable advancement of ML. Additionally, the study also fosters a deeper understanding of the broader implications of ML models [2].

The development of natural language processing tools involves various aspects, including datasets, models, individual skills, and the crucial task of following guidelines for documentation for potential future use. The adoption of standards for the documentation process enhances accessibility across the research community. This study presents two case studies to showcase reusable documentation templates on Hugging Face. It describes the key processes of developing templates, focusing on relevant stakeholders and entities [9].

Recent studies describe how transformer architectures have significantly empowered natural language processing. Additionally, enhanced pre-training processes have effectively broaden the applications of these models to address diverse scientific tasks across research domains. Furthermore, the study also highlights the usefulness of transformers and underscore the value of research repositories, emphasizing the impact of these advancements on scientific communities [14].

A recent study focused on the usability of pre-trained models (PTMs), specifically within the Hugging Face ecosystem. The study presents an empirical investigation of PTM reuse by interviewing twelve practitioners from Hugging Face to understand best practices and challenges in PTM re-usability. Additionally, the study enables researchers to model the decision-making process by identifying key attributes that contribute to the re-usability of models in the Hugging Face ecosystem [6].

A recent study analyzed the 1,417 hugging face machine learning models and their respective dataset for carbon footprint. The study aimed to provide insights and recommendations to optimize the carbon efficiency of these models. Focusing on the Hugging Face Hub API, the study addressed two research questions regarding the measurement and reporting of carbon emissions. The findings indicate a slight decrease in reported carbon footprint over the past two years and reveal correlations between carbon emissions and attributes such as model size, dataset size, ML application domains, and performance metrics [4].

The number of research papers across various domains is increasing due to advancements in science and technology, making it challenging for researchers to find relevant literature. This study proposes a novel approach based on Heterogeneous Information Networks (HIN) that targets potential relationships such as citation links, author collaborations, and research areas. By constructing network and using a random walk strategy to simulate natural sentences, the approach effectively discovers relevance between papers [5]. With the rapid growth of digital publishing, efficiently visualizing scholarly data has become increasingly demanding. This data includes millions of raw data points such as authors, papers, citations, and scholarly networks. Various visualization techniques can be applied to better represent data structures and uncover hidden patterns. The study introduces the basic concepts and collection methods for scholarly data and provides a comprehensive overview of related visualization tools and techniques [8].

Recently, many works have incorporated auxiliary information into recommender systems to address data sparsity. Heterogeneous information network (HIN) based recommender systems provide a unified approach to integrate various auxiliary information, enhancing the performance and Interpretability. This study delves into the network-based recommender systems, covering concepts, methods, applications, and resources. Finally, the study discusses the potential research directions incorporating Heterogeneous Information Network [7].

3 Proposed Exploration

This sections describes the overall exploration approach being adopted in this study. The Fig. 1 showcase the block diagram of tasks by highlighting the top-down approach being followed for the Hugging Face exploration.

The exploration commence with data retrieval by the use of Hugging Face API, targeting models to fetch unstructured data in JSON format. This extracted data includes numerous available metadata features for each model, such as architecture, pipelines, likes, downloads, tags and more. The initial processing ensures that the data is structured and ready for subsequent tasks.

The acquired data is further passed to the information parsing module, which segregates and extracts the relevant feature set for each of the Hugging Face model. However, the feature engineering is applied to transform these features into a meaningful set of interesting features, omitting irrelevant features such as URLs and local IDs. Furthermore, pre-processing techniques are applied to handle missing values, encoding issues, and remove redundant information, ensuring the data is clean, consistent, and ready for further analysis. It has been observed that some models have multilingual tags and duplicate entries, which were removed.

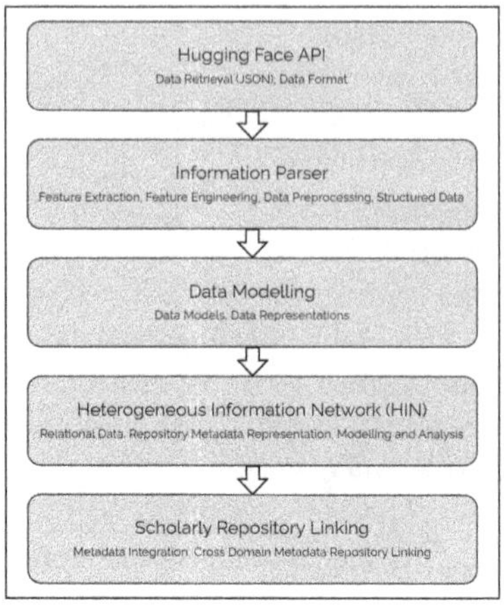

Fig. 1. Block Diagram of Hugging Face Exploration Pipeline

By extensively studying the diverse feature set, we elaborate the diversified data models being proposed for the insightful exploration of Hugging Face machine learning models as shown in Fig. 2 and Fig. 3 respectively. These data models exploit the feature set for each model to comprehensively explore and group the relevant models. These data models are designed to support the proposed research questions and capture the complexities and relationships within the feature set of Hugging Face models.

The first data model, as shown in the Fig. 2, focuses on the relational aspects of models by highlighting connections between models and their relevant features such as models being linked to specific pipeline and libraries respectively. In addition it also indicate the connections between models and their available tags, downloads, likes and citation tag available for each model.

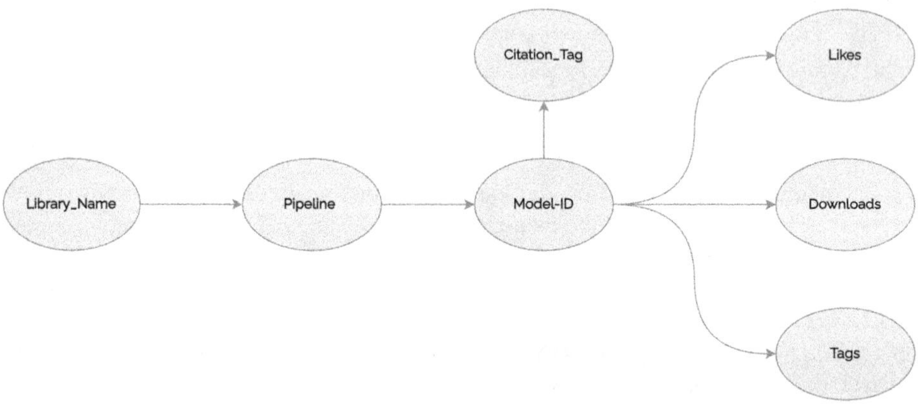

Fig. 2. Data Model for Hugging Face Exploration

The second data model primarily aim to support the cross-domain repository linking between Hugging Face and metadata repository such as arXiv. The feature representations help in understanding the potential patterns and enhance the interpretability and effectiveness of Hugging Face models across metadata repository. The models having a citation tags are considered at this stage for exploration. Beside, scholarly metadata features available against each model, important features such as library and pipelines are also incorporated to explore the cross disciplinary repository linking effectively.

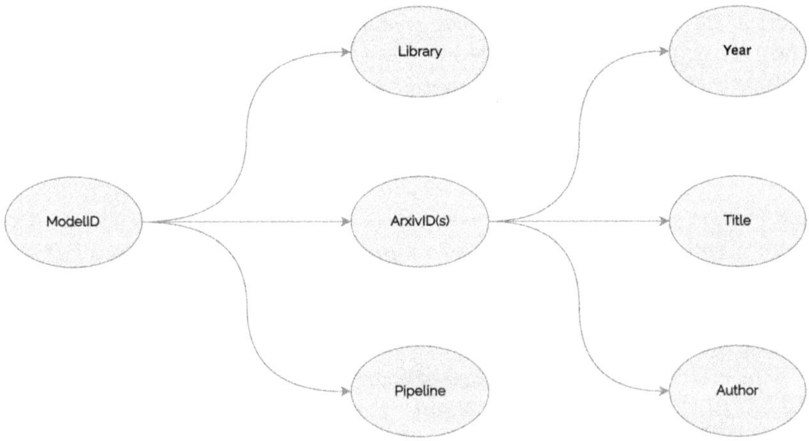

Fig. 3. Data Mode for Repository Linking having Citation Tag

Heterogeneous Information Networks (HIN) are advanced models used to represent and analyze complex data comprising multiple types of entities and relationships. Unlike traditional homogeneous networks that consist of a single type of node and edge, HINs incorporate diverse node and edge types, capturing the multifaceted nature of diverse data. HIN can be defined as a graph $G = (V, E)$ where V denotes the set of nodes (or vertices) and E represents the set of edges.

This diversity allows for a comprehensive representation and understanding of interconnected data, making HINs particularly useful in domains such as social network analysis, bioinformatics, and recommendation systems. The ability to model complex interactions and dependencies within the data provides a powerful tool for addressing intricate research questions, enabling more precise and insightful analyses [7]. So, HINs provide a powerful framework for modeling and analyzing complex relationships. In this problem setting, HINs are constructed by leveraging the proposed data models, which integrate the relational data and metadata of the models and perform comprehensive analyses to explore the structure and dynamics of the models.

Finally, the HINs are used to explore the Hugging Face models within the repository, additionally highlighting potential insights and patterns. Moreover, the second HIN provides the potential linking to external scholarly repositories, enabling metadata integration and cross-domain analysis. In general HIN explorations allows for seamless linking of data across repositories, enhancing the richness and utility of the information. This broadens the scope of analysis and facilitates comprehensive insights by leveraging data from diverse sources, leading to repository harmonization.

4 Results and Discussions

This section offers the comprehensive insight being collected over the course of this study by providing the detailed exploration of Hugging Face models for the selected feature set considering proposed data models.

4.1 Data Description

Hugging Face is the one of the biggest repository which facilitate the hosting and sharing of machine learning models to support the scientific communities. Over the course of this study we collected the metadata concerning hosted models of Hugging Face. The data considered from the exploration is till April 2024 and the dataset is being collected from the Hugging Face API. The dataset in this regard comprise of various potential features. But in this study we focused on *Model-IDs*, *Pipeline tag*, *Library Name*, *Downloads*, *Likes* and *tags*. Moreover, *tags* features carry a wide range of textual keywords attached to the models. These tags usually carry information about potential citation tags, language supports, pipeline, licences and many more and the study aim to consider wide range of these available tags as potential features.

4.2 Descriptive Exploration of Models

The potential feature set for the exploration of Hugging Face are quite diverse. We presents the segregation of Hugging Face models by two quantitative features such as downloads and likes. The Fig. 4 presents the distribution of downloads of hugging face models which clearly states the drops in number of models as the number of download increases. Likewise the Fig. 5 showcases the distribution of likes of hugging face models, which clearly states the huge drops in the number of hugging face models as number of like increases.

However, to further explore the downloads and likes features, we created various ranges for each of the feature and plot the findings in Fig. 4 and Fig. 5 respectively. The Fig. 4 shows the distribution of models for the various downloads ranges. It is evident that a substantial number of models lies in downloads range from 0 to 50 and further decreases as the value of ranges increases. Moreover, Fig. 5 also exhibits similar trends and indicate that a large number of models having likes from 0 to 50 and only one model have more than 10000 likes.

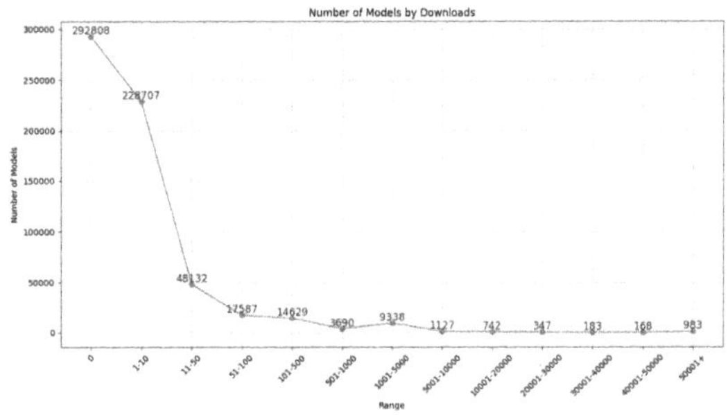

Fig. 4. Downloads Distribution of Hugging Face Models

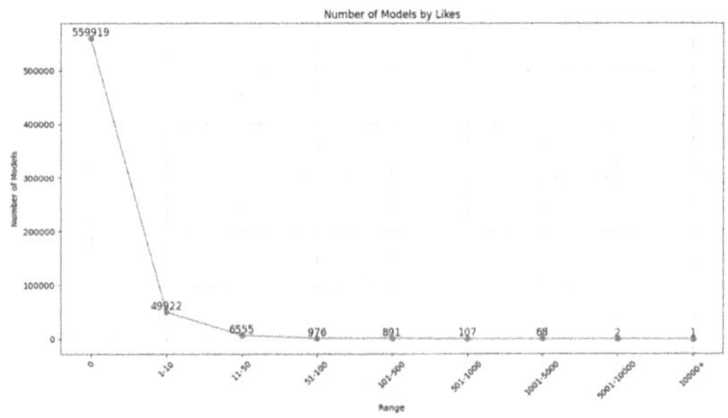

Fig. 5. Likes Distribution of Hugging Face Models

In addition for deeper insights, the most downloaded and liked models also corresponds to scholarly metadata available in the tags features. Thus it further emphasise to broaden the scope of exploration to have more interesting and useful insights. Furthermore, the information is also organized considering downloads and likes and presented in Table 1.

The overall findings demonstrate that downloads and likes tends to be the features which cannot be considered solely to explore the Hugging Face models. But could be important features. Hence indicating that for comprehensive exploration it is necessary to consider more features. To support that we collected the tags considering both downloads and likes and processed them against the defined ranges and created word-clouds to anticipate the usefulness of modelling features beyond downloads and likes. The word-clouds for downloads and likes are shown in Fig. 6 and Fig. 7 respectively.

As these representations, showcase the need of further exploration to have a deep insight about the models. As in both word clouds the frequent words are names of the pipelines or libraries and also the languages being supported by the model. In addition, there is also an considerable representation of citation tags which could lead to scholarly metadata repository. Finally, this exploration also supports the network modelling direction by keeping the downloads and likes ranges.

Table 1. Hugging-Face model Segregation

Segregation Criteria	Model Count
Downloads > 0	325633
Downloads = 0	292808
Likes > 0	58522
Likes = 0	559919
Both Downloads and Likes > 0	44822
Both Downloads and Likes = 0	279108

Fig. 6. Word Cloud of Tags for Downloads Ranges of Hugging Face Models

4.3 Heterogeneous Information Network Exploration

To further explore the feature set of Hugging Face models in Heterogeneous Information Network (HIN), the previously applied ranges for downloads and likes are taken into account alongside proposed data models. Hence, for simplicity we only showcase the one with best presentable network each for downloads and likes ranges. The criteria adopted for the presentation of the networks are the visibility of diverse nodes and their connections. Here to further enhance the interoperability of network, each type of nodes are color-coded.

The Fig. 8 showcased the information network presenting the diverse nodes and their interconnections as in the Fig. 8 the red colored nodes indicate the citation tags are available for the models and light blue colored nodes indicate the models and grey nodes are the generic tags for each model. The pipeline nodes and library nodes are represented in green and yellow respectively. For this specific network we picked the last download range i.e. 50001 and plus as it carries 983 models. The number of red colored nodes showcase the potential links of machine learning models to publications over specific download range.

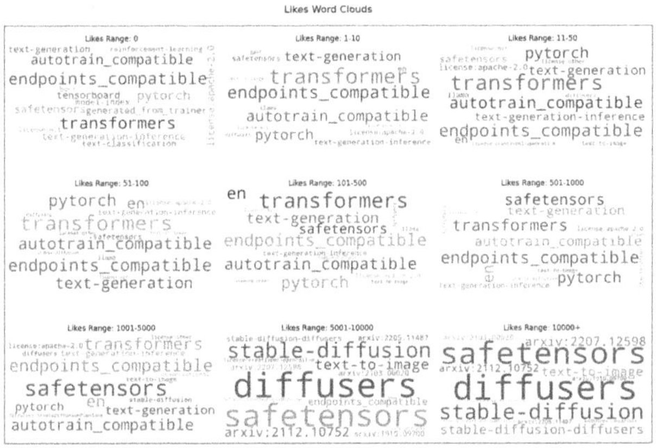

Fig. 7. Word Cloud of Tags for Likes Ranges of Hugging Face Models

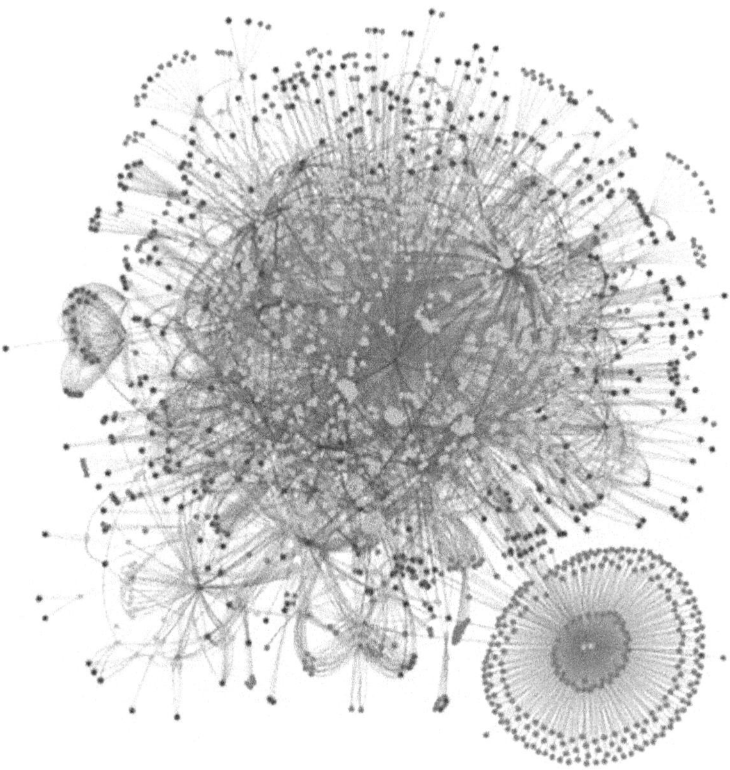

Fig. 8. Network of Download Range (50001 and Plus) using First Data Model (Color figure online)

Moreover, a network was also populated following the similar criteria for likes ranges and the exploration is quite evident that there exist a potential exploratory link to scholarly repository as shown in Fig. 9. For the likes, we picked the similar range picked earlier i.e. 501 to 1000 which carries 107 models.

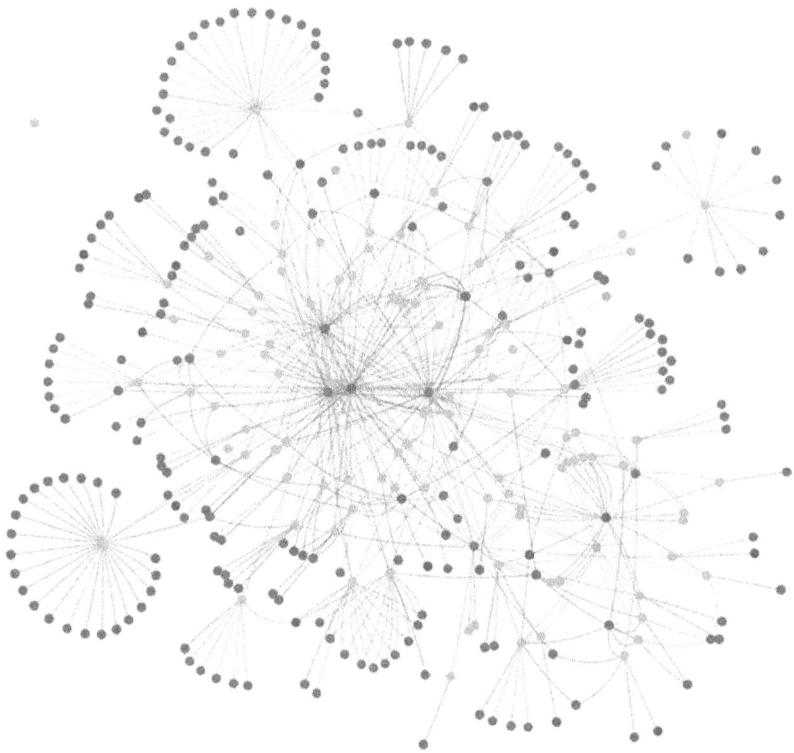

Fig. 9. Network of Likes Range (501-1000) using First Data Model

In addition to support the argument of linking information to scholarly repositories we processed tags features of models to have arXiv ids, we used the Python wrapper for arXiv[4] to collect the metadata. In addition, to further emphasize on the linking between scholarly repositories by considering second data model, we created the networks for both downloads and likes ranges as adopted earlier.

The Fig. 10 and Fig. 11 illustrates information networks by incorporating the metadata from arXiv and Hugging Face model features exploiting data model from Fig. 3. Each node in the network represents a different entity and corresponding link to other types of nodes. The red colored nodes denote arXiv ids, green nodes represent Hugging Face model ids, and blue nodes are the authors and publication dates. However, the connections between these nodes indicate

[4] https://github.com/lukasschwab/arxiv.py.

potential relationships, such as authorship or the presence of an arXiv paper id within tags of a model.

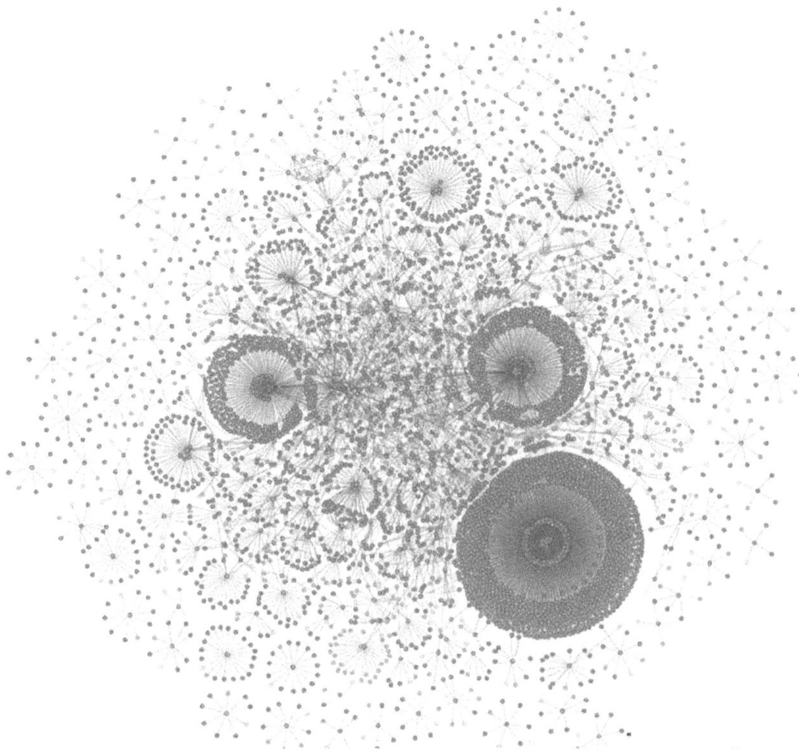

Fig. 10. Network of Downloads range (50001 and plus) models with arXiv tags

These networks demonstrate the direct and indirect inter-connectivity of research publications with respective machine learning models, highlighting how specific papers influence and contribute to the development of these models. In these network representations, the largest portion represents Gemini multi-model publication [13] and second largest representation is from the large language model article [11] which are connected to numerous Hugging Face models considering the picked ranges.

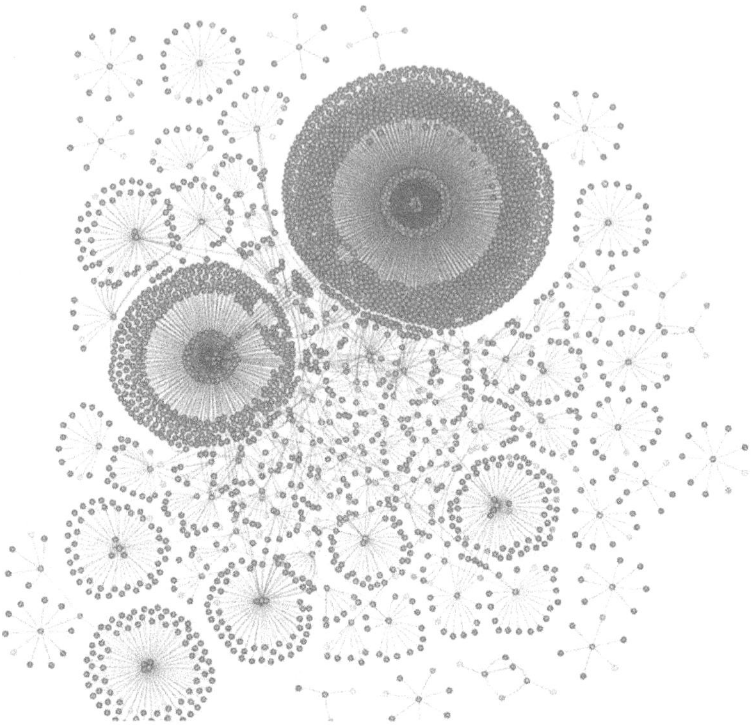

Fig. 11. Network of Likes range (501-1000) models with arXiv tags

5 Conclusion

This work present an insightful exploration of Hugging Face repository by structuring diverse features of machine learning models. As these machine learning models are crucial for the research communities and thus nourish the interdisciplinary research. However, modelling of heterogeneous features into information network will provide the potential linking among features and also pave to way for explore inter-repository scholarly information. The exploration also indicates the coverage of various computing libraries in interdisciplinary research activities. Furthermore the linkage to scholarly resources from arXiv is the key aspects and could be beneficial to compile available information against the models. As these cross repositories linkages are essential in exploiting the potential links of models to their corresponding research papers or potential references. In this AI driven era, this could also provide the potential reference to context to researchers for a deep insightful.

This work could also be enhanced in future to incorporate the other scholarly repositories and also encapsulation of traditional metadata could be beneficial for communities and contribute towards effective implications of research data management and open science. Moreover, a potential direction could also be to incorporate the broader scholarly metadata features for a comprehensive network representation of metadata and their corresponding AI models which could be lead to recommender systems and relevant research community detection. The addition of spatial features will also provide the location based concept network which could provide region based publications focusing specific machine learning models.

Acknowledgments. This work has been partially funded by the Deutsche Forschungsgemeinschaft (DFG, German Research Foundation), NFDI4DS (Grant number 460234259) and SmartER (Grant number 515537520). Authors also acknowledge the Hugging Face as data sources and also thanks the individuals involved in this research.

References

1. Ait, A., Izquierdo, J.L.C., Cabot, J.: Hfcommunity: an extraction process and relational database to analyze hugging face hub data. Sci. Comput. Program. **234**, 103079 (2024)
2. Castaño, J., Martínez-Fernández, S., Franch, X.: Lessons learned from mining the hugging face repository. arXiv preprint arXiv:2402.07323 (2024)
3. Castaño, J., Martínez-Fernández, S., Franch, X., Bogner, J.: Analyzing the evolution and maintenance of ml models on hugging face. arXiv preprint arXiv:2311.13380 (2023)
4. Castaño, J., Martínez-Fernández, S., Franch, X., Bogner, J.: Exploring the carbon footprint of hugging face's ml models: a repository mining study. In: 2023 ACM/IEEE International Symposium on Empirical Software Engineering and Measurement (ESEM), pp. 1–12. IEEE (2023)
5. Du, N., Guo, J., Wu, C.Q., Hou, A., Zhao, Z., Gan, D.: Recommendation of academic papers based on heterogeneous information networks. In: 2020 IEEE/ACS 17th International Conference on Computer Systems and Applications (AICCSA), pp. 1–6. IEEE (2020)
6. Jiang, W., et al.: An empirical study of pre-trained model reuse in the hugging face deep learning model registry. In: 2023 IEEE/ACM 45th International Conference on Software Engineering (ICSE), pp. 2463–2475. IEEE (2023)
7. Liu, J., Shi, C., Yang, C., Lu, Z., Philip, S.Y.: A survey on heterogeneous information network based recommender systems: concepts, methods, applications and resources. AI Open **3**, 40–57 (2022)
8. Liu, J., Tang, T., Wang, W., Xu, B., Kong, X., Xia, F.: A survey of scholarly data visualization. IEEE Access **6**, 19205–19221 (2018)
9. McMillan-Major, A., Osei, S., Rodriguez, J.D., Ammanamanchi, P.S., Gehrmann, S., Jernite, Y.: Reusable templates and guides for documenting datasets and models for natural language processing and generation: a case study of the huggingface and gem data and model cards. arXiv preprint arXiv:2108.07374 (2021)
10. McQuilton, P., et al.: Helping the consumers and producers of standards, repositories and policies to enable fair data. Data Intell. **2**(1–2), 151–157 (2020)
11. Srivastava, A., et al.: Beyond the imitation game: quantifying and extrapolating the capabilities of language models. arXiv preprint arXiv:2206.04615 (2022)
12. Sun, Y., Han, J., Yan, X., Yu, P.S., Wu, T.: Heterogeneous information networks: the past, the present, and the future. Proc. VLDB Endow. **15**(12) (2022)

13. Team, G., et al.: Gemini: a family of highly capable multimodal models. arXiv preprint arXiv:2312.11805 (2023)
14. Wolf, T., et al.: Transformers: state-of-the-art natural language processing. In: Proceedings of the 2020 Conference on Empirical Methods in Natural Language Processing: System Demonstrations, pp. 38–45 (2020)

Author Index

A
Abuhamad, Mohammed 355
Abushofa, Mohamad 131
Alam, Md Jahangir 79
Alkulaib, Lulwah 262
Almutairi, Mariam 262
Alvaro-Montes, Oscar 103
Amar, Mohamed Mahmoud 197
Arminio, Luigi 164
Atapour-Abarghouei, Amir 131

B
Bertaglia, Thales 3
Bertet, Karell 23
Berthet, Antoine O. 58
Bouguessa, Mohamed 197

C
Carlsson, Niklas 36
Cavalini, Athus 181, 307
Cheng, Lu 230
Clements, Nicholas 48

D
De Vinco, Daniele 148
de Vries, Sybe 3
Demko, Christophe 23
Diallo, Abdoulaye Banire 197
Dillon, S. J. 322
Dutta, Sujan 89

F
Ferro, Victoria 307
Forshaw, Matthew 131
Furman, Patrick 355

G
Ginde, Gouri 338
Goanta, Catalina 3
Goveia, Fabio 181, 307
Gui, Haoyang 3
Guillaume, Jean-Loup 23

H
Hall, Deborah L. 230, 355
Hammad, Omar 48
Haq, Ehsan-Ul 13, 69
Hayel, Yezekael 58
He, Jiahui 69
He, Yupeng 69
Hemery, Baptiste 117
Herkenhoff, Gabriel 181
Hollo, Aniya Khalili 48
Homann, Laura 103
Hossain, Ismail 79
Hui, Pan 13

J
Jeanne, Fabrice 117

K
Karmakar, Saurav 371
Kekriwal, Mayank 278
Kemgne, Martin Waffo 23
Khatri, Aadyant 262
KhudaBukhsh, Ashiqur R. 89
Kravitzch, Emmanuel 58
Krishnan, Siddharth 245
Kumar, Sumeet 89

L
Liang, Yueqing 322
Liu, Huan 230
Lu, Chang-Tien 262

M
Malini, Fabio 181, 307
Mathiak, Brigitte 371
McGough, A. Stephen 131
Melton, Joshua 245

L. M. Aiello et al. (Eds.): ASONAM 2024, LNCS 15213, pp. 387–388, 2025.
https://doi.org/10.1007/978-3-031-78548-1

Micallef, Nicholas 211
Miller, Shelly 48
Mishra, Shivakant 48
Mohammadinodooshan, Alireza 36
Morin, Maxence 117

N
Nananukul, Navapat 278
Nazari, Mujtaba 355

P
Pawlowski-Cherrier, Estelle 117
Puppala, Sai 79

R
Ravi, Rahul 338
Reid, Shannon 245
Rokne, Jon 338
Rossi, Luca 164
Russell Bernard, H. 322

S
Sandoval, Manuel 230, 355
Sha, Yuze 211
Shu, Kai 322
Silva, Yasin N. 230, 355
Sodré, Francis 181
Spanakis, Gerasimos 3
Sullivan, Esther 48
Suryani, Muhammad Asif 371

T
Tahir, Anique 230
Talukder, Sajedul 79
Tamura, Yuto 294
Terejanu, Gabriel 245
Tsugawa, Sho 294
Tupakula, Mallikarjuna 89
Tyson, Gareth 13, 69

V
Varma, Vineeth S. 58

W
Watabe, Kohei 294
Wu, Yan 211

Y
Yildiz Aktas, Melike 262

Z
Zhang, Peixian 13, 69
Zhu, Yiming 13

www.ingramcontent.com/pod-product-compliance
Lightning Source LLC
Chambersburg PA
CBHW070135030225
21297CB00004B/53
9783031785474